Prions and Brain Diseases
in Animals and Humans

NATO ASI Series

Advanced Science Institutes Series

A series presenting the results of activities sponsored by the NATO Science Committee, which aims at the dissemination of advanced scientific and technological knowledge, with a view to strengthening links between scientific communities.

The series is published by an international board of publishers in conjunction with the NATO Scientific Affairs Division

A	**Life Sciences**	Plenum Publishing Corporation
B	**Physics**	New York and London
C	**Mathematical**	Kluwer Academic Publishers
	and Physical Sciences	Dordrecht, Boston, and London
D	**Behavioral and Social Sciences**	
E	**Applied Sciences**	
F	**Computer and Systems Sciences**	Springer-Verlag
G	**Ecological Sciences**	Berlin, Heidelberg, New York, London,
H	**Cell Biology**	Paris, Tokyo, Hong Kong, and Barcelona
I	**Global Environmental Change**	

PARTNERSHIP SUB-SERIES

1. **Disarmament Technologies**	Kluwer Academic Publishers
2. **Environment**	Springer-Verlag
3. **High Technology**	Kluwer Academic Publishers
4. **Science and Technology Policy**	Kluwer Academic Publishers
5. **Computer Networking**	Kluwer Academic Publishers

The Partnership Sub-Series incorporates activities undertaken in collaboration with NATO's Cooperation Partners, the countries of the CIS and Central and Eastern Europe, in Priority Areas of concern to those countries.

Recent Volumes in this Series:

Series A: Life Sciences

Prions and Brain Diseases in Animals and Humans

Edited by

Douglas R. O. Morrison

Coppet, Switzerland

Plenum Press
New York and London
Published in cooperation with NATO Scientific Affairs Division

Proceedings of a NATO Advanced Research Workshop on
Prions and Brain Diseases in Animals and Humans,
held August 19–23, 1996,
in Erice, Italy

NATO-PCO-DATA BASE

The electronic index to the NATO ASI Series provides full bibliographical references (with keywords and/or abstracts) to about 50,000 contributions from international scientists published in all sections of the NATO ASI Series. Access to the NATO-PCO-DATA BASE is possible via a CD-ROM "NATO Science and Technology Disk" with user-friendly retrieval software in English, French, and German (©WTV GmbH and DATAWARE Technologies, Inc. 1989). The CD-ROM also contains the AGARD Aerospace Database.

The CD-ROM can be ordered through any member of the Board of Publishers or through NATO-PCO, Overijse, Belgium.

Library of Congress Cataloging-in-Publication Data

Prions and brain diseases in animals and humans / edited by Douglas
   R.O. Morrison.
      p.   cm. -- (NATO ASI series. Series A, Life sciences ; v.
   295)
      "Proceedings of a NATO Advanced Research Workshop on Prions and
   Brain Diseases in Animals and Humans, held August 19-23, 1996, in
   Erice, Italy"--T.p. verso.
      Includes bibliographical references and index.
      ISBN 0-306-45825-X
      1. Prion diseases--Congresses. 2. Prion diseases in animals-
   -Congresses. 3. Prions--Congresses.   I. Morrison, Douglas R. O.
   II. North Atlantic Treaty Organization. Scientific Affairs
   Division.   III. NATO Advanced Research Workshop on Prions and Brain
   Diseases in Animals and Humans (1996 : Erice, Italy) IV. Series.
      [DNLM: 1. Prion Diseases--congresses. 2. Prions--congresses.
   3. Brain Diseases--congresses.   WL 300 P9587 1998]
   QR201.P737P76 1998
   616.8'047--dc21
   DNLM/DLC
   for Library of Congress   98-12133
                                                          CIP

ISBN 0-306-45825-X

PREFACE

Most of the world's experts on prions met for a workshop in Erice in August 1996. The aim of the workshop was to discuss the fundamentals of the science of prions. It was fortunate that so many could be present given the pressure that they were under because of the data presented in March 1996, indicating that Bovine Spongiform Encephalopathy, BSE or Mad Cow Disease, had penetrated the species barrier and was beginning to cause a new disease in humans—the new variant of Creutzfeldt-Jakob Disease, nvCJD. This important and urgent subject became an additional major topic at the workshop.

This is a book containing most of the talks plus the abstracts of those unable to find time to write up their talks. Almost all papers were written in the spring and summer of 1997 and contain material added after the workshop; thus Bob Will's paper on the new variant of CJD contains data up to July 1997 and four contributions arrived in October 1997.

In addition to the talks given at our workshop, there was a special joint session with the Planetary Emergencies Workshop where many distinguished scientists, including three Nobel laureates, discussed major issues affecting our planet. Six talks were given by us to this other workshop, five about prions, BSE, and nvCJD, and one on the broader issue of new epidemics by Luc Montagnier.

Although most of the talks concerned research issues, there were a few special talks. Stan Prusiner described the early history of the Protein Only hypothesis and the initial skepticism—now overcome with the award of the 1997 Nobel prize for this and his work on Alzheimer's Disease. Paul Brown gave a fascinating account of the stone-age Fore tribe in Papua New Guinea who had indulged in ritual cannibalism causing the spread of the prion disease, kuru, which became the leading cause of death among women. He showed some of Carleton Gajdusek's films of the tribe and of kuru—these were later used by BBC television. David Westaway presented an additional talk where he gave an interesting review of Alzheimer's Disease, which he compared with prion diseases, bringing out the similarities and the differences. David found the subject so interesting that for this book he has done an appreciable amount of additional work so that the article published here is considerably more comprehensive than his original talk. This has also happened with other contributors whose papers printed here are more complete than their oral presentations.

Although BSE and its probable corresponding human version, the new variant of Creutzfeldt-Jakob Disease, nvCJD, were not the main direction of the workshop, the subject was discussed by John Wilesmith of the Central Veterinary Laboratory and by Bob Will of the CJD Surveillance Centre in Edinburgh. In addition many leaders in this matter from various countries were present, including Charles Weissmann and John Pattison who are Chairmen of the European Union and British committees, respectively, on the subject.

The workshop had been proposed as a quiet meeting in the beautiful surroundings of the Science and Cultural Centre of Erice, where we could discuss peacefully, hence I deliberately did not advertise it. However an unforeseen consequence of the great public interest in the subjects of the workshop was that teams from six TV companies in France, Italy, Switzerland, and the UK turned up. They were excluded from the workshop sessions but they filmed during the Joint session and interviewed outside of the meeting. Some excellent films were shown later on television. Several newspaper articles appeared, those in Le Monde by Dr. Nau being of particularly high standard.

Great thanks are due to the many speakers who despite great pressure of work, wrote up their talks for this volume which represents a rather complete summary of research work on prions in 1997. A few speakers were unable to submit papers because of time pressure and some wrote extra long abstracts which are printed here.

An Introduction has been written which serves as a guide to the talks. The main subjects and questions are stated and the papers are summarised under these headings.

Many have asked how this workshop came to be initiated and organised by someone who normally works on particle physics and astrophysics. A few years ago, at dinner during an astrophysics conference in Erice, Antonio Zichichi, who is the Director of the Scientific and Cultural Centre, said that we needed new subjects. I suggested prions which is pronounced "preeons". This caused some confusion as preons are the name given to hypothetical particles which are the sub-structure of quarks which are the sub-structure of nucleons which are the sub-structure of nuclei which are the centre of atoms. When I explained about Mad Cows and Cannibals, Nino asked if I was serious. Finally he said "Great idea—you organise it." A friend who is a professor of mathematics put me in touch with James Armour, a distinguished veterinary expert who told me who to contact. The International Organising Committee of John Collinge of London, Pawel Liberski from Lodz, and Stan Prusiner of San Francisco (and myself) encouraged many prion experts to attend. The invaluable Book of Abstracts was established by John Collinge's team at St Mary's.

The financial help that permitted the conference to take place was due to the generosity of NATO, the "Ettore Majorana" Centre for Scientific Cultural at Erice, Nestle, and Biogen and it is a pleasure to thank them. The kindness of the staff of the EMCSC made our stay in Erice particularly agreeable and smoothed all the arrangements. Great and patient help which made this book possible was given by Joanna Lawrence of Plenum Press.

An outstanding feature of the workshop was the friendly spirit among the participants despite the strong disagreements on some subjects during the meeting. The entire atmosphere of being in the ancient city of Erice, perched on a mountain top 700 metres above the hot coastal plain of Northwestern Sicily, was conducive to considering one's own work in a large historical perspective. An afternoon excursion to a Greek temple with a nearby beach reinforced the cohesive spirit of the workshop.

Douglas R.O. Morrison

Figure 1. Charles Weissmann (left) and John Collinge in front of the entrance to the Dirac lecture hall.

Figure 2. Stanley Prusiner. A plain clothes guard is on the left.

Figure 3. Luc Montagnier (left) and Douglas Morrison. The Toulouse-Lautrec book has a poster of "La Vache Enragee" - a hundred years ago there were processions of artists through Montemartre protesting that all they could afford to eat were "mad cows".

Figure 4. In successive rows are Charles Weissmann, Adriano Aguzzi, Dominique Dormont, Harriet Coles, Moira Bruce, and Andrea LeBlanc.

Figure 5. Paul Brown. Monte Calfano is in the peak on the right.

Figure 6. Adriano Aguzzi (left) and Reed Wickner.

Figure 7. The Dirac lecture hall with Andrea LeBlanc answering questions. Others visible are Detlev Riesner, Antonio Zichichi, and Douglas Morrison.

Figure 8. Moira Bruce and David Westaway in discussion during coffee.

SPONSORS OF THE ERICE WORKSHOP ON
PRIONS AND BRAIN DISEASES IN ANIMAL AND HUMANS

We are very pleased to express our gratitude to the following for their generous sponsorship of this workshop which could not have been held without their help.

The principal sponsor was **NATO** who considered it as part of their **International Scientific Exchange Programme** designed to assist Eastern European countries. This was particularly important as relatively little work is being performed there so that this workshop acted as a catalyst to spread knowledge of the subject of prions. This is particularly urgent as the Ministries of Health and of Agriculture of all European countries, East and West, are concerned.

We are indebted to **Prof. Antonio Zichichi**, the Director of the **"Ettore Majorana" Centre for Scientific Culture in Erice** for his generous support and for the staff of the Centre for unsparing help and kindness.

We are also indepted for major financial help to

Nestle S.A., Vevey, CH-1800

Biogen, Cambridge, MA 02142

CONTENTS

INTRODUCTION - GUIDE TO THE TALKS

Douglas R.O. Morrison

CH-1296 Coppet
Switzerland

This introduction is intended as a guide to the talks given at the Erice Workshop on Prions It will also serve as a summary for those who wish information on a specific topic but reference should be made to the full paper.

SUBJECTS
1. "Protein Only" Hypothesis
2. What is a Prion?
    2.1 Genes that Produce Prions
    2.2 Three-Dimensional shape
    2.3 Size of a Prion and of Prion Aggregates
    2.4 Life History of a Prion
    2.5 What do Prion Proteins Do?
3. Aggregates - Tubulovesicular Structures - Vacuoles
4. Strains
5. Polymorphisms and Mutations
6. Species Specificity or "Species Barrier"
7. Mechanism of Conversion of PrP^c to PrP^{sc}
8. Transgenic Mice - Importance and Results
9. Routes of Infection
10. Human Prion Diseases
    10.1 Creutzfeldt-Jakob Disease, CJD
    10.2 Gerstmann-Straussler-Scheinker Disease, GSS
    10.3 Fatal Familial Insomnia, FFI
    10.4 New Variant of Creutzfeldt-Jakob Disease, nvCJD
    10.5 Possible new Human Disease?
    10.6 General - Linkage of Clinical Symptoms and Structure of Prions
11. Yeast - Evidence for Non-Nucleic Acid Transmission
12. Purification of Samples
13. Early Warning Test for Animals
14. Drugs to Cure or Postpone Prion Diseases
15. Comparison of Prion Diseases with Alzheimer's Disease

1. "Protein Only" hypothesis
Stan Prusiner describes his journey from "Heresy to Orthodoxy" in a historical account of the "Prion Only" hypothesis that he presented in 1982.

Many other talks also give a good introduction with a historical content, e.g. **Charles Weissmann**. Those who have reservations about the "Protein Only" hypothesis, such as **Byron Caughey, Heino Diringer, Pawel Liberski,** and **Maurizio Pocchiari,** give different introductions and tend to use different names. The "Protein Only" (or "Prion Only" as it is often called), hypothesis uses PrP^c for the normal cellular prion protein and PrP^{sc} for the abnormal disease-causing prion protein where the "sc" refers to scrapie which is used as a general name for the infective agent that converts PrP^c to PrP^{sc}. **Byron Caughey** and **Bruce Chesebro** prefer more phenomenological names and use the fact that the detergent, proteinase K, (PK), destroys the normal form, PrP^c, but some part of the abnormal form, PrP^{sc}, is resistant and leaves an insoluble deposit which has been shown to be infectious. They use PrP-sen for the PK sensitive prion protein and PrP-res for the resistive prion protein or residue. This has advantages as scrapie is only one form of prion disease and the bovine form is turning out to be more widespread.

It has been suggested that since the prion samples are imperfectly purified, there could be an unknown and undetected virus which could be responsible, but **Reed Wickner** showed that there are two proteins in yeast which have fundamental prion qualities and it is believed that all viruses in yeast have already been discovered.

Viktor Zouev reported that the histopathological effects of viruses such as influenza A, were similar to those caused by prions.

Many arguments were presented and will be described below as they occur.

There was one experimental result that convinced many that the agent was not a virus but was the prion protein. **Charles Weissmann** and his group in Zurich, created transgenic mice which had no prion proteins. When these mice were injected with an infectious agent - scrapie - they did not become ill. This is as predicted by the "Prion Only" hypothesis.

However one recurring theme was that there were such an extensive variety of arguments in favour of the "Protein Only" hypothesis that this was convincing even although there were loopholes and problems.

In the Concluding talk, **Douglas Morrison** tries to summarise and comment on, the arguments for and against the Protein Only hypothesis. The results are shown of a vote taken which gave a large majority in favour of the Protein Only hypothesis.

2. What is a Prion?
2.1 Genes that Produce Prions
The prion protein is produced by a gene, i.e. it is encoded by a single-copy chromosomal gene. **David Westaway**, in his first paper, has studied these prion genes in humans and sheep starting from 145 kb of DNA, and also the mouse gene in 44 strains. The comparison gives many interesting results, two unexpected, including the indication that the ancestral PrP gene that predated the specification of animals, most likely exhibited a three exon structure. Thus it appears that all mammals have prions.

An important result is that they do not find any retroviral insertions in human or sheep PrP genes so that this work cannot support the idea that a retrovirus or retrovirus-like particle could cause prion diseases.

2.2 Three-Dimensional Shape
It is established that the prion has a chain of about 250 amino acids, (253 for the human prion). In first approximation the PrP^{sc} and PrP^c proteins are chemically identical but have different shapes e.g. have folded differently. If

the protein were to be stretched out in a line - Open Reading Frame, ORF - then each position (codon) would be linked only to its neighbour except for two sulphur ions at codons 179 and 217 which are joined.

Prions have two terminations, N and C. The most critical part appears to lie between codons of about 100 and 231, judging by the mutations observed though there can be many insertions (or deletions) between codons 51 and 91. Previous studies had suggested that PrP contains several helixes. The difference in shape between the abnormal and normal forms, was that there were more α-helixes in PrPc and more β-sheets in PrPsc.

Rudi Glockshuber described a ground-breaking experiment where the detailed shape of the mouse prion protein was established from codon 121 to codon 231. This is an intrinsically stable domain which is monomeric. It contains both glycosylation sites (where two sugars can be added) at Asn181 and Asn197, and codon 231 where is the GPI anchor which attaches the prion to the outer surface of the cell membrane.

Nuclear Magnetic Resonance, NMR, techniques were used with samples labelled with ^{15}N and secondly with ^{15}N/^{13}C labels. This incredible work (to a physicist) deserves description. Firstly the basic NMR structure was determined with 93% completeness using sequence-based resonance assignments. Then the ^{1}H, ^{13}C and ^{15}N resonances of the backbone were assigned from the intraresidual and sequential correlations of the amide ^{1}H and ^{15}N resonances with C$^\alpha$, C$^\beta$ and H$^\alpha$ signals using three-dimensional, 3D, triple resonance experiments and 3D ^{15}N-resolved (^{1}H,^{1}H)-NOESY (nuclear Overhauser enhancement spectroscopy) experiments. The side-chain signals were assigned from 3D through-bond correlation NMR experiments. The calculation was made with 1368 NOE distance constraints and 227 dihedral angle constraints. The most uncertain 20 conformers were energy-minimised. For these 20 conformers, the root mean square distance to the mean structure was 1.4 Angstroms for the N, C$^\alpha$ and C$^\beta$ atoms and 2.0 A for all the heavy atoms.

The resultant structure is different from previous suggestions. It consists of three α-helices and a β-sheet which has two strands of amino acids which are antiparallel.

The exceptional stability and resistance to attack of the prion can be understood as the prion is very tightly bound and the S-S bond is shielded near the centre of the structure.

It is again impressive that the position of electrons was added and the electrostatic surface potential found, as shown in the colour photographs in Riek et al., Nature 382 (1996) 180, the surface has pronounced regions which are electrically negative and others which are positive. The region which probably binds to the surface membrane is positively charged and contains the β-sheet while the glycosylation sites are exposed to the outside.

There are many other important results that can be derived from this new 3-D structure.

Studies of the structure of the human prion protein were reported by **Graham Jackson**. In view of the low abundance of soluble PrPc from endogenous sources and problems of purification, they employed an E. coli expression system to produce recombinant PrP. A fragment containing residues 89 to 230 was studied which is equivalent in amino acid sequence to PrP27-30, the protease resistant core of PrPsc that remains after digestion of the N-terminal 89 residues. Finally recombinant human prion protein was studied biophysically and its structure determined by NMR. The results are discussed in section 7.

2.3 Size of Prion and Prion Aggregates

The mass of a prion is about 35 kDaltons (where a Dalton is about one atomic mass unit, thus oxygen is 16 Da). Stan Prusiner et al. in 1982 showed that after treatment with detergents, generally an insoluble fragment of 27 to 30 kDa is left, and most importantly, this fragment is infective.

Detlev Riesner presented results on the study of this PrP 27-30 and its aggregated form. One infectious unit of scrapie contains 10^5 to 10^6 PrP molecules. There were prion rods of 100 to 300 nm length which were infectious. The prion rods were treated to produce a soluble fraction and an insoluble residue. He found that the soluble part did not show infectivity and had a high a-helix content. The insoluble part gave spheres of 10 nm diameter of a molecular weight of 117 kDa so that they could be considered as consisting of 4 to 6 molecules of PrP 27-30, i.e. pentameric 27-30; they were not infectious.

Detlev said in connection with the suggestion that a nucleic acid could be responsible for infectivity rather than prions, that no such acid has been found and with a quantitative analysis, it was found that they had eliminated oligonucleotides greater than about 80 nucleotides.

Rudi Glockshuber (priv. comm) said that the size of the 121-231 part is about 2 to 3 nm.

2.4 Life History of a Prion

This subject was little discussed in Erice, but essentially it is believed that the normal PrP is produced in the endoplasmic reticulum and passes via the Golgi apparatus and a secretory vesicle to the outside of the membrane of the cell - especially a nerve cell. Later it re-enters the cell. In vitro experiments, it lives about 4 to 6 hours. See for example Byron Caughey and Bruce Chesebro, Cell Biology 7(1997) 56-62.

The prion lifetime in vivo is unknown except that **Charles Weissmann** described an experiment with transgenic mice which are bred with no normal prions. When scrapie PrP^{sc} is injected in the brain, some PrP^{sc} is observed in the spleen after 4 days but not at the next measurement after 2 weeks. This would suggest that in vivo, prions can live days and not hours - further experiments on this, using Tg mice, would be of interest.

Randell Nixon describes experiments where scrapie alters the behaviour of the membrane, particularly in terms of Ca2++ channels. Further the claim is made that the plasma membrane fluidity decreases by a factor of seven, but this was questioned because of the large errors, the seven being the ratio of (8.04 +/- 6) to (1.13 +/- 1). He quotes previous work that 90% of the PrP^c are attached to the outer part of the cell membrane.

2.5 What Do Prion Proteins Do?

John Collinge found that PrP null mice (having no prions) demonstrated abnormalities of inhibitory neurotransmission (synaptic function and memory) raising the possibility that neurodegeneration caused by prion diseases is due, at least in part, from loss of the prion function. However a series of experiments presented by **Hans Kretzschmar**, give an entirely different reason for the existence of prions. Instead of using biochemistry, they study cellular pathogenesis and look at the processes in detail, in particular what causes the death of neuronal cells.

It is shown both in vitro and in vivo, that nerve cell death is due to an apostic (i.e. planned) mechanism rather than necrosis (severe and sudden injury). A synthetic peptide PrP 106-121 is found to have a toxic effect on neurones only if PrP^c is present and also, in culture, only if microglia are present. Further detailed experiments show that the microglia produce oxygen radicals. It appears that PrP 106-121 does not kill neurones directly, but rather cause microglia to emit substances which are toxic to the nerve cells.

The point is that nerve cells need a strong defence against oxidative stress and PrP^c appears to give this resistance. If an abnormal PrP^{sc} enters and converts or destroys PrP^c, then this defence is lost.

It may be noted that in terminally ill mice, only about 1% of the nerve cells were dead, but apoptosis can be completed in hours.

While some evidence had suggested that the prion protein was necessary for the normal synaptic function, but their extensive measurements showed no differences between normal and prion-free transgenic mice. The synaptic currents and voltages measured were in the range 20 pA to 5 nA and about 3 mV resp., while the average activation and decay time constants were 2.4 and 11.3 ms resp.

The question of the effect of oxidation of proteins in general, was discussed by **Luc Montagnier** who described its harmful effects, for example, in causing cancer.

3. AGGREGATES - TUBULOVESICULAR STRUCTURES - VACUOLES

Aggregates observed in TSE diseases have been partly discussed above in section 2.2.

Pawel Liberski described research with electron microscopes on Tubulovesicular structures, TVS, which are found in all TSE diseases. Using immunogold techniques, he reports that TVS are not stained with anti-sera and thus are not composed of PrP. He also reported on studies on primitive plaques in the scrapie-affected brains of hamsters. These amyloid plaques are different from plaques in kuru, CJD, GSS, and in mice. He writes that further experiments involving infectivity are required to ascertain if the infectious agent is PrP^{sc} or whether PrP is merely an amyloidic protein.

Reed Wickner reported that yeast contains two proteins that show prion-like characteristics. One of them has been shown to give aggregates and from further work, he considers that aggregation is the basic mechanism of abnormal prion formation.

Lajos Laszlo described the results of ultrastructural and immunocytochemical studies. As currently available PrP^{sc} has cell lines originating from the peripheral central nervous system, and does not show significant amounts of spongiform degeneration, a special hypothalmic cell line was established. These cells when infected, showed intense vacuolation and characteristic features of programmed cell death.

4. STRAINS

Moira Bruce (for a fuller account, see M. Bruce, Methods in Molecular Medicine: prion diseases, Eds. H. Baker and R.M. Ridley. Humana Press Totowa (1996)) described the work of the Edinburgh group which had shown that scrapie in sheep has some 20 strains each of which had distinct characteristics measurably clinically and neuropathologically. In particular, the technique of lesion patterns has been established where mice are infected and the way in which the brains are infected is studied. An important result is that when the agent transmitting the disease is passed through several hosts, it is found that it must contain an informational component which is independent of the host.

She reported that only one strain of BSE was observed even in many different species such as sheep, cows, cats, antelopes, etc. BSE was transmitted to various types of sheep both intercerebrally and orally. It was interesting to note that the most susceptible genotype was a "negative line" of Cheviot sheep in which natural scrapie has not been found. The spleen was also found to be infected.

Stan Prusiner reported on experiments which showed that prion strains seem to be generated and maintained by the sequence of the PrP substrates that are converted into PrP^{sc}.

Byron Caughey noted that the most remarkable case of strains was with hamster-adapted TSE from mink, where there are two major strains, the hyper, HY, and drowsy, DY, types - this is remarkable because they have identical amino acid sequences so appear to differ only in their 3D structure. They used a cell-free conversion reaction to show that the HY and DY PrP-res molecules give their own strain to PrP-sen while PrP-res is being formed. This means that there is a non-genetic propagation of these strain types. Further the distinctive 3D structure is transmitted by self-propagation of the PrP-res polymers.

5. POLYMORPHISMS AND MUTATIONS

It is a general characteristic that mutations occur all the time. Sometimes a mutation results in two forms which occur naturally as alternatives. A well-known example in the human prion is that at codon 129 either Methione, M, or Valine, V, can occur. Thus from the maternal and paternal alleles, people are found to have MM, MV and VV with probabilities of 37%, 52% and 11% resp.

Many papers were presented at Erice relating the effect of polymorphisms and mutations to diseases, especially human diseases. See papers by **Paul Brown, Pierluigi Gambetti, Bernardino Ghetti**, and **Bob Will** which will be discussed below under human diseases. **Maurizio Pocchiari** shows a table of 11 mutations or polymorphisms in human diseases and gives the corresponding amyloid protein allotypes.

6. SPECIES SPECIFICITY OR "SPECIES BARRIER"

It is found that a TSE can be transmitted most easily within the same species. It has been shown early that it is possible to transmit the disease from one species to another but the efficiency varies greatly. Sometimes the phrase "species barrier" is used but this is rather inappropriate and it is more scientific to talk of the efficiency of transmission. Many papers at the workshop discussed this. **Byron Caughey** discussed species specificity and strains generally. He described their experiments where they combined PrP-sen and PrP-res from several different species - mice, hamsters, sheep, cattle, and human - in cell-free conversion reactions. They found that the efficiencies found in these in vitro studies agreed closely with the efficiency found with real animals. For example, the BSE agent, PrP^{BSE}, readily converted PrP-sen from cows, mice and two different types of goats as happens in nature, but where the efficiency was negligible in nature, then the reaction was also not found in the lab. This also applied to sheep with high and zero susceptibility to PrP^{BSE}.

7. MECHANISM OF CONVERSION OF PRP^C TO PRP^{SC}

Stan Prusiner argues that since PrP^c is mainly α-helices while PrP^{sc} has approximately equal parts of α-helixes and β-sheets, then the underlying mechanism changing one into the other is a conversion of α-helices into β-sheets. Experiments were reported on this and they also indicate that the PrP^c/PrP^{sc} complex binds to a macromolecule, provisionally designated "protein X".

With transgenic mice with human and mice prions, it was found that amino acid mismatches at codons 102 and 129 resulted in delayed onset of Central Nervous System, CNS, disfunction while a mismatch at positions 178 or 200 did not. Further experiments suggested that prion diversity may be limited.

Randell Nixon described experiments made in the San Francisco and the Rocky Mountain labs, using radioactive PrP^c The radioactivity picked up by PrP^{sc} is delayed by about an hour and then increases to a maximum as the PrP^c loses its tracer. The PrP^{sc} keeps its radioactivity suggesting that the PrP^{sc} is not degraded but stays inside the cell. Some is found outside the cell perhaps from the subsequent death of the cell. Blocking the passage of the normal PrP^c from leaving the cell, also stops the formation of PrP^{sc} indicating that the conversion of normal to abnormal prion proteins must take place outside the cell, presumably mainly on the membrane. He also discusses the possibility that another particle, a chaperone, may be involved in the conversion mechanism.

Tony Clarke discussed protein folding and misfolding as he said there is the possibility that misfolding of the normal PrP produces the abnormal scrapie agent and such folding could be important generally. Spontaneous folding in vitro, of some model proteins have been studied and attempts have been made to define the mechanism of action of chaperonins which assist in selecting a structure among the very large number possible. The competition between folding and misfolding of a chaperone-dependent imported mitochondrial protein was discussed.

The detailed mechanism by which the abnormal PrP^{sc} converts the normal PrP^c to itself or a closely related infectious form, is not established. Two main theories are discussed by **Byron Caughey**. Firstly the dimerization model and secondly, the nucleated (or seeded) polymerisation model.

In the first model it is assumed that a PrP-res monomer binds to a monomer of PrP-sen to form a heterodimer. This then splits to give two PrP-res. However no monomer of PrP has been found which is resistant to proteinase K. Their results show that ordered aggregates of PrP-res can convert the normal PrP to the abnormal form, and that a free PrP-res monomer is not needed.

The nucleated polymerisation model proposes that PrP-res polymers could seed a polymerisation reaction and these polymers could range from small stable oligomers to large plaques.

Maurizio Pocchiari discussed the mechanism from studies of the amyloid deposits formed in the brain, in relation to the mutations in PrP-res. The formation of amyloid deposits is considered to be a major pathogenetic event since it occurs before histological lesions and clinical symptoms. Both human patients and cell cultures from transgenic hamsters were studied and also synthetic peptides with residues at for 90-145, 106-147, and 202-218.

A great variety of results for different mutations and polymorphisms, were obtained. For GSS, FFI, Famial and sporadic CJD, a table shows that in some cases the amyloid protein was determined only from the mutant allotype while in other cases the amyloid was from both the mutant and wild type. While the results were not decisive, they tend to favour the nucleation theory, Different strains (from different mutations) then give different mechanisms for forming the amyloid.

Reed Wickner from his studies in yeast, gives a model of aggregate formation which depends on the geometrical shape of the PrP-res.

Graham Jackson reported the results on the structure of recombinant human prion protein containing residues 89 to 230. There appear to be two distinct transitions during the unfolding of PrP^c. This region contains a large proportion of secondary structure and a well-defined tertiary structure. The overall folding energy found of -3.5 kcal/mole is low and implies that one molecule in 1600 is unfolded - such molecules would be available for conversion to PrP^{sc}.

Near residue tryptophan 99, the stability is still lower and Graham suggests that this could be an initiation site for conversion. Work is continuing and site-directed mutations should give fundamental information.

8. TRANSGENIC MICE - IMPORTANCE AND RESULTS

Charles Weissmann and his group were the first to develop a breed of transgenic mice which had no normal prion proteins. The main aim was to test the major prediction of the "Protein Only" hypothesis that such mice should be resistant to prion diseases - and indeed that was what was found. This is a remarkable achievement and the strongest single piece of evidence in favour of the hypothesis.

This work with transgenic mice has led to many more experiments. In breeding these Tg mice - called null or knockout mice- another breed was established which had half the normal amount of PrP, showed enhanced resistance to scrapie diseases even at high levels of infection - this surprising result still needs an explanation.

In another type of experiment, mouse PrP genes were introduced to the null mice, the mice became susceptible to infectious mice prions but not to hamster prions. But the opposite was the case when hamster PrP genes were introduced into the mice, which then became susceptible to hamster prions but to a much lesser extent to mouse prions. Thus there must be an important factor(s) with in vivo experiments not found in the lab.

This proof that introduction of PrP transgenes into PrP null mice again renders the mice susceptible to scrapie, enables a further class of experiments called Reverse Genetics. Here PrP genes which are modified by mutations or deletions, are introduced into PrP null mice. and then these Tg mice can be tested for their resistance to scrapie.

In an example of Reverse Genetics, three types of mice were bred with normal PrP, and PrP missing 26 or 49 proximal amino acids which in PrPsc can be cleaved off with protease. When inoculated with mouse scrapie, the mice with normal and truncated prions all were fatally infected, and prion propagation and PrPsc accumulation was observed. It was concluded that the segment of prions that were deleted and which contain 3.5 of its 5 octa repeats, plays no significant role in the conversion mechanism nor in the generation of PrPsc.

Adriano Aguzzi has extended the transgenic mice technique by making grafts containing an infective agent, on to knockout mice. While grafts have been used previously, this is the first time that the effect of an individual molecule can be studied. In addition, the grafts can be used to study what happens locally at long times, greater than the lifetime which normal mice with these fatal diseases would have, that is, mice can live with a lethal disease at the implant, but not elsewhere.

After an interesting introduction on grafting and on the Blood Brain Barrier, BBB, he showed that their grafting technique does not cause permanent leakage through the BBB.

Applying a graft to the brain containing mouse scrapie, to knockout mice appeared to give no general noticeable effect on the mice. At the graft site, after a normal incubation period, there were local signs of scrapie such as spongiosis but the effect of the infectious tissue in the graft did not spread to the rest of the brain. Yet PrPsc that was produced in the graft moved into the brain where it formed fine granules along white matter tracts; and also some PrPsc deposits were observed, for example in the hippocampus. This suggests that diffusion of PrPsc occurs via the extracellular space rather than by axonal transport.

The fact that PrPsc can move in the brain without any disease being produced, suggests that PrPsc is inherently non-toxic. It becomes toxic only after its interaction with the normal PrPc.

Bruce Chesebro reported on experiments using two lines of transgenic mice. The Tg52NSE line contained the neuron-specific enolase, NSE promotor plus only 1 kb of cDNA while the Tg10 mice had a cosmid H-PrP transgene with more(40 kb) flanking DNA. In the NSE mice HPrP was detected in the brain but not in six other tissues as might be expected with the neuron-specific agent - the Purkinje cells were most affected. When injected with a hamster scrapie agent, mice of both lines died more quickly than controls but the symptoms and histopathological findings were different. Previously mice expressing HPrP were made with 40 kb cosmid clone, but the Tg52NSE mice were highly susceptible to hamster scrapie with only 1 Kb of cDNA showing that this HPrP minigene is the critical element which induce susceptibility to the hamster scrapie agent in vivo.

While these results appear to support the "Protein Only" hypothesis, the authors warn that some of the results could be the consequence of abnormal over-expression of PrP. In support of this they quote experiments (D. Westaway et al, Cell, 76(1994)117-29 and K.K. Hsiao et al. Science

250(1990)1587-90) where spontaneous scrapie-like diseases have occurred without any scrapie. However as written for example in the concluding talk, if it is assumed that spontaneous mutations can cause disease e.g. with CJD, then the rate will be increased by increasing the concentration of normal PrP. Many other reports described work using the transgenic mice that Charles Weissmann developed and these are described elsewhere.

9. ROUTES OF INFECTION

After an infection at one point, it is important for understanding, to know the routes that the infection takes to reach the brain. This is of particular interest for the new variant of CJD where the PrP^{BSE} probably entered by people eating infected cow products.

Heino Diringer reported on the spread of infection in hamsters fed scrapie. The disease was first observed in the spinal chord between vertebrae T4 and T9 - this excited interest as it is the region where nerves enter the Central Nervous System, CNS, which enervate the gastrointestinal tract. The pathogenic process was observed almost simultaneously in the brain and the spinal chord - sometimes earlier in the brain which suggests that the infection must reach the brain by at least two routes, one of which is the spinal chord.

Charles Weissmann gave measurements of infectivity in the brain and in the spleen of knockout and normal mice at various times after inoculation in the brain of PrP^{sc}. With normal mice, substantial infectivity (about a million infectious units, IU, per ml) appeared in the spleen after 4 days but did not appear in the brain until several weeks where it rose to over 100 million IU per ml. This may be interpreted as showing that prions are synthesised in the spleen.

In contrast, low infectivity (100 IU per ml) was detected after 4 days in both the spleen and the brain of knockout mice but not at longer times. This would suggest that the abnormal injected prions are transported quickly to the spleen and then die or are dispersed.

As mentioned above, **Adriano Aguzzi** using grafts of scrapie applied to the brains of null mice, found that the PrP^{sc} moved in extracellular space rather than by axonal transport.

10. HUMAN PRION DISEASES

10.1 Creutzfeldt-Jakob Disease

Paul Brown studied two polymorphisms that occur on the chromosome 20 gene which encodes the prion protein. Codon 129 can have either methione, M, or valine, V, with an allelic ratio of 0.62 to 0.38 in Caucasians, and there can be a 24 base pair deletion between codons 51 and 91 which occurs with a 2% frequency in Caucasians. Cases are rare of human to human horizontal transmission where both the donor and receiver are known - results are given of three cases of surgical transmission and one case where both husband and wife died of CJD. The statistics were too small for definite conclusions. However the fact that in 2 out of 7 cases the recipient was M/V at codon 129 and M/M in the other 5 cases, shows that it is not necessary to be homologous (that is M/M or V/V). The results are consistent with previous work showing that in transmission from humans to non-human primates, it is not necessary for them to be homologous.

Ruth Gabizon reported on Libyan Jews who have an abnormally high rate of CJD and where all cases are found to have a point mutation at codon 200. She said that the E200K mutation is not enough to cause CJD - some unknown other factor which is age-related is needed. It was found that mutant PrP converts into PrP^{sc} normally, but wild-type PrP (no mutation) in the brains of heterozygous E200K CJD remains sensitive to PK digestion. It is suggested that there is some type of species barrier which inhibits the conversion of normal PrP^c to PrP^{sc} by interaction with mutant PrP.

Maurizio Pocchiari analysed the PrP-res found in the brain of patients with familial CJD who carried the (Val210Ile) mutation and of a patient with no mutation but who had a methione/valine combination at codon 129. He reported that both allotypes of PrP-sen contributed to the PrP-res. The presence of this (Val210Ile) mutation and the M/V combination at codon 129, seemed to have no direct effect on the pathological conversion of PrP-sen into PrP-res, i.e. it was secondary.

Eva Mitrova discussed 104 cases in Slovakia of whom 65 had a mutation at codon 200, to check whether it was familial CJD. The patients' profession was studied to investigate possible professional risk factors.

10.2 Gerstmann-Straussler-Scheinker Disease, GSS

Bernardino Ghetti described the characteristics of the GSS prion disease as observed in a family that was first studied in Indiana. This Indiana kindred, GSS-IK, now spans 8 generations and has over 3000 members of whom at least 57 members are known to be affected and more are at risk. There is some resemblance to Alzheimer's disease particularly in the presence of neurofibrillary tangles (although the age of onset, late-30's to mid-60's, is much earlier). The duration after onset, is 5 to 7 years on average which is longer than for other prion diseases.

At least five mutations are associated with GSS. In the Indiana kindred, the dominant mutation is a Serine (S) for a Phenylalanine (F) at codon 198. What is interesting is that this F198S mutation is coupled with a codon for Valine at position 129, that is all family patients have Serine at 198 and Valine at 129. All patients to date have been heterozygous for the F198S mutation, that is, each patient has one PRNP allele that produces mutant PrP and one allele that produces normal PrP. While the mutant allele is always (129V and 198S), the other (normal) allele is variable. There is a difference in the age of onset in the two cases being 44.5 years for V/V at codon 129 and 59.8 for M/V at codon 129 - this has since been confirmed with more cases (Bernardino Ghetti, priv. comm).

Neurofibrillary tangles, NFT, similar to those observed in AD, are a feature of GSS when there are mutations at codons 198 and 217. The main constituent of NFT are paired helical filaments each of which has a diameter of about 10 nm and the pair have a diameter of about 22 to 24 nm. The period of the helical twist is about 70 to 80 nm. The tau protein is the main component.

The PrP-res has bands of 27-29, 18-19 and 8 kDa. The PrP-sen had a prominent band at 33 to 35 kDa before digestion with proteinase-K. The amyloid contained two major peptides of about 11 and 7 kDa which spanned residues 58-150 and 81-150 resp.

Fabrizio Tagliavini used synthetic peptides corresponding to different fragments of human PrP, to study the pathogenesis of tissue changes in GSS.

GSS has been found to have 5 mutations (P102L, P105L, A117V, F198S and Q217R) and two polymorphisms at codon 129(M/V) and at codon 219 (E/K). Different clinical symptoms are found for different mutations. Analysis of the amyloid molecule showed that it contained only V129 showing that amyloid formation was due to the mutant PrP and not the normal one.

To find the PrP residues important for conversion of PrP^c to PrP^{sc} and for aggregation, synthetic peptides were made for the octapeptide repeat region and for residues 89-106, 106-126 and 127-147. Peptides for the 106-147 region could made fibrils but those of the N-terminal region did not. The peptide 106-126 quickly made fibrils like those of GSS, and other tests indicted that the 106-126 region is involved in the PrP^c to PrP^{sc} transition and in clinical changes.

10.3 Fatal Familial Insomnia, FFI

Fatal Familial Insomnia is the latest of the human prion diseases to be identified being first reported by **Elio Lugaresi** and colleagues in 1986. He said that the early symptoms are different from the other human diseases as the patients become apathetic and drowsy due to lack of sleep, so that it is frequently mis-diagnosed. Elio noted that the thalmus is the site in the brain most affected with severe neuronal loss - over 70%. It is in the thalamus that the bio-electric activities appear which are the characteristic of the onset of sleep. An explanation of this is given.

The symptoms are similar to those of Tg mice with no prions - that is, the circadian rhythm and sleep patterns are altered. This suggests that the protease-resistant protein disturbs the same anatomic circuits as the prion protein interacts with.

Some of the indicative features of FFI were discussed by **Katie Sidle**. The mutation at codon 178 where an aspartic acid is changed into asparagine, is a characteristic feature and FFI can be re-classified as an inherited disease. What is unusual is that the missense mutation at 178 can give two diseases - CJD and FFI. The difference has been found to lie in the polymorphism at codon 129 on the same allele. If there is Valine at 129, then CJD results; if methione at 129, the FFI is found.

Secondly, several attempts to transmit FFI have been unsuccessful, even to 18 non-human primates. It would be an important finding if FFI were a non-transmissible prion disease, but Katie reported that they had used transgenic mice with human PrP, and did obtain transmission.

A third unusual feature of FFI was discovered - PrP^{sc} was barely detectable. This was despite the observation of spongiform change and severe astrocytosis. This finding of such small amounts of PrP^{sc} may explain the transmission difficulties. Katie raised the question of whether neurodegeneration comes from loss rather than gain of function of PrP.

10.4 New Variant of Creutzfeldt-Jakob Disease, nvCJD

John Wilesmith gave an account of the Bovine Spongiform Encephalopathy, BSE, epidemic which began about 1985. Epidemiological studies indicated that the source was meat and bone meal which had been used as a protein supplement in cattle feedstuffs. There is evidence that the ban in July 1988, which prohibited the feeding of ruminant feedstuffs to ruminants, has not been completely followed, but the rate of new cases of BSE is declining and it is expected to be insignificant at the turn of the millennium. The question of transmission from mother to calf has been studied and a low risk has been found.

There was a discussion whether BSE originated with sheep as **John Wilesmith** contended or with cows (with "staggers") as **Douglas Morrison** proposed. Finally a **friendly bet** was made of a symbolic one dollar.

Bob Will has updated his report on the new variant of Creutzfeldt Jakob Disease to July 1997. There are now 20 cases of nvCJD in the United Kingdom and one in France. The most outstanding difference from classical CJD, is the age distribution. Normally CJD is a disease of late middle age but 18 of the 20 the UK cases were aged 35 or less at onset and their duration of illness averaged 16 months (range 9 to 38 months). The neuropathological phenotype was novel with the widespread occurrence of florid plaques in addition to the normal spongiform change, astrocytic gliosis and neuronal loss. A search of European files has not found any comparable case.

Bovine offals certainly entered the human food chain in the mid to late 1980's and could be the source of nvCJD. The incubation period for cow to human transmission is unknown. For kuru as an example of oral exposure, times range from 4.5 to 30 years. For iatrogenic CJD by human growth hormone administered by peripheral injection, the incubation periods range from 5 to over 25 years with an average of 13 years. Thus for exposure to BSE-contamination in the mid-eighties, the appearance of cases in the mid-nineties is consistent with the beginning of an epidemic.

Since there were substantial numbers of British cattle and meat and bone meal sent to France in the 1980's, the occurrence of one case of nvCJD in that country, is not inconsistent.

Among the 20 cases of nvCJD, no consistent dietary characteristic has been observed. Dietary exposure in the past is very hard to establish reliably. There is no satisfactory data of the amount of brain and spinal chord that entered the food chain in the UK, however in New Zealand, detailed investigation has shown that some cattle remains, including vertebral column, are used in the production of Mechanically Recovered Meat, MRM. In the UK, MRM is said to have been used in a variety of meat products, including burgers, pies, sausages and pate but there is no hard data available on quantities. He concluded that while no direct evidence links BSE and nvCJD, there is no other reasonable explanation.

John Collinge assessed the risks of the transmission of BSE to humans. He concluded that while the existence of a link between BSE and the new cases of CJD in young people was unproven, it is a likely explanation. In general the transmission of a prion disease between species is usually inefficient (species barrier) and results in extremely long incubation periods.

He said that molecular genetic analysis of PRNP coding and promotor sequence in the first 8 cases, showed no special feature except that all were homozygous for methione at codon 129 of the PrP. On the other hand, transgenic mice having human PrP but not mice PrP, were found not to have any apparent species barrier to human prions and gave short incubation periods irrespective of whether codon 129 had methione or valine.

Douglas Morrison showed a plot of the age distribution for the classical CJD showing a steeply rising curve while the age distribution for the first 8 cases of nvCJD peaked near 25 years and was entirely different with a probability of less than one in a million of being consistent with sporadic CJD. This plot has now been updated to 31 May 1997 with 18 cases.

10.5 Possible New Human Prion Disease?

Andrea LeBlanc reported on the discovery of a Brazilian family that had an atypical form of prion disease. The age of onset was relatively early, 44.8 years and lasted a further 4.2 years. The neuropathology was unusual - there were severe spongiform effects with an absence of gliosis but this was based on only three cases. A mutation was found at codon 183 which was transmitted in a Mendelian way to other family members. This mutation abolishes the first glycosylation site at codon 181. It was noted that all 14 members with the T183A mutation had at least one methione allele at codon 129.

The family history spans 5 generations and has 101 members. At present no member of the family is clinically ill.

From the unusual clinical and neuropathological symptoms plus the novel mutation at codon 183, this could be considered as a new variation of human prion diseases though no such claim was made.

Note added (September 1997). Andrea informs me that this new disease has been named **FASE** for Famial Atypical Spongiform Encephalopathy. The reference is; Nitrini, R., Rosemberg, S., Passos-Bueno, M.R., Lughetti, P., Papadopoulos, M., Carrilho, P.E., Caramelli, P., Albrecht, S., Zatz, M., and LeBlanc, A.C., Annals of Neurology, 42(2): 138-146, (1997).

Luc Montagnier gave a general talk describing how human activities were responsible for new pathogenic agents and he suggested that BSE by causing a new human disease, could be an example of this. Other examples were transmission by blood or sex, of HIV; diseases transmitted by air-borne pollution; and treatment causing selection of resistant micro-organisms. He recommended an increase of research on prions.

10.6 General - Linkage of Clinical Symptoms and Structure of Prions

Pierluigi Gambetti gave a talk of a broad nature where he noted that human prion diseases have been found to occur in many forms and there are many mutations and polymorphisms in the prion molecule. The important conclusion is that there is a measure of correlation between these and characteristics of the disease.

He classifies the diseases firstly into 3 types; sporadic, inherited and by infection; secondly each of these types can further present one or more of four basic clinico-pathological phenotypes. This classification is compared with the mutual and normal allele, the mutation and the nature of PrP-res. These give many clear effects whose nature needs to be understood. For example, with inherited diseases, the codon 129 has a dual effect on the nature of the disease according to whether there is a methione or a valine. For the mutant allele(e.g. with F198S for GSS) the clinical and neurological features change, whereas on the normal allele, according to whether it is a M or V determines the age of onset and the duration of the disease - which may then modify the lesions and also the type of PrP-res.

In conclusion, the mechanisms underlying the appearance of the disease, are multiple and interconnected. This is in contrast to Alzheimer's disease where there is more consistency.

11. YEAST- EVIDENCE FOR NON-NUCLEIC ACID TRANSMISSION

Two proteins in yeast replicate in a non-genetic (non-chromosomal) manner. **Reed Wickner** reported that they have the characteristics of prions and proposed the "Yeast Prion hypothesis". He described experiments supporting this. Further it appears that the yeast proteins give more fundamental understanding of prions than do mammalian prions. Also since it is probable that all viruses in yeast have already been identified and do not cause prion-like effects, it is unlikely that a undiscovered virus could be responsible for yeast prions.

In yeast, there are two non-Mendelian genetic elements, [PSI] and [URE3] whose characteristics can best be understood if they are considered as the abnormal (prion) form of the normal proteins Sup35p and ure2p respectively. This "Yeast Prion hypothesis" has now been supported by biochemical data that two normal proteins are altered to their corresponding abnormal strains. Further Reed Wickner et al. have defined a prion-producing domain in the normal Ure2p which mediates prion alteration.

Reed proposes 3 genetic characterisations for deciding if there is a Mendelian genetic element in a yeast protein or if there is nucleic acid replication,

(1) Reversible curability; if the protein is cured to remove the abnormal form, it can arise again at some low frequency

(2) overproduction of the normal form increases the frequency of abnormal prion production since there are more molecules available to undergo spontaneous transformation to the abnormal form. For example, overproduction of Ure2p results in a 20 to 100 fold increase in the rate of production of [URE3].

(3) phenotypes the same for (a) the presence of the prion form of the protein and (b) a defect in the chromosomal gene for the protein.

With mammals, treatment of abnormal prions with proteinase K leaves resistant residues and this is often considered a fundamental characteristic. However Reed reported that while the 40 kDa Ure2p is normally completely degraded by proteinase K, Ure2p in a [URE3] strain, gave relatively stable fragments of 30 and 32 kDa, and it was eventually degraded. Thus Reed indicates that protease resistance is not an essential feature of the "Prion Only" hypothesis.

Similarly Sup35p is easily digested by proteinase but in strains carrying [PSI] it is resistant and gives aggregates. These aggregates prevent Sup35p fulfilling its normal function. It has been found that the heat-shock protein Hsp104 can disaggregate [PSI] - this is taken as evidence that the mechanism of prion formation is aggregation and is not some covalent change in the Sup35 molecule.

When the C-terminal region of the normal ure2p protein was deleted, it was found that the remainder was 100 times more efficient at creating the prion [URE3]. The N-terminal 65 amino acid residues of ure2p was called the prion-inducing region as it gave the same high rate of [URE3] induction.

Since yeast mates frequently, the prion mechanism might be considered as a sexually-transmitted disease and hence it may be expected that more examples of prions will be found in other yeasts or fungi.

Sergei Inge-Vechtomov reported on studies of protein "prionization" (change from the normal to abnormal form) in the SUP35 - [PSI] system, it being noted that many experiments can be more simply studied in yeast where there is no problem of human infectivity. Two independent substitutions of the gene SUP35 in the same strain of Saccharomyces yeast, behave as recessive sup35 mutations of different suppressive efficiencies. Thus inheritance characteristics are not dependent on any other genetic factors besides the sup35 mutation. Further estimation of the specificity of the prionization process was studied. It was found that the quaternary structure influences the process of protein prionization.

12. PURIFICATION OF SAMPLES

Heino Diringer has repeatedly emphasised that the prion samples used are not completely purified and may contain significant amounts of nucleic acids which could be the transmissible agent rather than a prion protein. The practical problems of purification are discussed. His earlier work had suggested that the 12 nm diameter spheres found in the brains of CJD patients were the virus-like particles which were the causative agent of the CJD, but **Detlev Riesner** showed that they were oligomers of PrP 27-30 and definitely not "virus-like".

13. EARLY WARNING TEST FOR ANIMALS

While humans can indicate through speech etc., early symptoms of TSE infections before clinical symptoms appear, animals cannot do so. Using maze testing, **Boris Semenov** studied the early behavioural disorders of mice injected with several viruses (not TSE) which cause brain diseases. It was found that many mice showed abnormal behaviour in the maze, before clinical symptoms were detectable.

14. DRUGS TO CURE OR POSTPONE PRION DISEASES

At present prion diseases have been found to be invariably fatal but based on research, drugs are being looked for to prolong the incubation period or perhaps to cure the disease.

Dominique Dormont reported on studies of the effect of polyene antibiotics. Amphoterin B (Am B) had been found useful but had toxic effects so it and MS-8209, another antifungal and antiviral drug which was less toxic, were tried on scrapie infections. It was found that if increased doses of MS-8209 were given from the time of infection continuously until death, then the hamster and mice survival times were increased by as much as 100% although the period with clinical symptoms was unchanged. It was found that the rate of PrP-res accumulation was decreased proportionally to the dose. However when BSE was introduced the increase in survival time was only 5 to 15% showing that efficiency of the drugs varies with the strain. When the two drugs were given at a later preclinical stage, the increases in survival times were less, and they were greater for higher doses. Further work is needed particularly as it is interesting to note that Am B concentrates in the peripheral organs (liver, spleen, and kidney) and very little in the brain.

Hans Kretzschmar has experimental results which suggest that prions are toxic because of oxidative stress and hence the use of anti-oxidants such as vitamin E may block the effect of prions on neurones.

Luc Montagnier described in a more general manner, the problems caused by oxidation of proteins and recommended taking anti-oxidants.

Reed Wickner has pointed out that while there is no cure for scrapie-type illnesses, there are cures for [URE3] and [PSI] in yeast. These techniques are being tried for mammalian prion diseases.

15. COMPARISON OF PRION DISEASES WITH ALZHEIMER'S DISEASE

David Westaway gave a second talk on Alzheimer's Disease, AD. He first described it, gave a thought-provoking comparison with some prion diseases, and reported on experiments using transgenic mice which had presenilin 1, PS1.

So far factors causing AD are a) missense mutations in the presenilin 1, PSI, gene on chromosome 14, b) missense mutations in the presenilin 2, PS2, gene on chromosome 1, and c) rare mutations on the APP gene on chromosome 21. Also loci where AD is favoured, d) the e4 allele on the apoE gene, and e) chromosome 14.

The most frequent causes of early onset familial Alzheimer's disease, FAD, are PS1 and PS2 which cause long-tailed amyloid β-peptides ending at residues 42 or 43 (Aβ42), These were studied using transgenic mice with human PS1. It was found that the mutant PS1 transgenes act to programme over-production of these Aβ42 peptides in the brain. Further this biochemical difference was present at 2 to 4 months age, well before any detectable neuropathologic lesions.

Two of the main features of AD are accumulations of extracellular amyloid deposits (mainly Aβ peptide) and accretion of neurofibrillary tangles which consist principally of modified tau proteins which hold microtubules in place. Supporters of the dominance of these are called baptists and tauists respectively. The baptists stress mutations in the amyloid precursor protein in familial AD and early amyloid deposits in Down's syndrome cases. The tauists note the poor correlation between amyloid deposits and cognitive decline whereas there is a good correlation with tangle formation. David considers that the debate is almost over with the baptists being preferred.

The comparison of Alzheimer's with human prion diseases is very informative. First it is emphasised that the crucial difference is that prion diseases are transmissible while Alzheimer's is not. There are 3 informative tables of comparisons;

a) Neuropathology and epidemiology of AD, CJD and GSS
b) Pathological features in familial prion diseases and some overlaps with FAD
c) Pathogenic proteins in AD and CJD.

It can be seen that GSS is intermediate between Alzheimer's Disease and CJD. Spongiform change, often considered a characteristic of prion diseases, may be absent or mild in certain mutations of GSS. It may also be an important clue that PS1, PS2, APP and PrP are all associated with the cell membrane.

Maurizio Pocchiari compared amyloidosis in prion diseases, Alzheimer's Disease, and Transthyretin, TTR. A table is given with various disease types of TTR, corresponding mutations (where known) and the composition of the amyloid protein. Similarities are noted and the possibility is explored that all three are "conformational disease" where folding plays a critical role. However he notes that only prion diseases are transmissible.

From experience with other similar scientific controversies, I would expect that both baptists and tauists are right and that they turn out to be different manifestations of AD. While there are strong resemblances between Alzheimer's and prion diseases, the fact that only prion diseases can be transmitted, may be fundamental.

Acknowledgements

It is a pleasure to thank several participants at the meeting - Paul Brown, Byron Caughey, Bruce Chesebro, Heino Diringer, Bernadino Ghetti, Andrea LeBlanc, and Charles Weissmann - for kindly reading part or all of the text and for their constructive comments. However the responsibility for errors is the author's alone.

POLYMORPHIC GENOTYPE MATCHING IN ACQUIRED CREUTZFELDT-JAKOB DISEASE: AN ANALYSIS OF DONOR/RECIPIENT CASE PAIRS

Paul Brown[1], Larisa Cervenáková[1], Lisa McShane[2],
Paul Kleihues[3], Jean-François Foncin[4], George Collins[5],
Frank Bastian[6], Lev G. Goldfarb[1], and D. Carleton Gajdusek[1]

[1]Laboratory of CNS Studies, NINDS, NIH, Bethesda MD
[2]Biometric Research Branch, NCI, NIH, Bethesda MD
[3]Institut für Neuropathologie, Universität Zurich, Zurich
[4]INSERM U-106, Hôpital de la Salpêtrière, Paris
[5]Department of Pathology, State University of New York
  Health Science Center, Syracuse NY
[6]Department of Pathology, University of South Alabama Medical
  Center, Mobile AL

INTRODUCTION

Two different expressed polymorphisms have been identified in the chomosome 20 (PRNP) gene encoding the amyloid precursor protein of infectious cerebral amyloidosis: codon 129 may specify either methionine or valine (at an allelic ratio of 0.62 to 0.38 in Caucasians), and a 24 base pair deletion may occur in the region of octapeptide coding repeats between codons 51 and 91 (at a frequency of about 2% in Caucasians). Although neither polymorphism is by itself pathogenic, codon 129 homozygosity is over-represented in both sporadic and iatrogenic forms of Creutzfeldt-Jakob disease (CJD) [1-4], and influences certain phenotypic features of sporadic and familial forms of disease [5-8], and octapeptide coding deletions can occur in association with CJD and other degenerative diseases [9, 10]. It therefore seemed of more than passing interest to learn whether polymorphic 'matching' could be demonstrated in those rare instances of proven or possible human-to-human contact infection in which tissue from identifiable donors and recipients was still available. We here report our findings in seven patients involved in three separate occurrences of surgically transmitted CJD, and in one conjugal pair in which both spouses died of CJD.

MATERIALS AND METHODS

Cases

The three published instances of iatrogenic surgical disease transmission for which we were able to retrieve tissue from both donors and recipients resulted from a corneal graft [11], stereotactic electroencephalographic (EEG) recordings [12], and a neurosurgical operation [13]. All seven cases had typical clinical and neuropathological features of CJD, and brain tissue from both EEG cases and the corneal transplant case transmitted disease to experimental primates [14]. None had a history of familial disease.

As the conjugal pair has not been previously reported, a summary account is provided here. During the Autumn of 1985, a 54 year-old husband suffered a gradual loss of memory and job capability, together with visual failure and ataxia, progressing within the next 12 months to global dementia, mutism, permanent myoclonus, and spasticity. An EEG at this time showed periodic triphasic slow wave activity. He died after a prolonged period of bedridden helplessness in June 1987, approximately 20 months after the onset of illness. A post-mortem biopsy of the frontal lobe revealed extensive spongiform changes, particularly in the mid-layers of the grey matter, associated with severe neuronal loss and gliosis.

In January 1992, the husband's then 55-year old widow experienced the sudden onset of confusion that rapidly progressed and became associated with ataxia, myoclonus, and primitive reflexes. An EEG showed periodic discharges, and a brain biopsy showed spongiform encephalopathy. The biopsy was complicated by a subarachnoid hemorrhage, and a week later she became comatose and died (one month after the onset of her illness). Autospy revealed a hematoma at the site of the biopsy and a diffuse subarachnoid hemorrhage. Microscopic examination revealed widespread spongiform changes throughout the brain, associated with astrocytosis and mild to moderate neuronal loss. Neither she nor her husband had a family history of neurological disease.

Molecular genetic analysis

DNA was extracted from either frozen or formalin-fixed, paraffin embedded brain tissue. The translated region of the PRNP gene was amplified in two overlapping fragments by PCR using *Taq* polymerase and oligonucleotide primers as previously described for frozen tissues [15]; for paraffin-embedded tissues, we used the same pair of primers for the 3' fragment, but a different pair of primers producing a slightly shorter 5'-fragment: 5'-TACTGAGCGGCCGCATGCTGGTTCTCTTTGT-3' and 5'-TACTGATCTAGAAATGTATGATGGGCCTGCTCA-3'. Specimens were examined for deletions and the codon 129 genotype by electrophoretic visualization of the PCR fragments with and without exposure to *Mae II* restriction endonuclease [16].

Statistical analysis

We evaluated four different types of possible CJD donor/recipient matches: 1) a homozygous allelic codon 129 match; 2) any type (homozygous or heterozygous) of codon 129 match; 3) deletion status (present or absent); and 4) a combined codon 129-deletion status match (overall polymorphic homology). We estimated prevalence rates for each polymorphism in the general population from pooled published [2, 3, 9, 17], and unpublished data from our own laboratory: 261 individuals tested for the codon 129 genotype; 135 (52%) methionine/valine, 97 (37%) methionine/ methionine, 29 (11%) valine/valine; and 624 individuals tested for the presence of deletions (12 = 1.9%).

The probability that two individuals randomly selected from the general population will match for a given genotype is calculated as the square of the prevalence rate for the genotype. The probability that they match with regard to any one of several specified mutually exclusive genotypes is equal to the sum of the match probabilities for the individual genotypes. Assuming that deletions occur independently of the codon 129 genotype, the probability of both a codon 129 match and a deletion status match is equal to the product of the two matching probabilities.

Using our estimated prevalence rates, the probability of a match for two randomly chosen individuals from the general population was calculated to be 0.150 for a homozygous codon 129 match, 0.418 for any type of codon 129 match, 0.962 for a deletion status match, and 0.402 for a combined codon 129 and deletion status match. Assuming that the five donor/recipient pairs can be treated as independent occurrences (even though the EEG two recipients shared the same donor) the numbers of matches follow binomial distributions, and allow probability values (p-values) to be calculated for tests of the null hypothesis that the proportion of matches in the CJD group is not greater than expected in the general population.

Because the prevalence of homozygous genotypes is higher among sporadic CJD cases (i.e., donors) than the general population [1, 18], an increased number of homozygous matches might simply reflect this increased prevalence rather than the effect of homozygous matches. To adjust for these different genotypic prevalence rates and specifically test for the effect of genetic homology, we also performed an exact conditional statistical analysis using a permutation test [19].

The idea behind a permutation test is to consider all possible rearrangements of subjects into groups, and then to determine if the arrangement actually observed in the data set is unusual in the sense of suggesting group effects. In our genetic matching example, we considered all possible rearrangements of subjects into donor/recipient pairs, and under each of these rearrangments we computed the number of pairs with genetic matches. The p-value was calculated from the porportion of rearrangments yielding an equal or greater number of matches than we observed in our data set.

RESULTS

Codon 129 genotypes and the presence or absence of deletions for each of the nine patients are shown in Table 1. Three of five possible donor-recipient matches had identical codon 129 genotypes (all were homozygous); however, the corneal graft recipient and one spouse of the conjugal pair also had a 24 base pair deletion, so that an identical combination of polymorphic genotypes was present in only one of the five possible matchings.

The proportions of genotypic matches for CJD pairs did not significantly differ from those expected for a pair of individuals chosen at random from the general population for three of the four tested types of matches: any codon 129 match (p = 0.349); deletion status match (p = 0.999); and a combined codon 129 and deletion status match (p = 0.924).

The nonsignificant statistical findings for these three types of matches using the binomial test could be the result of its diminished power when applied to such small data sets. In situations where data is very difficult to obtain, one may be willing to accept an 80% chance (power) to detect a statistically significant difference. As an example, the observed proportion of any type of codon 129 match in our data set was 3/5, which is approximately 50% higher than that expected in the general population. However, to have an 80% power of detecting a statistically significant 50% increase, we would still need 34 CJD donor/recipient pairs, and even higher numbers of pairs for deletion status or combined codon 129/deletion status matches.

Table 1. PRNP polymorphic genotypes of donor and recipient CJD patients in three iatrogenic surgical incidents, and in a CJD conjugal pair.

Case Identification	Age/ Sex	Incubation interval (months)	Duration of illness (months)	Codon 129	24 bp deletion
Corneal graft					
Donor	55 M		2	Met/Met	No
Recipient	55 F	18	8	Met/Met	Yes
Stereotactic EEG					
Donor	69 F		7	Met/Met	No
Recipient #1	23 F	20	23	Met/Val	No
Recipient #2	17 M	16	8	Met/Met	No
Neurosurgery					
Donor	59 F		2	Met/Met	No
Recipient	46 M	28	6	Met/Val	No
Conjugal pair					
Husband	54 M		9	Met/Met	No
Wife	55 F	$\geq55^1$	1	Met/Met	Yes

[1] Interval between husband's death and onset of wife's illness.

For the fourth type of match (identical homozygous codon 129 genotypes), the proportion of donor/recipient pairs was significantly higher than would be expected in the general population (p = 0.027), but when the permutation test which adjusts for the genotypic prevalence rates in sporadic (donor) CJD patients was applied to this statistic, the difference was not significant (p = 0.833). Thus, we do not have sufficient statistical evidence to claim that the increased proportion of homozygous codon 129 matches was due to an effect of genetic homology rather than to an increased prevalence of the homozygous genotype in the CJD group.

DISCUSSION

The corneal graft recipient and stereotactic EEG patients are the only known examples of experimentally proven iatrogenic CJD. The neuro-surgical case is a highly likely but unproven additional example of iatrogenic disease, having been operated in the same suite as a patient with CJD three days earlier. Histological brain sections from patients involved in two other neurosurgical incidents [20] yielded DNA that was too degraded to analyze, and the much larger group of iatrogenic cases due to contaminated pituitary hormones and dura mater grafts could not be used because the sources of infection were tissues pooled from multiple unidentified donors.

We also chose to include the only known instance of conjugal CJD, even though infection of one spouse by the other is conjectural, because the conjugal occurrence of a disease as rare as CJD mandates consideration as contact transmission, especially since the minimum five year interval between disease onsets is consistent with incubation periods in both experimentally transmitted and iatrogenic CJD [14, 21]. The Austrian spouses reported in 1972 to have died of CJD [22], have recently been shown not to have had CJD on clinical, pathological, and immunohistological grounds [23].

This small group of patients thus comprises all known instances of proven or possible horizontal transmission of CJD between humans for whom the disease donor could be precisely identified, and for whom tissue was available for molecular genetic analysis. Our results (one of five possible identical donor-recipient genotypes) suggest that overall genotypic homology is not of overriding importance to horizontal disease transmission in humans, and is consistent with our observation that the ease with which CJD can be experimentally transmitted from humans to non-human primates does not correlate with the degree of PRNP homology between humans and other primate species [24]. The occurrence of homozygous methionine codon 129 genotypes in three of five donor/recipient pairs of iatrogenic and conjugal CJD appears to result from the increased prevalence of this genotype in both sporadic and iatrogenic cases of CJD. It is possible that direct introduction of the infectious agent into the nervous system in our iatrogenic cases may have overcome the influence of host susceptibility genotypes that affect peripheral routes of infection.

Because the primary structure of the amyloid precursor protein is identical in normal and diseased brain tissue, it has been concluded that the difference between the two isoforms must result from post-translational events influencing secondary, tertiary, or even quaternary protein patterning and aggregation. The consequence of deletions remains problematic, but the polymorphism at position 129 clearly influences both susceptibility to disease and its phenotypic features. We need to learn much more about the actual (as opposed to predicted) conformational changes resulting from amino acid alterations if we are to make sense of the molecular genetic component of CJD susceptibility.

REFERENCES

1. M.S. Palmer, A.J. Dryden, J.T. Hughes and J. Collinge, Homozygous prion protein genotype predisposes to sporadic Creutzfeldt-Jakob disease, *Nature* 352:340-342 (1991).

2. J. Collinge, M.S. Palmer and A.J. Dryden, Genetic predisposition to iatrogenic Creutzfeldt-Jakob disease, *Lancet* 337:1441-1442 (1991).

3. J.P. Deslys, D. Marcé and D. Dormont, Similar genetic susceptibility in iatrogenic and sporadic Creutzfeldt-Jakob disease, *J Gen Virol* 75:23-27 (1994).

4. P. Brown, L. Cervenáková, L.G. Goldfarb, W.R. McCombie, R. Rubenstein, R.G. Will, M. Pocchiari, J.F. Martinez-Lage, C. Scalici, C. Masullo, G. Graupera, J. Ligan and D.C. Gajdusek, Iatrogenic Creutzfeldt-Jakob disease: an example of the interplay between ancient genes and modern medicine, *Neurology* 44:291-293 (1994).

5. K. Doh-Ura, T. Kitamoto, Y. Sakaki and J. Tateishi, CJD discrepancy, *Nature* 353: 802-803 (1991).

6. M. Poulter, H.F. Baker, C.D. Frith, M. Leach, R. Lofthouse, R.M. Ridley, T. Shah, F. Owen, J. collinge, J. Brown, J. Hardy, M.J. Mullan, A.E. Harding, C. Bennett, R. Doshi and T.J. Crow, Inherited prion disease with 144 base pair gene insertion. 1. Genealogical and molecular studies, *Brain* 115:675-685 (1992).

7. S.F. Dlouhy, K. Hsiao, M.R. Farlow, T. Foroud, P.M. Conneally, P. Johnson, S.B. Prusiner, M.E. Hodes and B. Ghetti, Linkage of the Indiana kindred of Gerstmann-Sträussler-Scheinker disease to theprion protein gene, *Nature Genet* 1:64-67 (1992).

8. L.G. Goldfarb, R.B. Petersen, M. Tabaton, P. Brown, A.C. LeBlanc, P. Montagna, P. Cortelli, J. Julien, C. Vital, W.W. Pendelbury, M. Haltia, P.R. Wills, J.J. Hauw, P.E. McKeever, L. Monari, B. Schrank, G.D. Swergold, L. Autilio-Gambetti, D.C. Gajdusek, E. Lugaresi, Fatal familial insomnia and familial Creutzfeldt-

Jakob disease: disease phenotype determined by a DNA polymorphism, *Science* 258:806-808 (1992).

9. M.S. Palmer, S.P. Mahal, T.A. Campbell, A.F. Hill, C.L. Sidle, J.-L. Laplanche and J. Collinge, Deletions in the prion protein gene are not associated with CJD, *Hum Molec Genet* 2:541-544 (1993).

10. L. Cervenáková, P. Brown, P. Piccardo, J.L. Cummings, J. Nagle, H.V. Vinters, P. Kaur, B. Ghetti, D.C. Gajdusek and L.G. Goldfarb, 24-nucleotide deletion in the PRNP gene: analysis of associated phenotypes, in *Transmissible Subacute Spongiform Encephalopathies: Prion Diseases*, L .Court and B. Dodet, eds., Elsevier, Amsterdam-Paris (1996).

11. P. Duffy, J. Wolf, G. Collins, A.G. DeVoe, B. Streeten and D. Cowen, Possible person-to-person transmission of Creutzfeldt-Jakob disease, *New Engl J Med* 290:692 (1974).

12. C. Bernoulli, J. Siegfried, G. Baumgartner, F. Regli, T. Rabinowics, D.C. Gajdusek and C.J. Gibbs Jr, Danger of accidental person-to-person transmission of Creutzfeldt-Jakob disease by surgery, *Lancet* i:478-479 (1977).

13. J. Foncin, J. Gaches, F. Cathala, E. El Sherif and J. Le Beau, Transmission iatrogène interhumaine possible de maladie de Creutzfeldt-Jakob avec atteinte des grains du cervelet, *Rev Neurol (Paris)* 136:280 (1980).

14. P. Brown, C.J. Gibbs Jr, P. Rodgers-Johnson, D.M. Asher, M.P. Sulima, A. Bacote, L.G. Goldfarb and D.C. Gajdusek, Human spongiform encephalopathy: the NIH series of 300 cases of experimentally transmitted disease, *Ann Neurol* 35:513-529 (1994).

15. L.G. Goldfarb, P. Brown, W.R. McCombie, D. Goldgaber, G.D. Swergold, P.R. Wills, L. Cervenáková, H. Baron, C.J. Gibbs Jr and D.C. Gajdusek, Transmissible familial Creutzfeldt-Jakob disease associated with five, seven, and eight extra octapeptide coding repeats in the *PRNP* gene, *Proc Natl Acad Sci (USA)* 88:10926-10930 (1991).

16. L.G. Goldfarb, P. Brown, D. Goldgaber, D.M. Asher, N. Strass, G. Graupera, P. Piccardo, W.T. Brown, R. Rubenstein, J.W. Boellaard and D.C. Gajdusek, Patients with Creutzfeldt-Jakob disease and kuru lack the mutation in the PRIP gene found in Gerstmann-Sträussler syndrome, but they show a different double-allele mutation in the same gene, *Am J Hum Genet* 45 (suppl):A189 (1989).

17. C.L. Vnencak-Jones and J.A.I. Phillips, Identification of heterogeneous PrP gene deletions in controls by detection of allele-specific heteroduplexes (DASH), *Am J Hum Genet* 50:871-872 (1992).

18. P. Brown, K. Kenney, B. Little, J. Ironside, W. R, L. Cervenáková, A. San Marin, J. Safar, R. Roos, S. Harris, M. Haltia, C.J. Gibbs Jr and D.C. Gajdusek, Intracerebral distribution of infectious amyloid protein in spongiform encephalopathy, *Ann Neurol* 38:245-253 (1995).

19. P. Good. *Permutation Tests - A Practical Guide to Resampling Method for Testing Hypotheses*, Springer Verlag, New York (1994).

20. R.G. Will and W.B. Matthews, Evidence for case-to-case transmission of Creutzfeldt-Jakob disease, *J Neurol Neurosurg Psychiat* 45:235-238 (1982).

21. P. Brown, M.A. Preece and R.G. Will, 'Friendly fire' in medicine: hormones, homografts, and Creutzfeldt-Jakob disease, *Lancet* 340:24-27 (1992).

22. V.K. Jellinger, F. Seitelberger, W.D. Heiss and W. Holezabek, Konjugale form der subakuten spongiösen enzphalopathie, *Wiener Klin Wochenschr* 84:245-249 (1972).

23. J.A. Hainfellner, K. Jellinger and H. Budka, Testing for prion protein does not confirm previously reported conjugal CJD, *Lancet* 347:616-617 (1996).

24. L. Cervenáková, P. Brown, L.G. Goldfarb, J. Nagle, K. Pettrone, R. Rubenstein, C.J. Gibbs Jr and D.C. Gajdusek, Infectious amyloid precursor gene sequences in primates used for experimental transmission of human spongiform encephalopathy, *Proc Natl Acad Sci (USA)* 91:12159-12162 (1994).

25. S.B. Prusiner, Transgenetic investigations of prion diseases of humans and animals, *Phil Trans R Soc Lond B* 339:239-254 (1993).

HUMAN PRION PROTEIN GENE MUTATION AT CODON 183 ASSOCIATED WITH AN ATYPICAL FORM OF PRION DISEASE

Ricardo Nitrini, M.D.[1] Sergio Rosemberg, M.D.[2],
Maria Rita Passos-Bueno, Ph.D.[3],
Luis S. Texeira da Silva, M.D.,[4] Paula Iughetti[3],
Maria Papadopoulos, B.Sc.[5,6], P.M. Carrilho[1],
Paulo Caramelli, M.D.[1], Steffen Albrecht, M.D.[7],
Mayana Zatz, Ph.D.[3], and Andréa LeBlanc, Ph.D.[5,6]

[1]Departments of Neurology and [2]Pathology of the Faculty
of Medicine, and at the [3]Department of Biology,
University of Sao Paulo, Brazil.[4] Dept. of Internal
Medicine, State U. of Londrina, Londrina, Brazil.
[5]Bloomfield Center for Research in Aging, Sir Mortimer B.
Davis JGH, [6]Departments of Neurology and [7]Pathology,
McGill University, Montreal Canada

ABSTRACT

We identified a Brazilian family with an autosomal dominant
fronto-temporal form of dementia. The disease had a mean age of
onset of 44.8 ± 3.8 years and a duration of 4.2 ± 2.4 years.
Neuropathological examination of three cases showed neuronal loss
and severe spongiform change in the deep layers of the cortex and
in the putamen in absence of the usual gliotic reaction associated
with most neurodegenerative diseases. Prion protein
immunoreactivity was detected as diffuse staining in the otherwise
normal cerebellum, and as prion plaque-like deposits in the
putamen. A mutation at codon 183 of the prion protein gene was
identified in two affected individuals and transmitted in a
Mendelian fashion to 12 family members. This new inherited variant
of human prion diseases represents an unusual clinico-pathological
entity of prion diseases.

INTRODUCTION

Inherited human prion diseases, Creutzfeldt-Jakob disease (CJD),
Gerstmann-Straussler-Scheinker disease (GSS) and Fatal familial
insomnia (FFI) are distinguished by particular genetic mutations,
and clinical and pathological characteristics [1]. Clinically,
CJD is identified by extrapyramidal, pyramidal and cerebellar
signs associated with rapidly progressing dementia [2]. GSS
presents as either cerebellar ataxia or gradual dementia [3], and
FFI with sleep, autonomic, endocrine and motor problems [4,5]. The
pathological hallmarks of prion diseases comprise spongiform
change, neuronal loss and gliosis [6]. Biochemically, the prion

protein from brains affected by prion protein diseases becomes proteinase K resistant. In absence of available fresh or frozen tissue, resistant prion protein immunoreactivity after formic acid treatment and hydrolytic autoclaving also represents a major marker of prion protein diseases [6].

CASE REPORT

A Brazilian patient of 47 years initially developed progressive apathy and memory problems and within three years presented with severe dementia, slurred speech, Parkinsonian syndrome, and frontal release signs while his general physical condition, coordination and limb reflexes, electroencephalogram (EEG), and cerebro-spinal fluid analysis remained normal. Diffuse cortical atrophy and moderate ventricular enlargement were observed by brain computed tomography (CT). He became bedridden five years after the onset of symptoms and after eight years was mute, unable to obey simple commands, and suffered frequent myoclonic jerks of both hands and arms. He died one year later, nine years after the onset of the disease. His family history indicated an inherited disease and pathological examination of the brain showed the classical sign of spongiform change and neuronal loss normally associated with prion diseases. Therefore, we proceeded to study if this family suffered from an inherited form of prion disease.

RESULTS

The information on the family spans five generations consisting of 101 individuals. The maternal grandfather of the propositus also reported to be affected came to Brazil in 1912 from the province of Malaga, Spain. The father of the propositus is Italian. In total, nine individuals are clinically confirmed affected of which three were further examined pathologically. Eight other individuals were reported affected by family members. To date, there is no clinically affected individual alive.

Within the clinically confirmed cases, there are 7 males and 2 females. The mean age of onset is 44.8 ± 3.8 years; the mean age of death is 49.0 ± 5.1 years and the mean duration of the disease is relatively long at 4.2 ± 2.4 years. Personality changes are the

Table I.   Clinical manifestation of affected individuals

Symptoms	# patients	Manifestation
Personality changes	9/9	early
Memory problem	4/9	early
Progressive Dementia	9/9	later
Aggressivity	4/9	later
Hyperorality	5/9	later
Verbal Stereotypies	5/9	later
Parkinsonian Syndrome	6/9	later

prominent early manifestation reported clinically and four of the nine cases also had memory problems (Table I). After the initial onset, rapidly progressive dementia occurred in all cases. Four of these also showed aggressivity, five showed hyperorality, five had verbal stereotypies, six showed Parkinsonian syndrome characterized by low amplitude and high frequency tremors of both hands at rest, reduced swing of the arms and walking with a slight flexion of the trunk.

Although no other live family member suffered from this disease, pathological material was available from the mother of the patient and a cousin, both affected by the disease (Fig 1).

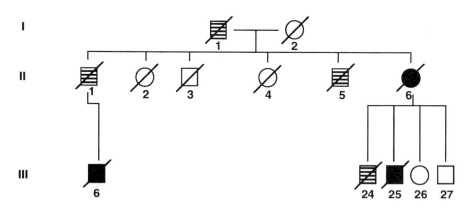

Fig1. Partial pedigree of the family. Solid symbols are pathologically confirmed cases, and symbols containing horizontal lines indicate subjects reported affected by family members. Diagonal line indicates deceased individuals.

In the three cases pathologically examined, III-6 was from a biopsy and II-6 and III-25 from autopsy material. Each displayed spongiform change in layers IV, V, VI as well as in layer II in the most affected areas of the cortex. Neuronal loss was evident in these areas. A schematic diagram of the areas affected by spongiform change in case II-6, the propositus' mother who lived two years after the onset of the disease, and her son, case III-25, who lived 9 years after the onset of the disease is shown in Fig. 2. There is a severe involvement of the frontal and temporal cortex in both cases. In case II-6, there is moderate spongiform change of the striatum whereas III-25 showed severe spongiform change of the putamen, caudate nucleus and claustrum and moderate spongiform change in the globus pallidus and entorhinal cortex. The more severe pathology of case III-25 may be a direct consequence of the extended duration of his disease. Although normal glial fibrillary acidic protein positive astrocytes were detected in tissue sections of case III-25, both

27

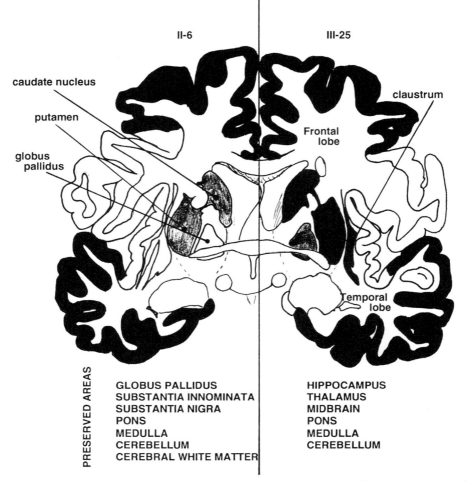

II-6   III-25

caudate nucleus

putamen

globus
pallidus

claustrum

Frontal
lobe

Temporal
lobe

PRESERVED AREAS

GLOBUS PALLIDUS   HIPPOCAMPUS
SUBSTANTIA INNOMINATA   THALAMUS
SUBSTANTIA NIGRA   MIDBRAIN
PONS   PONS
MEDULLA   MEDULLA
CEREBELLUM   CEREBELLUM
CEREBRAL WHITE MATTER

Fig. 2. Schematic diagram of spongiform change. Areas affected by spongiform
change in II-6 and III-25. Dark areas indicate severe and shade areas moderate
spongiform change.

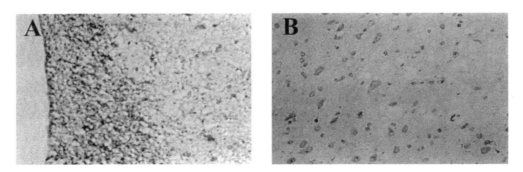

Figure 3. Glial fibrillary acidic protein immunocytochemistry of subependymal
region (Fig. 3A) and putamen (Fig. 3B). GFAP immunoreactivity was detected
with a Vectastain peroxidase kit.

28

in the subependymal and subpial region, there is a remarkable absence of gliosis in the affected areas of the brain (Figure 3). Whereas astrocytes were clearly detected in areas where they are normally present such as the subependymal region (Fig. 3A), the astrocytes were absent in areas strongly affected by spongiform change such as the putamen (Fig. 3B). A few astrocytes were detected in the cortical area. These stained only weakly for GFAP, had few arborizations and did not display the usual robust reactivity to GFAP. None were detected in the striatum. Therefore, we conclude that minimal gliosis accompanies the severely affected areas of case III-25, a surprising finding considering that this individual lived 9 years with his disease.

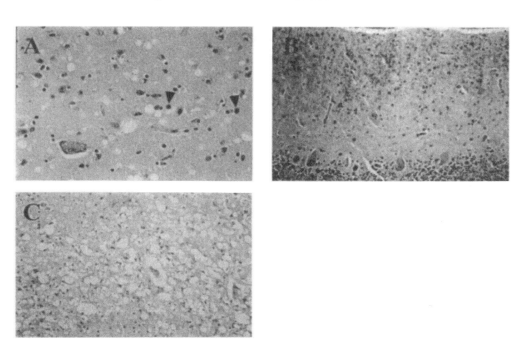

Figure 4. Immunoreactivity of prion protein using 3F4 antibody. Tissue sections were deparafinized, treated with formic acid for one hour followed by 10 minutes of hydrolytic autoclaving. Sections were blocked with 10% fetal goat serum and incubated with 3F4 monoclonal antibody overnight. Immunoreactivity was detected using Vectastain ABC immunoperoxidase kit. Fig 4A shows plaque-like deposits (arrow heads) in putamen, 4B shows diffuse cerebellum staining and Fig. 4C, the absence of PrP immunoreactivity in area severely affected by spongiform change.

Prion protein immunoreactivity was observed with the monoclonal 3F4 antibody after submitting the sections to a pre-treatment of one hour with formic acid followed by hydrolytic autoclaving (protocol kindly provided by P. Gambetti). Plaque-like deposits were detected but only in one area affected by spongiform change, the putamen (Fig. 4A); other strongly affected areas were negative (Fig 4B). In the cerebellum which seemed normal when examined by H&E, a strong and diffuse immunoreactivity was similar to that depicted in other prion diseases (Fig. 4C). Immunostaining of sections of the hippocampus, frontal, temporal, occipital, and parietal lobes and midbrain did not show any prion protein immunoreactivity (not shown). Unfortunately, we do not have frozen tissue to study proteinase K resistance of the prion protein.

Sequencing the prion protein coding exon after PCR amplification of DNA extracted from cases II-6 and III-25 identified a mutation at codon 183 of the prion protein gene in both these individuals (Fig 5). The codon 183 mutation results in a non-conservative substitution of a threonine for alanine (T183A) and abolishes the first of two N-linked glycosylation (N-X-T) sites of the prion protein (Fig. 6).

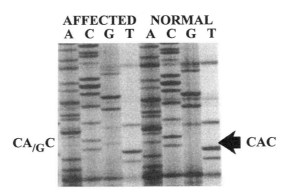

Figure 5.   Autoradiogram of direct PCR sequencing of PrP gene from case III-25 and a normal control shows heterozygosity for codon 183 in III-25.

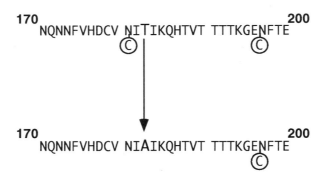

Figure 6.   T183A eliminates N-linked glycosylation site at codon 181. N-linked glycosylation site N-X-T is transformed into N-X-A by codon 183 mutation.

Since the T183A mutation did not create or eliminate a restriction site, we used the single strand conformational polymorphism (SSCP) method to screen 57 members of the family. We identified the T183A SSCP pattern in 12 "at risk " individuals of which nine were confirmed by direct PCR sequencing. To ensure that this mutation was not a rare polymorphism of the prion protein gene, we also screened 107 unrelated Brazilian individuals, none of whom showed the mutant SSCP pattern.

Codon 129 of the prion protein gene is polymorphic and can encode either a methionine or a valine. Since homozygosity at codon 129 is more common in sporadic CJD cases [7], and dictates the phenotype of the prion diseases associated with prion protein codon 178 mutation [8], we investigated the nature of codon 129 in individuals with the T183A mutation, using the restriction enzyme NspH1 digest to distinguish between the 129Met and 129 Val alleles. Case III-25 was heterozygous at codon 129 with methionine being on the mutant allele. Case II-6 was homozygous for methionine. We found that all other individuals carrying T183A also have at least one methionine allele. Subcloning must be done to clearly show that the 129Met co-segregates with the T183A mutation in all individuals. However, our results suggest that the codon 129 may explain the difference in duration between the mother and son affected by this disease.

DISCUSSION

The clinical, pathological and genetic features of this prion disease although similar to those in other prion disease are, as a whole, distinct from those reported to date. Clinically, the early manifestation of cognitive deficits, the frontal release signs and the protracted course of the disease resemble more Pick's disease or frontal-temporal dementia than the usual prion disease. Although the dementia is consistant with CJD, the longer duration of the disease is not. Typical CJD usually have a rapid course of 3-4 months. The new CJD variant which also presents with behavioral and psychiatric problems has a duration of the disease between 8-22 months [9]. On the other hand, GSS cases can live up to 12 years after the onset of the disease [10]. There are two forms of GSS, one presents with ataxia, the other as a dementing illness. Therefore, the duration of the disease and the initial dementia could classify this disease clinically as GSS. However, GSS is defined pathologically by the presence of cerebellar or cortical plaques. Although we have detected a few plaques in the putamen, these are so sparse that it is difficult to compare them with the classical GSS case. The later clinical symptoms such as the Parkinsonian Syndrome are the result of the development of sub-cortical involvment of the disease. These signs occur in several neurodegenerative diseases of longer duration and as such are not useful as markers for a particular disease.

Pathologically, the lack of significant gliosis is a most striking feature. Robust gliosis is usually associated with neurodegeneration and represents one of the common pathological markers of prion and other neurodegenerative diseases. Only few weakly reactive astrocytes were detected in cortical areas displaying mild, moderate or severe spongiform change and none were observed in the sub-cortical areas. The reason for this lack of gliosis is not clear at this time. One possible explanation is that the prion protein mutation directly affects astrocytic survival or function.

The T183A mutation was previously studied in T183A-PrP transfected cells and resulted in reduced glycosylation and accumulation of the mutant protein in endoplasmic reticulum and cis-Golgi compartments rather than the normal GPI-anchored cell surface localization [11]. It is highly likely that the human T183A PrP also accumulates in the endoplasmic reticulum. If so, this mutant protein may disrupt normal translational processes and contribute to neural cell death. The absence of prion protein immunoreactivity in most severely affected areas supports the possibility that cells expressing high levels of T183A or without a mechanism of clearance for the mutant protein have undergone cell death.

In conclusion, we believe to have identified a clinico-pathological variant of prion disease associated with a novel mutation at codon 183 of the prion protein gene. This finding adds to the growing diversity of human prion protein diseases.

REFERENCES

1.   Prusiner, S., *Inherited prion diseases.* Proc. Natl. Acad. Sci. USA, 1994. **91**: p. 4611-4614.

2.   Brown, P., Cathala, F., Castaigne, P., and Gadjusek, D., *Creutzfeldt-Jakob disease; clinical analysis of consecutive series of 230 neuropathologically verified cases.* Ann. Neurol., 1986. **20**: p. 597-602.

3.   Hsiao, K. and Prusiner, S.B., *Molecular genetics and transgenetic model of Gerstmann-Straussler-Scheinker disease.* Alzheimer's disease and Associated diseases, 1991. **5**(3): p. 155-162.

4.   Lugaresi, E., Medori, R., Montagna, P., Baruzzi A., Cortelli P., Lugaresi A., Tinuper P., Zucconi M., Gambetti P., *Fatal familial insomnia and dysautonomia with selective degeneration of thalamic nuclei.* New Engl J Med, 1986. **315**: p. 997-1003.

5.   Medori, R., Tritschler, H.J., LeBlanc, A., Villare, F., Manetto, V., Chen, H.Y., Xue, R., Leal, S., Montagna, P., Cortelli, P., and et, a.l., *Fatal familial insomnia, a prion disease with a mutation at codon 178 of the prion protein gene [see comments].* N Engl J Med, 1992. **326**(7): p. 444-9.

6.   Budka, H., Aguzzi, A., Brown, P., Brucher, J.M., Bugiani, O., Gullota, F., Haltia, M., Hauw, J.-J., Ironside, J.W., Jellinger, K., Kretzschmar, H.A., Lantos, P.L., Masullo, C., Schlote, W., Tateishi, J., and Weller, R.O., *Neuropathological diagnosis criteria for Creutzfeldt-Jakob disease (CJD) and other human spongiform encephalopathies (prion diseases).* Brain Pathol., 1995. **5**: p. 459-466.

7.   Palmer, M.S., Dryden, A.J., Hughes, J.T., and Collinge, J., *Homozygous prion protein genotype predisposes to sporadic Creutzfeldt-Jakob disease [published erratum appears in Nature 1991 Aug 8;352(6335):547] [see comments].* Nature, 1991. **352**(6333): p. 340-2.

8.   Goldfarb, L.G., Petersen, R.B., Tabaton, M., Brown, P., LeBlanc, A.C., Montagna, P., Cortelli, P., Julien, J., Vital, C., Pendelbury, W.W., Haltia, M., Willis, P.R., Hauw, J.J., McKeever, P.E., Monari, L., Schrank, B., Swergold, G.D., Autilio-Gambetti, L., Gajdusek, C., Lugaresi, E., and Gambetti, P., *Fatal familial insomnia and familial Creutzfeldt Jakob Disease: Disease phenotype determined by a DNA polymorphism.* Science, 1992. **258**: p. 806-808.

9.   Will, R.G., Ironside, J.W., Zeidler, M., Cousens, S.N., Estibeiro, K., Alperovitch, A., Poser, S., Pocchiari, M., Hofman, A., and Smith, P.G., *A new variant of Creutzfeldt-Jakob disease in the UK.* The Lancet, 1996. **347**: p. 921-925.

10.   Amano, N., Yagishita, S., Yokoi, S., Itoh, Y., Kinoshita, J., Mizutani, T., and Matsuishi, T., *Gerstmann-Straussler syndrome - A variant type: amyloid plaques and Alzheimer's neurofibrillary tangles in cerebral cortex.* Acta Neuropathol., 1992. **84**: p. 15-23.

11.   Rogers, M., Taraboulos, A., Scott, M., Groth, D., and Prusiner, S.B., *Intracellular accumulation of the cellular prion protein after mutagenesis of its Asn-linked glycosylation sites.* Glycobiology, 1990. **1**(1): p. 101-9.

FATAL FAMILIAL INSOMNIA: A HUMAN MODEL OF PRION DISEASE

Elio Lugaresi, Pasquale Montagna, Pietro Cortelli, Paolo Tinuper,
Patrizia Avoni, and Giuseppe Plazzi

Institute of Clinical Neurology,
University of Bologna
40123 Bologna, Italy

Drowsiness has been reported to occur in goats, cats, minks and hamsters with prion encephalopathies.[1-3] Instrumental evaluation of sleep-wake abnormalities in these prion affected animals, however, has never been reported.

Very recently Tobler et al.[4] documented that mice devoid of prion protein manifest impaired sleep-wake and other circadian rhythms. A clinical picture characterized by disordered sleep and impaired circadian activities has been described in Fatal Familial Insomnia.[5]

FFI is an autosomal dominant prion disease[6] presenting with loss of sleep, sympathetic hyperactivity and progressive attenuation of autonomic and endocrine circadian oscillations. These cardinal features of the disease are associated with selective thalamic degeneration. Therefore experimental and clinical data suggest that the prion protein plays a major role in physiological mechanisms regulating the rest-activity cycles of living organisms. FFI represents a human model to study the anatomic and molecular basis of sleep-wake and other circadian rhythms.

FFI patients appear apathetic and drowsy: neuropsychological tests demonstrate a progressive impairment of attention and vigilance.[7] Behaviorally, these patients appear to be in an ongoing transition between sleep and wakefulness. 24h polygraphic recordings demonstrate a progressive inability to generate physiological EEG sleep patterns.[8]

Sleep spindles, EEG activities of thalamic origin, which characterize physiological sleep onset and the shift from NREM to REM sleep and vice versa, disappear very early in the course of the disease and only brief episodes of REM sleep or delta sleep emerge, directly and abnormally, from wakefulness.

In other words, in FFI patients the disappearance of spindling activity impairs the onset and maintenance of sleep. 24 h sleep time is radically shortened and any cyclic organization of sleep is lacking. Disappearance of sleep is associated with sympathetic hyperactivity with tachycardia, tachypnea, hypertension, hyperthermia, sweating, impotence and constipation.

24 h recordings demonstrate that body temperature, and heart and breathing rates are abnormally high and that circadian oscillations are severely reduced. Blood pressure also increases, showing ever more reduced circadian oscillations with the progression of the disease. 24 h determinations of hormonal secretion document an abnormal pattern; cortisol, prolactin, catecholamines display increased concentrations; ACTH and GH lack physiological peaks of secretion.[9] In addition, the amplitude of circadian oscillations of all

hormones is reduced. Nocturnal melatonin increase, physiologically associated with darkness (and not with sleep), is greatly reduced from the early stages of the disease and disappears in the advanced stages.[10] PET scans reveal that the thalamus is the only or the most severely affected cerebral region.[11]

The medio-dorsal and antero-ventral thalamic nuclei are invariably and severely affected (with a neuronal loss of over 70%). Other thalamic nuclei (pulvinar, central-median and others) are less consistently and less severely involved. The pathology of the cerebral cortex is very mild in short-lasting cases, more pronounced in long duration cases.[12] The lesions are more severe in some limbic and paralimbic cortical regions (i.e., the rostral part of the cingulate gyrus, orbital and entorhinal cortex). Therefore FFI can be defined as a preferential thalamic degeneration prevalently involving the dorso-median and anteroventral nuclei (i.e., the visceral or limbic portion of the thalamus). The DM nucleus is closely connected with the prefrontal cortex on the one side and the anterior hypothalamus and basal forebrain on the other. The AV nucleus is strictly connected with the rostral part of the cingulate gyrus on the one hand and the posterior hypothalamus and midbrain on the other. The anterior hypothalamus and the basal forebrain contain sleep-inducing structures, whereas the posterior hypothalamus and the midbrain incorporate activating neuronal structures which trigger wakefulness.[13] Atrophy of the DM and AV nuclei of the thalamus could therefore disturb both the circuits activating wakefulness and those promoting sleep.

The hypersomnolent-insomniac condition characteristic of FFI and also typical of paramedian thalamic strokes,[14-15] could be due to the impairment of two different neuronal circuits controlling vigilance level and sleep respectively.

According to Moruzzi[16] and many others, sleep is an instinctive behaviour which is triggered by a low level of vigilance (sub-wakefulness or drowsiness). Recalling this, it becomes clear that the two neuronal circuits, one regulating vigilance and the other inducing sleep, are different.

Sleep onset is marked by the appearance of characteristic bioelectric activities (the sleep spindles) which originate in the thalamus and from there irradiate to the prefrontal cerebral cortex and other neuronal brain structures.

Some experimental studies have demonstrated that atrophy of the DM nuclei prevents spindling spreading to the prefrontal cortex.[17] This could be why patients with FFI have difficulty initiating and maintaining sleep despite being in a condition of persistent somnolence. In other words, degeneration of the limbic portion of the thalamus, disturbing both the circuits controlling wakefulness and those which induce sleep leads to hypersomnolence associated with the inability to sleep: the typical behaviour pattern of FFI patients. Loss of sleep is associated with persistent sympathetic hyperactivation and progressive attenuation of autonomic and endocrine circadian oscillations.[18] A tentative explanation for this complex autonomic and endocrine impairment is the following: atrophy of DM and AV thalamic nuclei (i.e.: the limbic portion of the thalamus) interrupts the most important cortical control over the hypothalamus. The impaired hypothalamic regulation of body homeostasis results in an activated autonomic and endocrine pattern stemming from this functional imbalance.[19]

The progressive attenuation of circadian and endocrine oscillations is an expression of ever increasing prevalence of activating system: if a pendulum swinging between two extremes tilts its axis towards one end, the greater the shift, the more limited its oscillations will be. FFI anatomoclinical findings demonstrate that the selective degeneration of the limbic thalamus impairs sleep and wakefulness and, more generally, endocrine and autonomic homeostasis. This is probably due to the fact that the hypothalamus is released from major cortico-limbic control.

The symptoms encountered in FFI are in some way similar to those observed in mice devoid of prion protein.[4] This suggests that the anatomic damage caused by the protease resistant prion protein disturbs the same circuits in which the physiological prion protein exerts its action. The key feature of FFI is a functional imbalance with the activating system prevailing over the deactivating system. All this implies that the prion protein is somehow involved in regulating the rest-activity cycle in living organisms. Our findings, however, do not establish whether this action involves gabaergic transmission, as has been suggested, or originates from other mechanisms. In FFI the system regulating homeostasis does appear to be extensively impaired involving some crucial circuits interconnecting the cerebral cortex, hypothalamus and other subcortical neuronal structures.

References

1. I.H. Pattison and G.C. Millson, Scrapie produced experimentally in goats with special reference to clinical syndrome,. *J Comp Pathol* 71:101-108 (1961).
2. P. Gourmelon, H.L. Amyx, H. Baron, G. Lemercier, L. Court and C.J. Gibbs, Electrophysiological findings in experimental Creutzfeldt-Jakob disease in cats, in: *Unconventional Virus Diseases of the Central Nervous System*, L. Court, D. Dormont, P. Brown and D.T. Kingsbury, eds., CEA Diffusion, Fontenay aux Roses, pp. 231-249 (1989).
3. R.A. Bessen and R.F. Marsh, Biochemical and physical properties of the prion protein from two strains of the transmissible mink encephalopathy agent, *J Virol* 66:2096-2101 (1992).
4. I. Tobler, S.E. Gaus, T. Deboer, P. Achermann, M. Fischer, T. Rülicke, M. Moser, B. Oesch, P.A. McBride and J.C. Manson, Altered circadian activity rhythms and sleep in mice devoid of prion protein, *Nature* 380:639-642 (1996).
5. E. Lugaresi, R. Medori, P. Montagna, A. Baruzzi, P. Cortelli, A. Lugaresi, P. Tinuper, M. Zucconi, P. Gambetti, Fatal familial insomnia and dysautonomia with selective degeneration of thalamic nuclei, *N Engl J Med* 315:997-1003 (1986).
6. R. Medori, H.J. Tritschler, A. LeBlanc, F. Villare, V. Manetto, H.Y. Chen, R. Xue, S. Leal, P. Montagna, P. Cortelli, P. Tinuper, P. Avoni, M. Mochi, A. Baruzzi, J.J. Hauw, J. Ott, E. Lugaresi, L. Autilio-Gambetti and P. Gambetti, Fatal familial insomnia, a prion disease with a mutation at codon 178 of the prion protein gene, *N Engl J Med* 326:444-449 (1992).
7. Gallassi, A. Morreale, P. Montagna, P. Cortelli, P. Avoni, R. Castellani, P. Gambetti and E. Lugaresi, Fatal familial insomnia: behavioral and cognitive features, *Neurology*, 46:935-939 (1996).
8. E. Sforza, P. Montagna, P. Tinuper, P. Cortelli, P. Avoni, F. Ferrillo, R. Petersen, P. Gambetti, E. Lugaresi, Sleep-wake cycle abnormalities in fatal familial insomnia. Evidence of the role of the thalamus in sleep regulation, *Electroencephalogr clin Neurophysiol* 94:398-405 (1995).
9. A. Lugaresi, A. Baruzzi, E. Cacciari, P. Cortelli, R. Medori, P. Montagna, P. Tinuper, M. Zucconi, I. Roiter, E. Lugaresi, Lack of vegetative and endocrine circadian rhythms in fatal familial thalamic degeneration, *Clin Endocrinol* 26:573-580 (1987).
10. F. Portaluppi, P. Cortelli, P. Avoni, L. Vergnani, P. Maltoni, A. Pavani, E. Sforza, E.C. degli Uberti, P. Gambetti, E. Lugaresi, Progressive disruption of the circadian rhythm of melatonin in fatal familial insomnia, *J Clin Endocrinol Metab* 78(5):1075-1078 (1994).
11. D. Perani, P. Cortelli, G. Lucignani, P. Montagna, P. Tinuper, R. Gallassi, P. Gambetti, G.L. Lenzi, E. Lugaresi, and F. Fazio, [^{18}F] FDG PET in fatal familial insomnia: the functional effects of thalamic lesions, *Neurology* 43:2565-2569 (1993).
12. P. Montagna, P. Cortelli, P. Gambetti and E. Lugaresi, Fatal familial insomnia: sleep, neuroendocrine and vegetative alterations, *Adv Neuroimmunol.* 5:13-21 (1995).
13. M. Sallanon, C. Aubert, M. Denoyer, K. Kitahama, M. Jouvet, L'insomnie consecutive à la lesion de la region préoptique paramédiane est reversible par inactivation de l'hypothalamus postérieur chez le chat, *CR Acad Sci [III]* 305:561-567 (1987).
14. C. Guilleminault, M.A. Quera-Salva, M.P. Goldberg, Pseudo hypersomnia and pre-sleep behaviour with bilateral paramedian thalamic lesions, *Brain* 116:1549-1563 (1993).
15. C. Bassetti, J. Mathis, M. Gugger, K.O. Lövblad, and C.W. Hess, Hypersomnia following paramedian thalamic stroke: a report of 12 patients. *Ann Neurol* 39:471-480 (1996).
16. G. Moruzzi, Sleep and instictive behavior, *Arch Ital Biol* 108:175-216 (1969).
17. L. Imeri, M.E. Moneta, and M. Mancia, Changes in spontaneous activity of medialis dorsalis thalamic neurones during sleep and wakefulness, *Electroenceph clin Neurophysiol* 69:82-84 (1988).
18. P. Cortelli, P. Parchi, M. Contin, G. Pierangeli, P. Avoni, P. Tinuper, P. Montagna, A. Baruzzi, P.L. Gambetti, E. Lugaresi, Cardiovascular dysautonomia in fatal familial insomnia, *Clin Autonom Res* 1:15-21 (1991).
19. E. Lugaresi , The thalamus and insomnia, *Neurology* 42(6):28-33 (1992).

MECHANISMS OF PHENOTYPIC HETEROGENEITY IN HUMAN PRION DISEASES

Pierluigi Gambetti, Shu G. Chen, and Piero Parchi

Division of Neuropathology
Institute of Pathology
Case Western Reserve University
Cleveland, OH 44106
U.S.A.

At variance with other neurodegenerative diseases, such as Alzheimer disease, amyotrophic lateral sclerosis and Huntington's chorea, in which the disease phenotype is fairly homogeneous, human prion diseases occur in a variety of forms and display a large spectrum of disease phenotypes.

Human prion diseases are commonly divided into three forms: sporadic and inherited forms and the form acquired by infection. All these forms present one or more of four basic clinico-pathological phenotypes. The sporadic form, which is the most common and accounts for over 80% of all the cases, is associated with the phenotype of the Creutzfeldt-Jakob disease (CJD) which is characterized by a fairly rapid dementia with myoclonus associated histopathologically with widespread spongiform degeneration. The inherited form may have, in addition to the CJD phenotype, the Gerstmann-Sträussler-Scheinker disease (GSS) phenotype characterized by cerebellar signs of long duration and widespread presence of PrP amyloid plaques, and the fatal familial insomnia (FFI) phenotype characterized by the loss of the ability to sleep associated with a severe atrophy of thalamic nuclei. Moreover, a group of inherited prion diseases associated with insertion mutations in the prion protein gene (*PRNP*) shows an heterogeneous phenotype which includes subtypes similar to the CJD and GSS, as well as subtypes with histopathology that is not distinctive or minimal.

Known determinants of the phenotypic variability in prion diseases are the *PRNP* genotype, the type of the protease resistant prion protein (PrP^{res}) and the rate of conversion of the normal or cellular PrP (PrP^C). Other less known mechanisms, such as the rate and topography of PrP^{res} spreading within the central nervous system and the route of entry of the exogenous PrP^{res} in the forms acquired by the infection, are also likely to play a role.

THE ROLE OF THE GENOTYPE IN THE PHENOTYPIC HETEROGENEITY OF INHERITED PRION DISEASES

The first observation that the *PRNP* codon 129, the site of a common methionine (M)/valine (V) polymorphism[1], influences human prion diseases was made by Collinge and coworkers who found that the prevalence of sporadic CJD and of the CJD acquired by infection is higher in subjects who are homozygous than in those that are heterozygous at codon 129[2,3]. Subsequently, it was shown that in the GSS subtype with the F198S-129V haplotype, affected 129 homozygotes had an earlier onset and shorter course of the disease[4]. The first demonstration that the 129 polymorphism can act as a determinant of the basic features of the disease phenotype was made by the observation[5] that FFI and a familial CJD subtype (CJD[178]) share the same D178N mutation but the mutation is coupled to the 129M codon in FFI and to the 129V codon in CJD[178] on the mutant allele. FFI and CJD[178] have distinct clinical features and histopathologies, both for the type and the distribution of the lesions. Moreover, these two diseases are also associated with PrP[res] fragments that have two distinct sizes as determined by the different mobility on gel electrophoresis and a different ratio of the glycoforms[6]. We suggested[6] that the distinct size of the PrP[res] fragments generated as a result of the differential cleavage by the protease is likely to reflect a different conformation of the two PrP[res]. The prominent effect on the phenotype of codon 129 as well as of the codon 219, the site of a glutamic acid/lysine polymorphism in the Japanese population, when they are located on the mutant allele, has also been recently reported in other inherited prion diseases[7,8]. Moreover, as for the F198S-129V GSS subtype, it has been also observed that codon 129 on the normal allele which results in 129 homozygous and heterozygous FFI and CJD[178] subjects, plays a role in determining disease duration in FFI and age of onset and disease duration in CJD[178]. Therefore, the phenotypic heterogeneity of several inherited prion diseases is determined not only by the diversity of the *PRNP* mutations but also by the polymorphisms at codon 129, 219 and possibly other codons. On the mutant allele, these codons determine the type of disease linked to the pathogenic mutation, on the normal allele they modulate the severity of the disease[5].

THE ROLE OF THE GENOTYPE IN THE PHENOTYPIC HETEROGENEITY IN THE SPORADIC FORM

It has been shown that the sporadic form of CJD can be subdivided into at least four groups or variants that differ in various phenotypic features such as clinical and histopathological characteristics, pattern of immunostaining and distribution of PrP[res]. These four phenotypes codistribute with the *PRNP* genotype at codon 129 and with the type of PrP[res], as defined by the gel migration characteristics which, as mentioned above, probably reflect a different conformation, and the ratios of the glycoforms (see below)[9]. Thus, group 1, that includes the typical and sporadic CJD, is associated with the 129M/M genotype, and a type 1 PrP[res] fragment corresponding to about 20 kDa that is indistinguishable from that in CJD[178]; group 2, which is characterized by a much longer disease duration, has the same 129M/M genotype but is associated with a type 2 PrP[res] fragment of about 19 kDa indistinguishable from that in FFI, whereas groups 3 and 4 have the 129M/V and 129V/V genotypes respectively, and are both associated with the type 2 PrP[res] (ref. 9). The role of the 129 polymorphism and of the PrP[res] type in the determination of the disease phenotype is unclear. The finding that, group 1 and 2 have different disease phenotypes, different PrP[res] subtypes but have the same 129 genotype, suggests that the protein type itself may be a determinant of the phenotype.

PRP[RES] TYPES AND PHENOTYPIC HETEROGENEITY

We have recently confirmed[10] the two types of PrP[res] reported previously[9] in a large series of cases including sporadic CJD, CJD transmitted by infection, including the so-called iatrogenic subtypes acquired by administration of contaminated hormones, transplant of contaminated tissues or use of contaminated surgical instruments, kuru probably due to ingestion of contaminated food stuff, and the new variant of CJD recently reported in United Kingdom which is believed to be acquired from bovine spongiform encephalopathy (BSE) through consumption of contaminated meat products. In this large series of cases, type 1 and 2 PrP[res] were found in the sporadic CJD associated to all 129 genotypes 129M/M, 129M/V and 129V/V, although the prevalence of type 1 and type 2 appeared to be related to the dosage of the 129M and 129V codons, respectively[10]. Thus, the great majority of the subjects with 129M/M (54 of 58) had type 1 PrP[res], type 1 and 2 PrP[res] appeared to be almost evenly represented (5 and 7) in the 129M/V group while the great majority (15 of 18) of the 129V/V cases had type 2 PrP[res]. Although the number of cases was low, a similar codistribution of 129 genotype and PrP[res] type seems to be present in the forms acquired by infection since all four 129M/M cases had type 1 PrP[res] and all three cases with either 129M/V or 129V/V, whether iatrogenic or kuru, had type 2. Therefore, the analysis of this large series indicates that in sporadic CJD, and possibly in the form acquired by infection, the 129 *PRNP* genotype has some influence on the type of PrP[res] expressed, the 129M codon favoring the type 1 and the 129V codon favoring the type 2. However, this effect is partial as each of the two PrP[res] types can be associated with each of the three 129 genotypes (M/M, MV, and V/V). An apparent exception to this rule is the new variant CJD in which only the 129M/M genotype has been observed in all cases reported to date. At least in the case that we examined[10], the new variant CJD is associated with type 2 PrP[res]. The mechanisms for the association of type 2 PrP[res] with the 129M/M genotype in the new variant remains to be determined. It might be due to the putative mode of acquisition of the new variant from BSE in which the PrP[res] has been reported[11] to be of the same type and to the susceptibility of the 129M/M subjects to this CJD variant. Moreover, as previously reported[10,11], the new variant CJD has a distinctive pattern of PrP[res] glycosylation. PrP[res] commonly includes three forms that are differently glycosylated and display different mobilities on gels: a form of high molecular weight, thought to contain two glycan chains, an intermediate form with one glycan chain and a low molecular weight band with no glycans[6,12]. In all the groups of sporadic and acquired CJD associated either with type 1 and type 2 PrP[res], the intermediate glycoform was the dominant form; in the new variant the highly glycosylated was clearly the dominant form while the unglycosylated form was markedly underrepresented[10,11]. The mechanisms of this distinctive glycoform ratio also is unclear. Since the size of the PrP[res] but not the ratio of the glycoforms seems to be reproduced when PrP[res] are transmitted between species, the distinctive glycoform ratio in the new variant might be the result of preferential conversion of the highly glycosylated form over the other glycoforms to PrP[res].

ANOTHER MECHANISM OF PHENOTYPIC HETEROGENEITY: ALLELIC ORIGIN OF PRP[RES]

Inherited prion diseases are overwhelmingly heterozygous for the pathogenic mutation. Therefore, not only the ease with which the mutant PrP (PrP[M]) converts into the PrP[res] but also the capacity of the mutant PrP[res] to convert the PrP expressed by the normal or wild type allele (PrP[Wt]). These two variables influence the rate of production and, therefore, the amount and distribution of the PrP[res] formed, which, in turn, have been shown to be related to the severity and topography of the lesions in human and in animal prion diseases[9,13,14]. In the GSS subtypes associated with the A117V and the F198S mutations it has been shown that only the mutant PrP (PrP[M]) is present in the PrP amyloid deposits that characterize these diseases[15]. In the CJD subtype linked to the E200K mutation it was shown[16] that only PrP^M is resistant to proteases, but both PrP^M and PrP^Wt are insoluble in nonionic detergents indicating that wild type PrP although not converted to PrP^res is in an aggregated form and therefore is modified by PrP^M. We examined the detergent solubility and protease resistance of PrP^M and PrP^Wt in brain tissues from FFI and CJD[178] subjects, as well as from CJD subjects carrying insertional mutations. Taking advantage of polymorphisms in the mutant allele of heterozygous subjects, we were able to differentiation PrP^M from PrP^Wt on the basis of differences in size and susceptibility to endoproteinases. We observed[17] that PrP^res in FFI and CJD[178] derives exclusively from PrP^M, while PrP^res in CJD with insertional mutations is likely to include both PrP^M and PrP^Wt. It appears therefore that the *PRNP* mutation may regulate the amount of PrP^res present in inherited prion diseases not only by affecting the conversion of PrP^M to PrP^res but also through the capability of the mutant PrP^res to convert the PrP^Wt. This finding suggests two considerations. First, since FFI and CJD[178] can be transmitted to mice that do not carry *PRNP* mutations and that the affected mice produce PrP^res, D178N mutant PrP^res (or other molecules associated with it) must be capable to convert PrP^C into PrP^res under certain conditions, for example, following intracerebral inoculation which produces a high PrP^res concentration in one brain region. Second, the role of codon 129 in modulating the disease duration was previously explained as the result of the higher convertibility to PrP^res of PrP^Wt in 129 homozygotes than heterozygotes due to the more extensive homogeneity of PrP^M and PrP^Wt in the 129 homozygotes[18]. However, in FFI and CJD[178] the mechanism causing a shorter disease duration in the 129 homozygous subjects must be different.

CONCLUSIONS

The mechanisms underlying the phenotypic heterogeneity of prion diseases are multiple and probably interconnected.

The genotype may modify the disease phenotype in all forms of prion diseases. In the inherited forms, the pathogenic mutation affects the expression of the disease phenotype according to several parameters: 1, the alteration of the PrP^M metabolism caused by the *PRNP* mutation; 2, the convertibility to PrP^res of the PrP^M ; 3, pathogenic effect of the mutant PrP^res ; 4, convertibility to PrP^res of PrP^Wt by the mutant PrP^res. In addition, codon 129 and other codons may have profound effect on the disease phenotype linked to the pathogenic mutation when they are located on the mutant allele. On the normal allele, codon 129 may influence disease duration and age of onset and, probably secondarily, the distribution of the histopathology and of the PrP^res.

In the sporadic form, the genotype at codon 129 plays a role in determining the variants within this form. Moreover, codon 129 influences the type of PrP^res expressed which, in turn, modifies the phenotype. The mechanisms of these effects and the interaction between codon 129 and PrP^res type are unclear.

Finally, the phenotype of the form acquired by infection is modulated by the route of contamination, the type of the exogenous PrP^res and the codon 129.

Therefore several variables often in concert play a role in modulating the disease phenotype in all forms of prion diseases. In striking contrast to the phenotypic variability of prion diseases is the phenotypic homogeneity of Alzheimer's disease of which the inherited form is linked to a variety of mutations on several genes and at least one polymorphism. A challenge will be to understand the mechanisms underlying this diversity.

REFERENCES

1. Goldfarb, P. Brown, and D. Goldgaber, Patients with Creutzfeldt-Jakob disease and kuru lack the mutation in the PRNP gene found in Gerstmann-Sträussler-Scheinker syndrome, but they show a different double-allele mutation in the same gene. *Am J Human Genetics* 45: (Supplement): A189 (1989).
2. M.S. Palmer, A.J. Dryden, J.T. Hughes, and J. Collinge, Homozygous prion protein genotype predisposes to sporadic Creutzfeldt-Jakob disease, *Nature* 352:340 (1991).
3. J. Collinge, M.S. Palmer, and A.J. Dryden, Genetic predisposition to iatrogenic Creutzfeldt-Jakob disease, *Lancet* 337:1441 (1991).
4. K. Hsiao, S.R. Dloughy, M.R. Farlow, C. Cass, M. Da Costa, P.M. Conneally, M.E. Hooles, B. Ghetti, and S.B. Prusiner, Mutant prion proteins in Gerstmann-Sträussler-Scheinker disease with neurofibrillary tangles, *Nature Genetics* 1:68 (1992).
5. L.G. Goldfarb, R. B. Petersen, M. Tabaton, P. Brown, A.C. LeBlanc, P. Montagna, P. Cortelli, J. Julien, C. Vital, W.W. Pendlebury, M. Haltia, P.R. Willis, J.J. Hauw, P.E. McKeever, L. Monari, B. Schrank, G.D. Swergold, L. Autilio-Gambetti, C. Gajdusek, E. Lugaresi, and P. Gambetti, Fatal familial insomnia and familial Creutzfeldt Jakob disease: disease phenotype determined by a DNA polymorphism, *Science* 258:806 (1992).
6. L. Monari, S.G. Chen, P. Brown, P. Parchi, R.B. Petersen, J. Mikol, F. Gray, P. Cortelli, P. Montagna, B. Ghetti, L.G. Goldfarb, D.C. Gajdusek, E. Lugaresi, P. Gambetti, and L. Autilio-Gambetti, Fatal familial insomnia and familial Creutzfeldt-Jakob disease: different prion proteins determined by a DNA polymorphism, *Proc Natl Acad Sci USA* 91:2839 (1994).
7. K. Young, H.B. Clark, P. Piccardo, S.R. Dlouhy, and B. Ghetti, Gerstmann-Sträussler-Scheinker disease with the PRNP P102L mutation and valine at codon 129, *Mol Brain Res* 44:147 (1997).
8. H. Furukawa, T. Kitamoto, Y. Tanaka, and J. Tateishi, New variant prion protein in a Japanese family with Gerstmann-Sträussler syndrome, *Mol Brain Res* 30: 385 (1995).
9. P. Parchi, R. Castellani, S. Capellari, B. Ghetti, K. Young, S.G. Chen, M. Farlow, D.W. Dickson, A.A.F. Sima, J.Q. Trojanowski, R.B. Petersen, and P. Gambetti, Molecular basis of phenotypic variability in sporadic Creutzfeldt-Jakob disease, *Ann Neurol* 39:669 (1996).
10. P. Parchi, S. Capellari, S.G. Chen, R.B. Petersen, P. Gambetti, N. Kopp, P. Brown, T. Kitamoto, J. Tateishi, A. Giese, and H. Kretzschmar, Typing in prion isoforms, *Nature* 386: 232 (1997).
11. J. Collinge, K.C.L. Sidle, J. Meads, J. Ironside, and A.F. Hill. Molecular analysis of prion stain variation and the aetiology of "new variant" CJD, *Nature* 383:685 (1996).
12. B. Caughey, R.E. Race, D. Ernst, M.J. Buchmeier, and B. Chesebro, Prion protein biosynthesis in scrapie-infected and uninfected neuroblastoma cells, *J Virol* 63:175 (1989).
13. K. Jendroska, F.P. Heinzel, M. Torchia, L. Stowring, H.A. Kretzschmar, A. Kon, A. Stern, S.B. Prusiner, and S.J. DeArmond, Proteinase-resistant prion protein accumulation in Syrian hamster brain correlates with regional pathology and scrapie infectivity, *Neurology* 41: 1482 (1991).
14. P. Parchi, R. Castellani, P. Cortelli, P. Montagna, S.G. Chen, R.B. Petersen, E. Lugaresi, L. Autilio-Gambetti, and P. Gambetti, Regional distribution of protease-resistant prion protein in Fatal Familial Insomnia, *Ann Neurol* 38:21 (1995).
15. F. Tagliavini, F. Prelli, M. Porro, G. Rossi, G. Giaccone, M.R. Farlow, S.R. Dloughy, B. Ghetti, O. Bugiani, and B. Frangione, Amyloid fibrils in Gerstmann-Sträussler-Scheinker disease (Indiana and Swedish kindreds) express only PrP peptides encoded by the mutant allele, *Cell* 79:695 (1994).
16. R. Gabizon, G. Telling, Z. Meiner, M. Halimi, I. Kahana, and S.B. Prusiner, Insoluble wild-type and protease-resistant mutant prion protein in brains of patients with inherited prion diseases, *Nature Med* 2:59 (1996).
17. S.G. Chen, P. Parchi, P. Brown, S. Capellari, W. Zou, E.J. Cochran, C.L. Vnencak-Jones, J. Julien, C. Vital, J. Mikol, E. Lugaresi, L. Autilio-Gambetti, and P. Gambetti, Allelic origin of the abnormal prion proteins in familial prion diseases, *Nature Med* (in press)
18. M.S. Palmer, and J. Collinge, Human prion diseases, *Curr Opin Neurol Neurosurg* 5:895 (1992).

TRANSGENIC MICE WITH NEURON-SPECIFIC EXPRESSION OF A HAMSTER PRION PROTEIN MINIGENE ARE SUSCEPTIBLE TO HAMSTER SCRAPIE AGENT

Bruce Chesebro, Suzette A. Priola, and Richard E. Race

Laboratory of Persistent Viral Diseases, Rocky Mountain Laboratories, National Institute of Allergy and Infectious Diseases, Hamilton, Montana 59840

INTRODUCTION

The degenerative brain diseases known as transmissible spongiform encephalopathies (TSE) occur naturally in primates and ruminants, and include scrapie of sheep, bovine spongiform encephalopathy, and several human diseases such as Creutzfeldt-Jakob disease, Kuru, and Gerstmann-Sträussler-Scheinker syndrome. During the course of TSE diseases brain tissue accumulates a proteinase K-resistant protein, known as PrP-res or PrP^{Sc}, which is associated with the pathogenic process [1-3]. PrP-res is post-translationally derived from a normal host proteinase K-sensitive PrP molecule (PrP-sen) [4-6] by an as yet undefined mechanism. PrP may play a critical role in interspecies transmission of TSE diseases. Many species show resistance to disease induction by TSE agents derived from other species. This resistance or "species barrier" is manifested either by total lack of disease induction or by a prolonged incubation period prior to onset of clinical disease. Genetic studies have indicated that the PrP genotype strongly influences the host susceptibility to TSE agents [7-9]. Transgenic mice with a 40 kb transgene expressing high levels of hamster PrP (HPrP)-sen are susceptible to disease when inoculated with the hamster scrapie agent while normal mice are resistant [10,11]. This suggests that transmission of scrapie may be dependent on interactions between the host PrP-sen and the PrP-res derived from the donor of the inoculated agent. Interactions between PrP molecules from different species have been found to inhibit generation of PrP-res in both scrapie-infected cells [12] and in cell-free systems [13,14], and such inhibitory interactions may provide a biochemical explanation for the species barrier.

TSE agents are known to replicate in both lymphoreticular organs and brain, but the precise cell types involved are not known. PrP-sen expression has recently been demonstrated in astrocytes and oligodendrocytes [15], and neurons [16-19]. However, it is unclear which of these three cell types are

Prions and Brain Diseases in Animals and Humans
Edited by Morrison, Plenum Press, New York, 1998

critical to TSE agent replication in brain. The present work shows that transgenic mice expressing the hamster PrP gene under control of the neuron-specific enolase promoter [20,21] are highly susceptible to the hamster scrapie agent. In these mice HPrP expression was found exclusively in neurons and not in glial cells or cells within the spleen or lymph nodes. Thus, hamster PrP expression in neurons was sufficient to abrogate the TSE species barrier. Furthermore, because the transgene used in these experiments contained only 1 kb of hamster DNA including the open reading frame of hamster PrP, the susceptibility of these transgenic mice to hamster scrapie was mediated by the hamster PrP gene itself rather than the additional 39 kb of transgene DNA used in previous studies [10,11,22].

RESULTS AND DISCUSSION

The Tg52NSE line of transgenic mice was derived by inoculation of a construct containing the neuron-specific enolase (NSE) promotor plus a 1 kb cDNA containing the HPrP open reading frame as previously described [21]. Tg52NSE mice expressed a 1.6 kb PrP mRNA band corresponding to the predicted HPrP transgene in addition to the 2.3 kb PrP mRNA band expressed in nontransgenic mice. Using immunoblotting with monoclonal antibody 3F4, which has a strong reactivity for HPrP and no reactivity for mouse PrP [23], HPrP protein was detected in brain of Tg52NSE mice, but not in six other tissues studied. In contrast, Tg10 mice containing a cosmid H-PrP transgene plus 40kb of flanking DNA [10,21], and normal hamsters expressed HPrP protein in a variety of other tissues in addition to brain. Thus, the transgene in Tg52NSE mice showed a restricted pattern of expression as expected from the use of the neuron-specific enolase promoter [20].

In situ hybridization was used to demonstrate localization of HPrP mRNA in a variety of neuronal populations in Tg52NSE mouse brain. Highest expression was seen in Purkinje cells of the cerebellum, neurons of the dentate gyrus, and pyramidal neurons of the hippocampus, whereas granular layer neurons of the cerebellum, cells in various layers of the cerebral cortex and cells with large nuclei in the dorsal portion of the thalamus also expressed moderate amounts of HPrP mRNA [21]. These results were consistent with neuron-specific expression of the HPrP transgene in Tg52NSE mice.

To determine the influence of HPrP expression on susceptibility to hamster scrapie agent, animals were inoculated intracerebrally with hamster scrapie strain 263K. Tg52NSE mice all died between 70 and 118 days postinoculation while non-Tg littermates were clinically normal 400 days postinoculation. Affected mice exhibited a 1-3 day clinical course characterized by a "stilted" gait, mild ataxia and inactivity. Tg10 mice had a more variable and protracted clinical course lasting several weeks or even months and died between 160 and 405 days postinoculation. The clinical symptoms in Tg10 mice included whole body tremors, ataxia, and progression to paralysis and death. In hamsters clinical symptoms included ataxia, tremor and somnolence. Clinically ill Tg52NSE mice and Tg10 mice had histopathological findings typical of scrapie with astrocytosis and spongiosis being the most prominent findings and easily detectable proteinase K-resistant HPrP (HPrP-res). In summary, Tg52NSE and Tg10 mice were highly susceptible to hamster 263K scrapie agent, but both the tempo and symptoms of clinical disease were different in the two transgenic strains.

In order to determine if the expression of HPrP would modify the pathogenesis of mouse scrapie, Tg52NSE and Tg10 mice were also inoculated with the Chandler strain of mouse-adapted scrapie agent. Tg52NSE mice died 180±3 days postinoculation (n=8), while normal littermates died 160±3 days postinoculation (n=11). Tg10 mice died 201±4 days postinoculation (n=17), while their normal littermates died 164±4 days after inoculation (n=14). Thus, expression of HPrP in both Tg52NSE and Tg10 mice delayed the onset of clinical disease induced by the mouse scrapie agent suggesting that expression of HPrP could partially interfere with development of mouse scrapie.

Transgenic mice expressing HPrP were made previously using a 40 kb cosmid clone containing the HPrP gene [10,11], but because of the large amount of DNA in these transgenes it was not possible to prove that HPrP expression was the only genetic factor involved in the induction of susceptibility to hamster scrapie in these mice. In contrast, the present experiments succeeded in getting high levels of HPrP expression by using a transgene containing only 1 kb of hamster DNA including the 762 base pair open reading frame of HPrP together with the neuron-specific enolase promoter. Thus, the high susceptibility of Tg52NSE mice to hamster scrapie demonstrates that this HPrP minigene including the open reading frame itself is the critical element in inducing susceptibility to the hamster scrapie agent *in vivo*.

Several previous reports indicate that astrocytes and splenic follicular dendritic cells (FDC) may be the earliest sites of PrP-res accumulation following scrapie infection [15,24,25], and these sites might also be important in restriction of agent replication following interspecies transmission of TSE agents [25]. However, based on the present findings, HPrP expression in astrocytes or FDC was not required to mediate susceptibility of mice to intracerebral inoculation with hamster scrapie. Nevertheless, HPrP expression in cells such as astrocytes, FDC or even other cell types might also be sufficient to overcome the scrapie species barrier, and FDC in spleen and lymph nodes might be particularly involved in interspecies transmission following intraperitoneal inoculation of agent.

Although experiments with PrP transgenic mice have provided helpful insights into the importance of PrP in scrapie pathogenesis and interspecies transmission, it is possible that some phenomena observed in transgenic mice are the result of abnormal overexpression of PrP. For example, some transgenic mice overexpressing mouse or non-mouse PrP have developed a spontaneous degenerative brain disease in the absence of scrapie infection [22,26]. In other instances transgenic mice expressing a different mouse PrP gene usually associated with resistance to most mouse scrapie strains had increased sensitivity to mouse scrapie [27]. Both of these unexpected examples of neurodegenerative disease in PrP transgenic mice are believed to involve overexpression of PrP, but the detailed pathogenic mechanisms involved in each are not known. Thus, it will be important in the future to confirm conclusions derived from transgenic mice by using PrP null mice where the mouse PrP gene can be replaced in its normal context in the mouse genome by a single copy of a mutant or foreign PrP gene.

REFERENCES

1. Bolton DC, McKinley MP, Prusiner SB. Identification of a protein that purifies with the scrapie prion. *Science* 1982;218:1309-11.
2. Diringer H, Gelderblom H, Hilmert H, Ozel M, Edelbluth C, Kimberlin RH. Scrapie infectivity, fibrils and low molecular weight protein. *Nature* 1983;306:476-8.
3. Prusiner SB. Novel proteinaceous infectious particles cause scrapie. *Science* 1982;216:136-44.
4. Caughey B, Raymond GJ. The scrapie-associated form of PrP is made from a cell surface precursor that is both protease- and phospholipase-sensitive. *J Biol Chem* 1991;266:18217-23.
5. Stahl N, Baldwin MA, Teplow DB, Hood L, Gibson BW, Burlingame AL, Prusiner SB. Structural studies of the scrapie prion protein using mass spectrometry and amino acid sequencing. *Biochemistry* 1993;32:1991-2002.
6. Borchelt DR, Scott M, Taraboulos A, Stahl N, Prusiner SB. Scrapie and cellular prion proteins differ in the kinetics of synthesis and topology in cultured cells. *J Cell Biol* 1990;110:743-52.
7. Carlson GA, Kingsbury DT, Goodman PA, Coleman S, Marshall ST, DeArmond S, Westaway D, Prusiner SB. Linkage of prion protein and scrapie incubation time genes. *Cell* 1986;46:503-11.
8. Hunter N, Hope J, McConnell I, Dickinson AG. Linkage of the scrapie-associated fibril protein (PrP) gene and sinc using congenic mice and restriction fragment length polymorphism analysis. *J Gen Virol* 1987;68:2711-6.
9. Race RE, Graham K, Ernst D, Caughey B, Chesebro B. Analysis of linkage between scrapie incubation period and the prion protein gene in mice. *J Gen Virol* 1990;71:493-7.
10. Scott M, Foster D, Mirenda C, Serban D, Coufal F, Walchli M, Torchia M, Groth D, Carlson G, DeArmond SJ, et al. Transgenic mice expressing hamster prion protein produce species-specific scrapie infectivity and amyloid plaques. *Cell* 1989;59:847-57.
11. Prusiner SB, Scott M, Foster D, Pan KM, Groth D, Mirenda C, Torchia M, Yang SL, Serban D, Carlson GA, et al. Transgenetic studies implicate interactions between homologous PrP isoforms in scrapie prion replication. *Cell* 1990;63:673-86.
12. Priola SA, Caughey B, Race RE, Chesebro B. Heterologous PrP molecules interfere with accumulation of protease-resistant PrP in scrapie-infected murine neuroblastoma cells. *J Virol* 1994;68:4873-8.
13. Kocisko DA, Come JH, Priola SA, Chesebro B, Raymond GJ, Lansbury PT, Caughey B. Cell-free formation of protease-resistant prion protein. *Nature* 1994;370:471-4.
14. Kocisko DA, Priola SA, Raymond GJ, Chesebro B, Lansbury PT, Jr., Caughey B. Species specificity in the cell-free conversion of prion protein to protease-resistant forms: a model for the scrapie species barrier. *Proc Natl Acad Sci USA* 1995;92:3923-7.
15. Moser M, Colello RJ, Pott U, Oesch B. Developmental expression of the prion protein gene in glial cells. *Neuron* 1995;14:509-17.
16. Kretzschmar HA, Prusiner SB, Stowring LE, DeArmond SJ. Scrapie prion proteins are synthesized in neurons. *Am J Pathol* 1986;122:1-5.
17. Brown HR, Goller NL, Rudelli RD, Merz GS, Wisniewski HM, Robakis NK. The mRNA encoding the scrapie agent protein is present in a variety of non-neuronal cells. *Acta Neuropathol* 1990;80:1-6.
18. Manson J, West JD, Thomson V, McBride P, Kaufman MH, Hope J. The prion protein gene: a role in mouse embryogenesis? *Development* 1992;115:117-22.
19. Manson JC, Clarke AR, McBride PA, McConnell I, Hope J. PrP gene dosage determines the timing but not the final intensity or distribution of lesions in scrapie pathology. *Neurodegen* 1994;3:331-40.

20. Forss-Petter S, Danielson PE, Catsicas S, Battenberg E, Price J, Nerenberg M, Sutcliffe JG. Transgenic mice expressing beta-galactosidase in mature neurons under neuron-specific enolase promoter control. *Neuron* 1990;5:187-97.
21. Race RE, Priola SA, Bessen RA, Ernst D, Dockter J, Rall GF, Mucke L, Chesebro B, Oldstone MBA. Neuron-specific expression of a hamster prion protein minigene in transgenic mice induces susceptibility to hamster scrapie agent. *Neuron* 1995;15:1183-91.
22. Westaway D, DeArmond SJ, Cayetano-Canlas J, Groth D, Foster D, Yang S, Torchia M, Carlson GA, Prusiner SB. Degeneration of skeletal muscle, peripheral nerves, and the central nervous system in transgenic mice overexpressing wild-type prion proteins. *Cell* 1994;76:117-29.
23. Kascsak RJ, Rubenstein R, Merz PA, Tonna-DeMasi M, Fersko R, Carp RI, Wisniewski HM, Diringer H. Mouse polyclonal and monoclonal antibody to scrapie-associated fibril proteins. *J Virol* 1987;61:3688-93.
24. Diedrich JF, Bendheim PE, Kim YS, Carp RI, Haase AT. Scrapie-associated prion protein accumulates in astrocytes during scrapie infection. *Proc Natl Acad Sci USA* 1991;88:375-9.
25. Muramoto T, Kitamoto T, Hoque MZ, Tateishi J, Goto I. Species barrier prevents an abnormal isoform of prion protein from accumulating in follicular dendritic cells of mice with Creutzfeldt-Jakob disease. *J Virol* 1993;67:6808-10.
26. Hsiao KK, Scott M, Foster D, Groth DF, DeArmond SJ, Prusiner SB. Spontaneous neurodegeneration in transgenic mice with mutant prion protein. *Science* 1990;250:1587-90.
27. Westaway D, Mirenda CA, Foster D, Zebarjadian Y, Scott M, Torchia M, Yang S, Serban H, DeArmond SJ, Ebeling C, et al. Paradoxical shortening of scrapie incubation times by expression of prion protein transgenes derived from long incubation period mice. *Neuron* 1991;7:59-68.

THE USE OF TRANSGENIC MICE IN THE INVESTIGATION OF TRANSMISSIBLE SPONGIFORM ENCEPHALOPATHIES

Charles Weissmann, Marek Fischer, Alex Raeber,
Hans Ruedi Büeler, Andreas Sailer[+], Doron Shmerling,
Thomas Rülicke[+], Sebastian Brandner[#], and Adriano Aguzzi[#]

Institut für Molekularbiologie der Universität Zürich, Abteilung I,
Hönggerberg, 8093 Zürich, [+]Biologisches Zentrallabor and
[#]Institut für Neuropathologie, Universitätsspital Zürich, 8091
Zürich, Switzerland

[+]Current address: Salk Institute, La Jolla, CA 92037, USA

ABSTRACT   The prion, the transmissible agent that causes spongiform encephalopathies is believed to be devoid of nucleic acid and identical with a modified form of the normal host protein PrP^C. The *"protein only"* hypothesis predicts that an animal devoid of PrP^C should be resistant to prion diseases. We generated homozygous PrP null mice (*Prnp$^{o/o}$*) mice and showed that, after inoculation with prions, they remained free of scrapie for at least 2 years while wild-type controls all died within 6 months. There was no propagation of prions in the *Prnp$^{o/o}$* animals. Surprisingly, heterozygous *Prnp$^{o/+}$* mice, which express PrP^C at about half the normal level, also showed enhanced resistance to scrapie disease despite high levels of infectious agent and PrP^{Sc}, a protease-resistant form of PrP, in the brain early on. After introduction of murine PrP transgenes *Prnp$^{o/o}$* mice became highly susceptible to mouse but not to hamster prions, while the insertion of Syrian hamster PrP transgenes rendered them susceptible to hamster but to a much lesser extent to mouse prions. These complementation experiments paved the way to the application of reverse genetics. We have prepared animals transgenic for genes encoding PrP with amino terminal deletions of various lengths and have found that PrP lacking 48 amino proximal amino acids, which comprise four of the five octa repeats of PrP, is still capable of mediating susceptibility to scrapie to PrP null mice.

INTRODUCTION

The unusual properties of the scrapie agent early on gave rise to speculations that it might be devoid of nucleic acid (Alper et al., 1967) Currently, the most widely accepted proposal is the *'protein only' hypothesis*, first outlined in general terms by Griffith (Griffith, 1967) and enunciated in its updated and detailed form by Prusiner (Prusiner, 1989; Prusiner, 1993).

The 'protein only' hypothesis proposes that the prion contains no nucleic acid and is identical with a modified form of PrP^C (Prusiner, 1989). PrP^C is a normal host protein (Basler et al., 1986; Chesebro et al., 1985; Oesch et al., 1985) found predominantly on the outer surface of neurons. PrP^{Sc} is defined as a form of PrP^C that readily forms protease-resistant aggregates after treatment with detergents (McKinley et al., 1991; Oesch et al., 1985). Prusiner proposed that PrP^{Sc}, when introduced into a normal cell, causes the conversion of PrP^C or its precursor into PrP^{Sc}. The exact nature of the conversion is unknown but it is currently ascribed to conformational modification (Cohen et al., 1994); it has been determined that the ß-sheet content of PrP^{Sc} is high while that of PrP^C is low (Caughey et al., 1991; Pan et al., 1993). No chemical differences have so far been found between PrP^C and PrP^{Sc} (Stahl et al., 1993). However, because the ratio of infectious units to PrP^{Sc} molecules is only about 1:100'000 (Bolton et al., 1991), the structure of the PrP molecule actually associated with infectivity cannot be definitively inferred. For this reason and because specific infectivity can vary considerably, the PrP species responsible for infectivity is presently better designated as PrP* (Weissmann, 1991); it may or may not be identical with PrP^{Sc}, the major species that has been characterized chemically and physicochemically. If it is identical, the low specific activity could be due to a low efficiency of infection or to the infectious unit being an aggregate of a large number of PrP^{Sc} molecules. The conclusion that some form of PrP is the essential, perhaps only constituent of the infectious agent is based on biochemical and genetic evidence, as outlined below.

1. Genetic evidence linking the PrP gene with prion disease

Prions are transmitted from one species to another much less inefficiently, if at all, than within the same species and only after prolonged incubation times. In the case of prion transmission from hamsters to mice, this so-called species barrier was overcome by introducing hamster *Prnp* transgenes into recipient wild-type mice (Prusiner et al., 1990; Scott et al., 1989). Importantly, the properties of the prions produced in these transgenic mice corresponded to the prion species used for inoculation (Prusiner et al., 1990), that is, infection with hamster prions led to production of hamster prions but infection with mouse prions gave rise to mouse prions. Within the framework of the 'protein only' hypothesis this means that hamster PrP^C but not murine PrP^C (which differs from the former by 10 amino acids), is a suitable substrate for conversion to hamster PrP^{Sc} by hamster prions and vice versa.

Most, if not all familial forms of human spongiform encephalopathies are linked to one of a number of mutations in the PrP gene (Hsiao et al., 1990); for reviews see refs (Baker and Ridley, 1992; Goldfarb et al., 1994). Prusiner (Prusiner, 1989; Prusiner, 1991) proposed that the mutations allow spontaneous conversion of PrP^C into PrP^{Sc} with a frequency sufficient to allow expression of the disease within the lifetime of the individual. Sporadic CJD could be attributed to rare instances of spontaneous conversion of PrP^C into PrP^{Sc} or rare somatic mutations in the *Prnp* gene. In both cases the initial conversion is thought to be followed by autocatalytic propagation. Hsiao et al. (Hsiao et al., 1990) showed that mice overexpressing a murine PrP transgene with a mutation corresponding to the human GSS mutation Pro102-->Leu spontaneously contract a lethal scrapie-like disease. The brains of these animals contain low levels of infectious prions which can be detected in indicator mice expressing the same mutant transgene but at lower levels which do not lead to spontaneous disease (Hsiao et al., 1994).

2. Resistance to scrapie of mice devoid of PrP^C

The 'protein only' hypothesis predicts that in the absence of PrP^C mice should be resistant to scrapie and fail to propagate the infectious agent.

To generate mice devoid of PrP, we disrupted one *Prnp* allele of murine embryonic stem (ES) cells by homologous recombination with a recombinant DNA fragment in which two thirds of the 254-codon open reading frame were replaced by extraneous DNA. The ES cells were introduced into blastocysts, from which chimeric mice were generated. Appropriate breeding gave rise to offspring homozygous for the disrupted *Prnp* gene (*Prnp$^{o/o}$*). PrP was undetectable in *Prnp$^{o/o}$* brains and present at about half the normal level in the brains of heterozygous (*Prnp$^{o/+}$*) mice (Büeler et al., 1992). No abnormalities were noted in *Prnp$^{o/o}$* mice at the macroscopic, microscopic or behavioral levels (Büeler et al., 1992). The suggestion that there may be a synaptic deficiency in *Prnp$^{o/o}$* mice (Collinge et al., 1994; Whittington et al., 1995) has not been confirmed (Lledo et al., 1996). The claim that aged mice (with a mixed genetic background) develop ataxia and suffer a loss of Purkinje cells (Sakaguchi et al., 1996) as a consequence of PrP gene disruption is not consistent with previous investigations on independently generated *Prnp$^{o/o}$* mouse lines (Büeler et al., 1992; Manson et al., 1994). Because the phenotype might be due to the undefined, mixed genetic background of the knockout mice (Gerlai, 1996), it is necessary to show that complementation with a PrP transgene restores the normal phenotype.

When challenged with mouse prions, mice devoid of PrP were completely protected against scrapie disease (Fig.1a). Prions were not propagated in brains of *Prnp$^{o/o}$* mice at detectable levels, while in scrapie-inoculated *Prnp$^{+/+}$* animals infectious agent was absent up to 2 weeks after inoculation (p.i.) but was present at 8 weeks and increased to about 8.6 log LD$_{50}$ units/ml by 20 weeks p.i. (Table 1) (Büeler et al., 1993; Sailer et al., 1994). As opposed to brain, spleen of *Prnp$^{+/+}$* animals contained infectivity at the earliest time point tested, namely 2 days p.i. and increased thereafter to a level of about 7 log LD$_{50}$ units/ml. In contrast, spleen of knockout animals showed only a low prion level at 4 days p.i., which thereafter became undetectable, suggesting that prions are initially transported from the intracerebral injection site to the spleen, where they are soon degraded. It had previously not been clear whether infectivity in spleen of wild-type animals, particularly at early times, was due to transport from the site of inoculation or whether it was synthesized in the spleen itself. The fact that in spleen of wild-type animals the prion titer is high at 2 weeks, when no infectivity is found in the brain, coupled with the fact that in knockout animals inoculum-derived infectivity has disappeared by that time, strongly suggests that in wild-type animals prions are in fact synthesized in the spleen. It is, incidentally, quite puzzling that following intracerebral injection prion synthesis occurs so early in spleen and only after a long delay in brain.

Interestingly, even heterozygous *Prnp$^{o/+}$* mice were partially protected, inasmuch as they showed prolonged incubation times of about 290 days as compared to about 180 days in the case of the wild-type controls. Moreover, the disease progressed much more slowly in *Prnp$^{o/+}$* mice than in *Prnp$^{+/+}$* mice, the interval between first symptoms and death being about 13 days in the case of *Prnp$^{+/+}$* mice and 150 or more days in *Prnp$^{o/+}$* mice (Fig.2) (Büeler et al., 1994) These and other findings (see below and ref. (Prusiner et al., 1990)) show that susceptibility to scrapie is a function of PrP^C levels in the host.

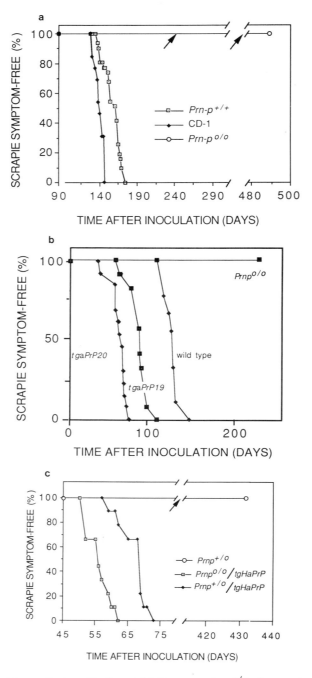

Figure 1. Scrapie resistance of mice with disrupted PrP genes **a.** *Prnp^(o/o)* mice remain symptom-free after inoculation with mouse scrapie prions. *Prnp^(+/+)* litter mates or wild-type CD-1 mice show scrapie symptoms at the times indicated. Arrows: Five mice were killed at various times; none had scrapie symptoms. Modified from ref. (Büeler et al., 1993). **b.** *Prnp^(o/o)* mice were rendered transgenic for *Prnp* genes. *tga19/+* mice had 3-4 times the normal PrP^C level; *tga20/+* mice had 6-7 times the normal level. From ref. (Fischer et al., 1996). **c.** *Prnp^(o/o)* and *Prnp^(o/+)* mice with hamster PrP transgenes at different times after inoculation with hamster scrapie prions. Groups of 9-11 mice of each genotype were inoculated with the Sc237 isolate of hamster prions. Arrow: One animal died spontaneously without scrapie symptoms and one was killed because of a tumor. Modified from Büeler et al. (1993).

Table 1. Prion titers in brain and spleen of $Prnp^{+/+}$ and $Prnp^{o/o}$ mice

Log LD50 units/ml

Time after inoculation	Brain		Spleen	
	$Prnp^{+/+}$	$Prnp^{o/o}$	$Prnp^{+/+}$	$Prnp^{o/o}$
4 days	< 1.5	2.0	5.7 ± 0.9	2.3
2 weeks	< 1.5	< 1.5	6.2 ± 0.8	< 1.5
8 weeks	5.4	< 1.5	6.9 ± 1.0	< 1.5
12 weeks	6.8	< 1.5	5.9 ± 0.6	< 1.5
20 weeks	8.6	< 1.5	6.9 ± 0.6	< 1.5
23/25 weeks	8.1 ± 0.8	< 1.5	n.d.	< 1.5

Mice with the genotype indicated were inoculated intracerebrally with mouse prions. Titers were determined by end point titration on homogenates of pooled organs from 4 mice after heating for 20 min at 80C. Data from ref. (Büeler et al., 1993).

It is evident that mice can carry high levels of scrapie infectivity and PrP^{Sc} in their brain without showing clinical disease; the same might be true for humans and other animals.

When a modified phenotype is observed following a targeted mutation or gene disruption, it is important to show that the original phenotype -in this case, susceptibility to scrapie- can be recovered by restoring wild-type function. We therefore introduced murine $Prnp$ transgenes into $Prnp^{o/o}$ mice and obtained several lines with varying expression levels of PrP^C. As shown in Fig.1b, knockout mice expressing $Prnp$ transgenes became susceptible to mouse prions; in fact, the higher the PrP^C content of the brain, the shorter the incubation times (Fischer et al., 1996). Even more interestingly, introduction of multiple hamster $Prnp$ transgenes into $Prnp^{o/o}$ mice rendered them very susceptible to hamster-derived prions (56 days incubation time) (Fig.1c) but much less so to mouse-derived prions (303 days incubation time) (Büeler et al., 1993), demonstrating the requirement for a homotypic relationship between incoming prion and resident PrP protein for optimal prion propagation and development of pathology, as foreshadowed by the results of Prusiner et al. (Prusiner et al., 1990) described above.

3. Reverse genetics

The demonstration that disruption of the PrP gene confers resistance to scrapie and reintroduction of a PrP-encoding transgene restores susceptibility to the disease opens up the possibility of practising reverse genetics on PrP, that is, introducing deletions or mutations into the $Prnp$ gene and determining the capacity of the modified gene to confer susceptibility to scrapie to a PrP knockout mouse.

As mentioned above, protease treatment of prion preparations cleaves off about 60 amino terminal residues of PrP^{Sc} (Hope et al., 1988) but does not abrogate infectivity (McKinley et al., 1983). We introduced into PrP knockout mice transgenes encoding wild-type PrP or PrP lacking 26 or 49 amino proximal amino acids which are protease-susceptible in PrP^{Sc}. Inoculation with prions led to fatal disease (Fig.3), prion propagation and accumulation of PrP^{Sc} both in mice expressing wild-type and truncated PrPs (Fischer et al., 1996). Within the framework of the "protein only" hypothesis this means that the amino proximal segment of PrP^C, which contains 3.5 of its 5 octa repeats,

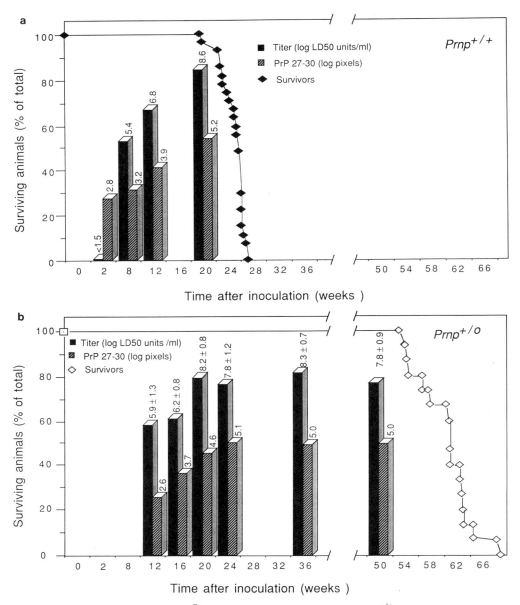

Figure 2. Survival, prion titers and PrP^Sc in brains of (**a**) wild type and (**b**) *Prnp^o/+* mice at various times after inoculation with mouse prions. From Büeler et al. (1994).

is required neither for its susceptibility to conversion into the pathogenic, infectious form of PrP nor for the generation of PrP^Sc.

4. Implications and outlook

While each individual piece of evidence described above could be explained in several ways, the conjunction of data strongly supports the proposal that the prion is composed partly or entirely of a PrP-derived molecule (PrP* or PrP^Sc), and that protein-

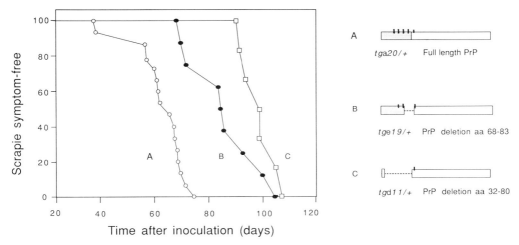

Figure 3. Susceptibility to scrapie of *Prnp* knockout mice carrying *Prnp* transgenes with amino proximal deletions. *Prnp* transgenes with the structures indicated in the right half of the panel were introduced into *Prnp*[o/o] mice. *tga20/+* (A, wild-type), *tge19/+* (B, a deletion removing 2 octa repeats) and *tgd11/+* (C, a deletion removing 3.5 octa repeats and additional N-proximal sequences) expressed PrP at levels 6-7, 3-4 and 6 times, respectively, those of wild-type. Mice were inoculated intracerebrally with mouse prions. Data from Fischer et al. (1996).

encoding nucleic acid is not an essential component. Probably the closest one could come to irrefutable proof for the 'protein only' hypothesis would be the demonstration that biosynthetic, pure PrP[C] can be converted not only into a protease-resistant form, but to infectious scrapie agent under defined conditions *in vitro*.

Because mice with disrupted *Prnp* genes are viable and resistant, it should be possible to breed sheep or cattle that are resistant to this disease; knockout methodology is in principle available for sheep (Campbell et al., 1996) but not yet for cattle. While it is hardly practical to consider complete replacement of conventional animals by PrP knockout counterparts, herds of BSE-resistant cattle would be useful as a source for products required for pharmaceutical purposes. Moreover, the fact that *Prnp*[o/+] heterozygous mice show prolonged scrapie incubation times argues that an (as yet conjectural) drug leading to moderate reduction of PrP[C] synthesis or one retarding the conversion of PrP[C] to PrP[Sc] (Demaimay et al., 1994) might substantially mitigate disease progression in incipient cases of human spongiform encephalopathies.

Finally one may raise the question whether prion-like agents cause other diseases or appear in non-vertebrate organisms. Although several human diseases accompanied by amyloid formation are known, none of them have been reproducibly transmitted. Interestingly, two yeast phenotypes are ascribed to "heritable protein conversion", namely the [URE3] and the [psi+] systems (Chernoff et al., 1995; Masison and Wickner, 1995), opening new perspectives for the elucidation of this phenomenon.

Acknowledgements

This work was supported by the Erziehungsdirektion of the Kanton of Zürich and grants of the Schweizerische Nationalfonds and the Human Frontier Science Program to C.W. I thank H.Büeler, M.Fischer, M.Aguet, A.Sailer, A.Raeber, A.Aguzzi, H.-P.Lipp, T.Rülicke, S.Brandner and D.Shmerling for their important contributions to the work carried out in Zürich.

REFERENCES,

Alper, T., Cramp, W. A., Haig, D. A. and Clarke, M. C. (1967). Does the agent of scrapie replicate without nucleic acid? *Nature 214*, 764-766.

Baker, H. F. and Ridley, R. M. (1992). The genetics and transmissibility of human spongiform encephalopathy. *Neurodegeneration 1*, 3-16.

Basler, K., Oesch, B., Scott, M., Westaway, D., Wälchli, M., Groth, D. F., McKinley, M. P., Prusiner, S. B. and Weissmann, C. (1986). Scrapie and cellular PrP isoforms are encoded by the same chromosomal gene. *Cell 46*, 417-428.

Bolton, D. C., Rudelli, R. D., Currie, J. R. and Bendheim, P. E. (1991). Copurification of Sp33-37 and scrapie agent from hamster brain prior to detectable histopathology and clinical disease. *J Gen Virol 72*, 2905-2913.

Büeler, H., Aguzzi, A., Sailer, A., Greiner, R. A., Autenried, P., Aguet, M. and Weissmann, C. (1993). Mice devoid of PrP are resistant to scrapie. *Cell 73*, 1339-1347.

Büeler, H., Fischer, M., Lang, Y., Bluethmann, H., Lipp, H.-P., DeArmond, S. J., Prusiner, S. B., Aguet, M. and Weissmann, C. (1992). Normal development and behaviour of mice lacking the neuronal cell- surface PrP protein. *Nature 356*, 577-582.

Büeler, H., Raeber, A., Sailer, A., Fischer, M., Aguzzi, A. and Weissmann, C. (1994). High prion and PrPSc levels but delayed onset of disease in scrapie-inoculated mice heterozygous for a disrupted PrP gene. *Molecular Medicine 1*, 19-30.

Campbell, K. H., McWhir, J., Ritchie, W. A. and Wilmut, I. (1996). Sheep cloned by nuclear transfer from a cultured cell line. *Nature 380*, 64-66.

Caughey, B. W., Dong, A., Bhat, K. S., Ernst, D., Hayes, S. F. and Caughey, W. S. (1991). Secondary structure analysis of the scrapie-associated protein PrP 27-30 in water by infrared spectroscopy. *Biochemistry 30*, 7672-7680.

Chernoff, Y. O., Lindquist, S. L., Ono, B., Inge-Vechtomov, S. G. and Liebman, S. W. (1995). Role of the chaperone protein Hsp104 in propagation of the yeast prion- like factor [psi+]. *Science 268*, 880-884.

Chesebro, B., Race, R., Wehrly, K., Nishio, J., Bloom, M., Lechner, D., Bergstrom, S., Robbins, K., Mayer, L., Keith, J. M., Garon, C. and Haase, A. (1985). Identification of scrapie prion protein-specific messenger RNA in scrapie-infected and uninfected brain. *Nature 315*, 331-333.

Cohen, F. E., Pan, K. M., Huang, Z., Baldwin, M., Fletterick, R. J. and Prusiner, S. B. (1994). Structural clues to prion replication (see comments). *Science 264*, 530-1.

Collinge, J., Whittington, M. A., Sidle, K. C. L., Smith, C. J., Palmer, M. S., Clarke, A. R. and Jefferys, J. G. R. (1994). Prion protein is necessary for normal synaptic function. *Nature 370*, 295-297.

Demaimay, R., Adjou, K., Lasmezas, C., Lazarini, F., Cherifi, K., Seman, M., Deslys, J. P. and Dormont, D. (1994). Pharmacological studies of a new derivative of amphotericin B, MS-8209, in mouse and hamster scrapie. *J Gen Virol 75*.

Fischer, M., Rülicke, T., Raeber, A., Sailer, A., Moser, M., Oesch, B., Brandner, S., Aguzzi, A. and Weissmann, C. (1996). Prion protein (PrP) with amino-proximal deletions restoring susceptibility of PrP knockout mice to scrapie. *EMBO J. 15*, 1255-1264.

Gerlai, R. (1996). Gene-targeting studies of mammalian behaviour: is it the mutation or the background genotype ? *Trends Neurosci. 19*, 177-181.

Goldfarb, L. G., Brown, P., Cervenakova, L. and Gajdusek, D. C. (1994). Genetic analysis of Creutzfeldt-Jakob disease and related disorders. *Philos Trans R Soc Lond B Biol Sci 343*, 379-84.

Griffith, J. S. (1967). Self-replication and scrapie. *Nature 215*, 1043-1044.

Hope, J., Multhaup, G., Reekie, L. J., Kimberlin, R. H. and Beyreuther, K. (1988). Molecular pathology of scrapie-associated fibril protein (PrP) in mouse brain affected by the ME7 strain of scrapie. *Eur J Biochem 172*, 271-277.

Hsiao, K. K., Groth, D., Scott, M., Yang, S. L., Serban, H., Rapp, D., Foster, D., Torchia, M., Dearmond, S. J. and Prusiner, S. B. (1994). Serial transmission in rodents of neurodegeneration from transgenic mice expressing mutant prion protein. *Proc Natl Acad Sci U S A 91*, 9126-30.

Hsiao, K. K., Scott, M., Foster, D., Groth, D. F., DeArmond, S. J. and Prusiner, S. B. (1990). Spontaneous neurodegeneration in transgenic mice with mutant prion protein. *Science 250*, 1587-1590.

Lledo, P.-M., Tremblay, P., DeArmond, S. J., Prusiner, S. B. and Nicoll, R. A. (1996). Mice deficient for prion protein exhibit normal neuronal excitability and synaptic transmission in the hippocampus. *Proc.Natl.Acad.Sci.U.S.A. 93*, 2403-2407.

Manson, J. C., Clarke, A. R., Hooper, M. L., Aitchison, L., McConnell, I. and Hope, J. (1994). 129/Ola mice carrying a null mutation in PrP that abolishes mRNA production are developmentally normal. *Mol Neurobiol 8*, 121-127.

Masison, D. C. and Wickner, R. B. (1995). Prion-inducing domain of yeast Ure2p and protease resistance of Ure2p in prion-containing cells. *Science 270*, 93-95.

McKinley, M. P., Bolton, D. C. and Prusiner, S. B. (1983). A protease-resistant protein is a structural component of the scrapie prion. *Cell 35*, 57-62.

McKinley, M. P., Meyer, R. K., Kenaga, L., Rahbar, F., Cotter, R., Serban, A. and Prusiner, S. B. (1991). Scrapie prion rod formation in vitro requires both detergent extraction and limited proteolysis. *J Virol 65*, 1340-1351.

Oesch, B., Westaway, D., Walchli, M., McKinley, M. P., Kent, S. B., Aebersold, R., Barry, R. A., Tempst, P., Teplow, D. B., Hood, L. E. and al., e. (1985). A cellular gene encodes scrapie PrP 27-30 protein. *Cell 40*, 735-746.

Pan, K. M., Baldwin, M., Nguyen, J., Gasset, M., Serban, A., Groth, D., Mehlhorn, I., Huang, Z., Fletterick, R. J., Cohen, F. E. and Prusiner, S. B. (1993). Conversion of alpha-helices into beta-sheets features in the formation of the scrapie prion proteins. *Proc Natl Acad Sci U S A 90*, 10962-10966.

Prusiner, S. B. (1989). Scrapie prions. *Annu Rev Microbiol 43*, 345-374.

Prusiner, S. B. (1991). Molecular biology of prion diseases. *Science 252*, 1515-1522.

Prusiner, S. B. (1993). Transgenetic investigations of prion diseases of humans and animals. *Philos Trans R Soc Lond Biol 339*, 239-254.

Prusiner, S. B., Scott, M., Foster, D., Pan, K. M., Groth, D., Mirenda, C., Torchia, M., Yang, S. L., Serban, D., Carlson, G. A., Hoppe, P. C., Westaway, D. and DeArmond, S. J. (1990). Transgenetic studies implicate interactions between homologous PrP isoforms in scrapie prion replication. *Cell 63*, 673-686.

Sailer, A., Büeler, H., Fischer, M., Aguzzi, A. and Weissmann, C. (1994). No propagation of prions in mice devoid of PrP. *Cell 77*, 967-968.

Sakaguchi, S., Katamine, S., Nishida, N., Moriuchi, R., Shigematsu, K., Sugimoto, T., Nakatani, A., Kataoka, Y., Houtani, T., Shirabe, S., Okada, H., Hasegawa, S., Miyamoto, T. and Noda, T. (1996). Loss of cerebellar Purkinje cells in aged mice homozygous for a disrupted PrP gene. *Nature 380*, 528-531.

Scott, M., Foster, D., Mirenda, C., Serban, D., Coufal, F., Wälchli, M., Torchia, M., Groth, D., Carlson, G., DeArmond, S. J., Westaway, D. and Prusiner, S. B. (1989). Transgenic mice expressing hamster prion protein produce species-specific scrapie infectivity and amyloid plaques. *Cell 59*, 847-857.

Stahl, N., Baldwin, M. A., Teplow, D. B., Hood, L., Gibson, B. W., Burlingame, A. L. and Prusiner, S. B. (1993). Structural studies of the scrapie prion protein using mass spectrometry and amino acid sequencing. *Biochemistry 32*, 1991-2002.

Weissmann, C. (1991). Spongiform encephalopathies. The prion's progress (news). *Nature 349*, 569-571.

Whittington, M. A., Sidle, K. C., Gowland, I., Meads, J., Hill, A. F., Palmer, M. S., Jefferys, J. G. and Collinge, J. (1995). Rescue of neurophysiological phenotype seen in PrP null mice by transgene encoding human prion protein. *Nat Genet 9*, 197-201.

LARGE-SCALE SEQUENCING OF HUMAN, MOUSE, AND SHEEP PRION PROTEIN GENES

Inyoul Lee[1], David Westaway[2], Arian Smit[1], Carol Cooper[3],
Hong Yao[2], Stanley B. Prusiner[3,4] and Leroy Hood[1]

1 Department of Molecular Biotechnology
University of Washington, Seattle, WA
2 Centre for Research in Neurodegenerative Diseases
and Department of Pathology, University of Toronto
Toronto, Ontario M5S3H2, Canada
3 Department of Neurology and of
4 Biochemistry and Biophysics
University of California, San Francisco, CA 94143

SUMMARY

The prion protein (PrP), first identified in scrapie-infected rodents, is encoded by a single-copy chromosomal gene. PrP genes are conserved and have been identified in 13 species of mammals. Cloning and analysis of the PrP gene has revealed that pathogenic (PrPSc) and cellular (PrPC) isoforms of the prion protein do not reflect alternative splicing events but instead share the same amino acid sequence: these data first suggested the hypothesis that prion isoforms are alternate conformers1. However, at this time the mechanism involved in the transition between PrPC and PrPSc is obscure, as is the normal function of PrPC. We have therefore sought to identify PrP-linked genes that might feature in conformational transitions, and conserved regulatory elements that might offer insights into the physiology of PrPC. Accordingly, 145 kb of DNA from human and sheep PrP genes, and two alleles of Prn-p has been analysed for conserved regions, ORFs, potential exons, repetitive sequences, putative CpG islands, and possible polymorphic motifs. These sequences reveal unexpected features within the wild-type PrP genes of each of these species. The predominant allele of the mouse gene, Prn-pa, found in 44 inbred laboratory strains contains a 6878 nucleotide retroviral genome inserted into the anticoding strand of intron 2. This intracisternal A particle (IAP) element is (i) flanked by duplications of a AAGCTT nucleotide motif found once in the *Prn-p*b gene, and (ii) highly related to a transpositionally-competent prototype IAP element, differing principally in small deletion of the pol gene: these data are indicative of a recent transpositional orgin. In the case of the sheep PrP gene, the unusually long 3' untranslated region (UTR) reflects in most part the presence of a "fossil" 1.2 kb mariner-like transposable element, which is absent from the human and mouse PrP genes. Chromosomal instability associated with mariner-like elements on chromosome 17 in humans suggests that DNA rearrangements may take place adjacent to the PrP gene in sheep, and also in cattle (which also exhibit an extended 3' UTR and presumably harbor a mariner-like element). Lastly, the large intron of the human PrP gene contains a sequence analogous to exon 2 of the mouse and sheep PrP genes, flanked by consensus splice acceptor and donor sites: while it remains to be established that "exon 2" sequences are included in a subset of human PrP mRNAs, these data indicate that an ancestral PrP gene most likely exhibited a three exon structure.

INTRODUCTION

Scrapie in sheep and goats, Creutzfeldt-Jakob Disease (CJD) in humans, and bovine spongiform encephalopathy (BSE), are caused by prions. Attempts to define a nucleic acid genome within these unusual infectious pathogens have been unsuccessful: instead, prions have been found to be composed largely or exclusively of an aberrant isoform (PrPSc) of a host-encoded protein, PrPC. Pulse-chase studies in prion-infected cells indicate that synthesis of PrPSc (which is protease-resistant and detergent-insoluble) from PrPC is a posttranslational event: since these protein isoforms have distinct contents of alpha-helical and beta-sheet structure it is likely that conformational changes play a fundamental role in the synthesis of PrPSc and prion infectivity(Cohen et al. 1994). Molecular genetic studies underscore a pivotal role for the prion protein as non-conservative amino acid substitutions in the human PrP gene are responsible for the dominant neurodegenerative disorders Gerstmann-Sträussler-Scheinker syndrome (GSS), familial CJD, and fatal familial insomnia (FFI)(Hsiao et al. 1992)(Medori et al. 1992), all of which are transmissible to experimental animals. In mice, PrP missense mutations in codons 108 and 189 co-segregate with a gene which profoundly affects susceptibility to experimental scrapie(Carlson et al. 1986).

In accord with the above findings, in experimental scrapie and CJD, the primary structure of host-encoded PrPC exerts a powerful influence over the ability of prions to cross from one species into another(Prusiner et al. 1990)(Telling et al. 1995). However it is also clear that factors besides PrP genotype can alter the course of prion infections. Description of distinct prion "strains" within the same inbred host genetic background suggests that prions contain either a second component(Bruce 1993), perhaps a peptide or lipid co-factor, or that "strains" comprise alternative, stable conformations of PrP distinct from that of PrPC(Bessen and Marsh 1992)(Bessen 1995). Some "strains" of prions are thought to transit more readily than others from one species to another. BSE, which appears to be comprised one strain of agent which has spread to domestic cats, exotic ungulates, ostriches, and (potentially) humans, all of which encode PrPs differing from bovine PrPC, is a particularly troubling case in point. Lastly, besides PrP genotype and "strain" type, modifier genes define a third variable, which affects disease penetrance and manifestation. Potential modifier genes include ApoE on chromosome 19(Amouyel 1995) (however see(Saunders et al. 1993)(Zerr et al. 1994)), the gene encoding protein X (which may interact with the C-terminal region of PrP(Telling et al. 1995)), and an unidentified locus on mouse chromosome 2 affecting the degree of CNS vacuolation(Carlson et al. 1994).

The biochemical function of PrPC itself is unclear. Syntenic localization between humans and mice argue that the PrP gene existed prior to the speciation of mammals (Sparkes et al. 1986). In mammals, the open reading frames (ORF) encoding PrP generally exhibit ~90% homology. The degree of homology rises to >95% when considering primates(Schätzl et al. 1994), is somewhat lower in marsupials (~80% homologous(Windl et al. 1995)), and is ~30% in the instance of chicken PrP(Gabriel et al. 1991). Attempts to find PrP-related genes in lower eukaryotes have to date been unsuccessful(Oesch et al. 1991). In all species studied, the PrP ORFs encode proteins of ~250 amino acids. All PrP molecules seem to be post-translationally modified by removal of an N-terminal signal peptides, cleavage of a C-terminal signal sequence upon additional of a GPI anchor and addition of 2 or 3 Asn-linked sugar chain(Stahl et al. 1987)(Gabriel et al. 1992). Since only subtle (and sometimes contradictory) phenotypic traits have been described in three independently-derived mouse lines homozygous for an ablation of the PrP gene(Bueler et al. 1992)(Manson et al. 1994)(Sakaguchi et al. 1996), it is formally possible that PrPC is functionless and that PrP genes have been retained in mammals by evolutionary inertia. More likely, PrPC serves a function that is duplicated by other (perhaps PrP-like) macromolecules.

Since functionally related mammalian genes may be organized into gene complexes, one means to address the issues raised above is to perform large-scale DNA sequencing. This task was undertaken on phage and cosmid clones encompassing the human, sheep, and mouse PrP genes. Primary objectives included the definition of genes adjacent to PrP which might encode PrP-like proteins or "private" chaperones that might interact with PrPC. The acquired sequences define two large transposon insertions within the transcribed regions of the mouse and sheep PrP genes, and a cryptic untranslated exon within the human PrP gene.

RESULTS

Principal Features of the nucleotide sequences

Physical maps of the three wild-type PrP genes (human, mouse "a" allele, and sheep) are presented in Figure 1. With sequences from two species it can be difficult to decipher whether a difference in size of a particular region affects a deletion in one gene, or an insertion in the other. The availability of three sequences is a great help in this respect, allowing the boundaries of many insertion events or complex rearrangements restricted to a single species to be assigned with reasonable accuracy. Insertion events are surprisingly common in gene evolution and the current estimate of sequences derived from mobile elements within the human genome is 30%. Our analyses reveal that at least 35% of the human *PRNP* locus seems to be derived from mobile elements, in close agreement with this figure. While most insertion events, typically involving the introduction of small elements such as Alu into flanking or intronic sequences will be phenotypically neutral, studies presented here reveal two unexpected and comparatively large sequence transpositions into the transcribed regions of the mouse and sheep PrP genes.

Mouse PrP gene alleles and an IAP insertion

To determine the complete structure of the mouse *Prnpa* region, two lambda clones, lambda 4 and lambda 7, and a plasmid subclone of a lambda clone 6 (all derived from a 129Sv mouse genomic DNA library(Westaway et al. 1994)) were sequenced and merged to a 38,761 bp long "contig" (Fig. 2).

Comparison of the prion gene non-coding exons regions.

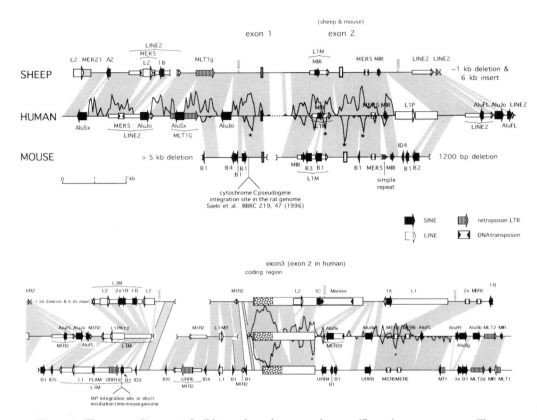

Figure 1. Three-way alignment of wild-type sheep, human, and mouse (*Prn-p*a) gene sequences. The complete sheep and human sequences are shown, whereas the mouse *Prn-p*a sequence extends 2.5 kb upstream and the *Prn-p*b mouse cosmid (not presented) 11 kb downstream. Common areas of homology are shaded. In two regions with deletions in the human sequence, homologous sequences between sheep and mouse are indicated by outlined shfding. Deletions in the genome larger than 250 bp are indicated by brackets. An approximately 350 bp region in in human intron 1 for which the sequence could not be obtained is shown with a dashed line. Repeats were identified as described. Some ancient elements were inferred to be present in the sheep and mouse from the identification of repeats at homologous postions in the the human DNA (the substitution level in the human lineage since the mammalian radiation has been lower than in other lineages, thus facilitating detection of ancient elements). identified interspersed repeats are represented by four differently shaded arrows for the four major classes of repeats: SINES, LINES, LTR elements, and DNA transposons. Wide arrows are used for elements that are absent from othologous sites the other two species and thus probably integrated after the mammalian radiation. Older elements are indicated with narrow arrows. A measure for similarity between the human and either sheep or mouse sequences, depicted by a graph above and below the human sequence, respectively, is derived from the cross_match (Smith-Waterman) scores of 100 bp fragments (overlapping by 50 bp) of human sequences with their orthologous sites in sheep and mouse. A cut-off score, the lowest above which all matches empirically were found to be significant (Smit and Green, unpublished data) was subtracted from the total score to form the baseline. Regions outside the coding region that are conserved in both sheep and mouse were studied in detail. Alignments of the most interesting regions are shown in figure 7.

Several restriction site polymorphisms have been mapped 5' to the mouse PrP coding region, within intron 2(Westaway et al. 1987)(Carlson et al. 1986): subsequent studies of cloned DNA revealed that *a* and *b* Prn-p alleles differed in size by 6.7 kb in the polymorphic region: since the *a* allele, found in 44 inbred mouse strains corresponds the larger size variant(Carlson et al. 1988), the smaller version of intron 2 present in *Prn-p*b mice such as I/LnJ mice was interpreted to represent a deletion event(Westaway et al. 1994). Sequencing defines this intron 2 size polymorphism (17,732 bp long in Prn-pa mice vs. 11,126 bp in *Prn-p*b mice) and reveals its molecular basis as an intracisternal A particle (IAP) element insertion in the wild-type *Prn-p*a allele (Figs. 1, 2, 3).

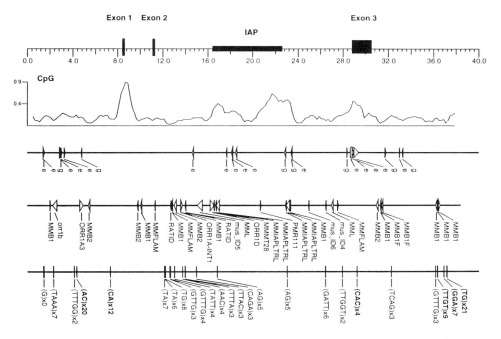

Figure 2. Structure of the mouse *Prn-p*a allele. Top-line physiacal map, second line, potential coding sequences revealed by the grail program are designated "e" excellent = 90 % probability or "g", good 60 % probability. Positions of repetitive elements and simple nucleotide repeats are shown on the third and fourth lines below the sequence.

IAPs are defective retroviruses, unable to spread by horizontal infection, and encoded by a gene family of about1,000 members present in the genome of many species of rodents reviewed in Kuff and Lueders, 1988(Kuff and Lueders 1988). The molecular clone MIA14 constitutes a prototype for a full-length, biologically active (i.e. transposition-competent) IAP genome: this retroposon is 7,095bp long and contains two identical 338 bp LTRs and 3 ORFs: the gag ORF with 827 aa for a major 73,000-molecular-weight gag-equivalent structural protein, the pol ORF with 867 aa for a reverse transcriptase, and the env ORF of 1100 bp with multiple stop codons. The IAP inserted in the large intron 2 of mouse *Prn-p*a is found ~5kb upstream of the translation start codon of exon 3, in the opposite transcriptional orientation. The nucleotide sequence of *Prn-p*a-IAP was compared with the full-size IAP genome of MIA14 using the "Inherit Analysis" program for a dot plot comparison. The results in

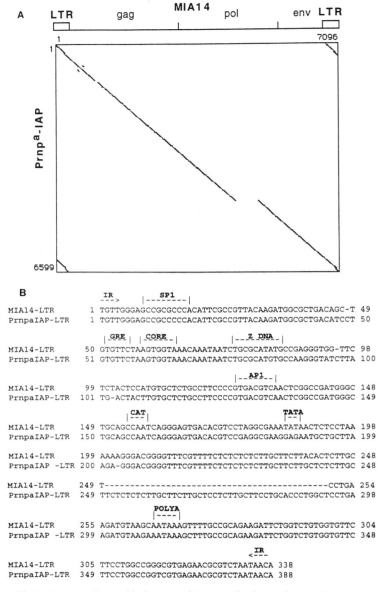

A

LTR   gag   MIA14   pol   env LTR

1   7096

Prnp^a-IAP

1

6599

B

```
                          IR,      |___SP1___|
                          --->
MIA14-LTR       1 TGTTGGGAGCCGCGCCCACATTCGCCGTTACAAGATGGCGCTGACAGC-T  49
PrnpaIAP-LTR    1 TGTTGGGAGCCGCCCCCACATTCGCCGTTACAAGATGGCGCTGACATCCT 50

                   |_GRE_| |_CORE_|            |___Z_DNA__|
MIA14-LTR      50 GTGTTCTAAGTGGTAAACAAATAATCTGCGCATATGCCGAGGGTGG-TTC  98
PrnpaIAP-LTR   51 GTGTTCTAAGTGGTAAACAAATAATCTGCGCATGTGCCAAGGGTATCTTA 100

                                        |__AP1__|
MIA14-LTR      99 TCTACTCCATGTGCTCTGCCTTCCCCGTGACGTCAACTCGGCCGATGGGC 148
PrnpaIAP-LTR  101 TG-ACTACTTGTGCTCTGCCTTCCCCGTGACGTCAACTCGGCCGATGGGC 149

                      |_CAT_|                        |TATA|
MIA14-LTR     149 TGCAGCCAATCAGGGAGTGACACGTCCTAGGCGAAATATAACTCTCCTAA 198
PrnpaIAP-LTR  150 TGCAGCCAATCAGGGAGTGACACGTCCGAGGCGAAGGAGAATGCTGCTTA 199

MIA14-LTR     199 AAAAGGGACGGGGTTTCGTTTTCTCTCTCTCTTGCTTCTTACACTCTTGC 248
PrnpaIAP -LTR 200 AGA-GGGACGGGGTTTCGTTTTCTCTCTCTCTTGCTTCTTGCTCTCTTGC 248

MIA14-LTR     249 T--------------------------------------------CCTGA 254
PrnpaIAP-LTR  249 TTCTCTCTCTTGCTTCTTGCTCCTCTTGCTTCCTGCACCCTGGCTCCTGA 298

                      |POLYA|
                      |----|
MIA14-LTR     255 AGATGTAAGCAATAAAGTTTTGCCGCAGAAGATTCTGGTCTGTGGTGTTC 304
PrnpaIAP -LTR 299 AGATGTAAGAAATAAAGCTTTGCCGCAGAAGATTCTGGTCTGTGGTGTTC 348

                                               IR
                                               <---
MIA14-LTR     305 TTCCTGGCCGGGCGTGAGAACGCGTCTAATAACA 338
PrnpaIAP-LTR  349 TTCCTGGCCGGTCGTGAGAACGCGTCTAATAACA 388
```

Figure 3. The IAP genome inserted in intron 2 of *Prn-p*a. The figure shows a dot-matrix alignment with the biologically-active MIA 14 prototype element.

Fig. 3 show a small internal deletion in *Prn-p*a-IAP (see below) and extensive homology over the remaining same regions of the IAP genomes. These region include the LTRs and the putative gag coding region with reading frame changes and env coding region with multiple stop codons, as observed in other integrated IAPs.

The IAP element is 6,878bp, in very close agreement with restriction mapping studies noted above. The element is flanked by a 6-bp duplication of cellular sequences, AAGGCT, present once in the corresponding region of the Prn-pb allele.

The 5' and 3' LTRs are 388 bp, and are identical: this identity is a special feature of *Prn-p*a-IAP genome, since in other inserted IAPs several differences have been found between the 5' and 3' LTRs(Kuff and Lueders 1988). While this identity suggests a relatively recent transposition into *Prn-p*a, it should be noted that the intron 2 IAP element differs from the MIA 14 prototype in that it bears a 217 bp deletion within the polymerase ("pol") gene (Fig. 2). Sequence comparisons were also undertaken with a cDNA of an IAP-related mRNA which is up-regulated in scrapie infected neuroblastoma cells: these analyses revealed minimal homology centred around the pol genes (not presented), indicating that the cDNA described by Doh-ura et. al. does not originate from within *Prn-p*(Doh-ura et al. 1995).

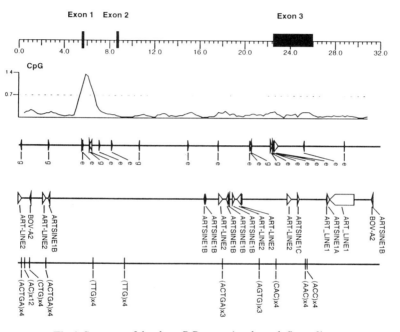

Fig 4. Structure of the sheep PrP gene. (see legend, figure 2)

A mariner-like transposon fossil in the Sheep PrP gene

The sequence of the sheep PrP cosmid from the scrapie-susceptible Suffolk breed confirms and extends earlier studies (Fig. 4). The size of intron 2 deduced from sequencing was 14,030 bp, though further studies may be needed to confirm this figure as instabilites were encountered propagating molecular clones in E. coli (unpublished data of Hong Yao and David Westaway). The sheep PrP gene has the largest 3' UTR described to date (3,246 bp). Surprisingly these untranslated sequences differed at 82 positions from those reported by Goldmann and co-workers. Although errors in our sequence cannot be excluded, at least one highly "polymorphic" region between nucleotides ~25800 and 26,200 was covered by between 6 and 8 independent M13 clones. Furthermore, our sequence is fully concordant with the partial 3' UTR sequence of Shinegawa et. al. (GENBANK accession D38179) in an area which includes some of the disparities with the sequence of Goldmann et. al. Thus, errors may exist in the earlier sequence for this region, or the 3'UTR, perhaps being under minimal selection pressure, may evolve rapidly.

A 1.2 kb mariner-like DNA transposable element was detected in the sheep PrP 3' UTR. "TBLASTN" searches of the human and mouse PrP genes, using a mariner consensus sequence (representing the conserved residues of four mariner subfamilies in insects), were negative. Many transposons of the mariner family described have been studied in mammals, invertebrates and in plant species(Auge-Gouillou et al. 1995)(Oosumi et al. 1995)(Robertson 1995). They possess inverted terminal repeats (ITRs of 8-40 bp) and encode a transposase. The transposase pseudogene in the sheep prion is interrupted by seven or eight frameshifts and five stop codons, indicative of an ancient rather than recent transpositional origin (Fig 5). The deduced amino-acid sequence of the transposase is similar to that of the mariner transposase from the Mellifera (honeybee) subfamily.

The Human PrP gene contains an exon 2-like sequence

The nucleotide sequence determined for the human PrP region from a cosmid clone(Puckett et al. 1991) was 35,204 bp in length and had a dG+dC content of 44.7 % (Fig. 6). Based on the analysis of 2 clones obtained from a retinal cDNA library

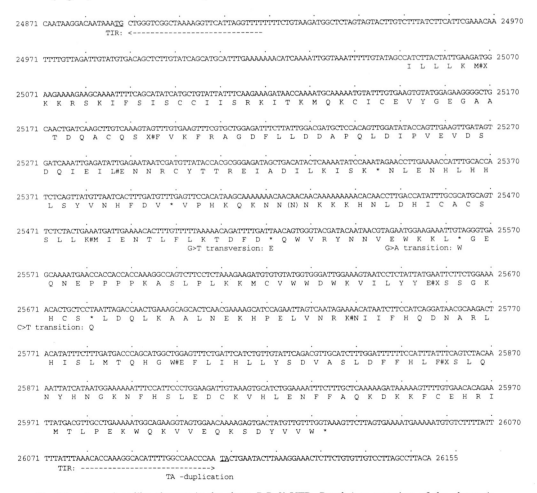

Fig 5A. A mariner-like element in the sheep PrP 3' UTR. Panel A = overview of the element's structure.

66

```
Sheep_mar.       KKKHNLDHICACSSLLKMIENTLFLKTDX--DEQWVRY---NNVEWKKLWGEQNEPPX--PKASLPLKK--MCVWW
Cat flea 10.6    TQKNLLDRINACDMLLKRNELDPFLKRMVTGDEKWISY---DNIKRKRSWSKAGESSQTVAKPGLMAREVLLCVWW
Deer fly 9.2     SVKNLMDRINICDALLKRNEIQPFLKQLITGDEKWITY---DNRKRKRSWIKDGEPSQVFAKPGLTFKKIMLCVWW
                  **  *      *   *       *              **                                *

Human_mar.       TANQKNRRFEVSSSLILRNNNEPFLDRIVTCDEKWILY---DNRRRPAQWLDREEAPKHFPKPNLHQKKVMVTVWW
House ant 16.2   NQKQMDNRFEACISLLSRHKSEPFLHRIVTCDEKWILF---DNRKRKSASWLDKNESPKHCPKQNIHQKKLMVSVWW
Carpelimus33.2   NENQKNRRFEVSSALLLRNNNDPFLDRIVTCDEKWILY---DNRRRSAQWLDRDQAPQHFPKPALHQKKVMVTVWW


Humar1           DQLQTXAELSMEILXKWNRDSEAFLRRIVTGDETWLYQYNPEDKAQSKQWLPRGGSGPVKAKVDRSRAKVMATAFL
Horn_fly         DQKQQRVDDSERCLQLLTRNTPEFFRRYVTMDETWLHHYTPESNRQSAQWTATGEPAPKRGKTQKSAGKVMASVFW
Silverfish 8.1   TQKQKRVESCQQLX-QYSENPTEFFERLVTVDETWFLYETPEKKRQSMEWRHTGSPRPKKARMGLFVRKEMATVFW


Sheep_mar.       DWKVILYYE-------XSSGKHCSQLDQLKAALNEKHPELVNRKNIIFHQDNARLHISLMTQH-----GWEFLIHLL
Cat flea 10.6    DWKGIIHYELLPYGQMLNSTIYCEQLDRLKQAIDQKRPELANGKGVVFHQDNARPHTSLMTRQKLRELGWEVLSHLP
Deer fly 9.2     DWKGIVHYELLPVGQTVDSQRYCEQLERLRQAIEKKRPELYNGKGVIFHHDNARPHTSLMTRQKLRELGWEVLMHPP
                  **   *          *         *          *        *          *           **

Human_mar.       SAAGLIHYSFLNPGETITSEKYAQQIDEMHRKLQRLQPALVNRKGPILLHDNARPHVAQPTLQKLNELGYKVLPHPF
House ant 16.2   TDSGIIYRTFLKPGESITAEIYCSQLDEMMIRLAIKKPRLINRDGPILLQDNARPHVAKNTLKLQSLHLETLLHPA
Carpelimus33.2   SVAGVIHHSFLNPGETITAEIYCQQIDEMHQKLQRMCPRLVNMKGPILLHDNARPHVAQPTLQKLNQLGYETLPHPA


Humar1           RHKGILLDSFPEXGRTITSAYYESVLRKLAKALAEKHPGKLHQQ-VLLHHDNAPAHSSHQTRAIS*EFLWEIIRHPP
Horn_fly         DAHGIIFIDYLEKGKTINSDYYMALLERLKVEIAAKRPHMKKKK-VLFHQDNAPCHKSLRTMAKIKELGFELLPHPF
Silverfish 8.1   DQEGILLVEWLPPNTSIDSESYCSSLHRLRRRIQQRRQGKWDRS-VLLQNDNARPHVSRQTIATVYELGCRVLPHPF
```

Fig 5B. A mariner-like element in the sheep PrP 3' UTR. Panel B = deduced sequence of the sheep PrP mariner transposase protein aligned with analogous sequences from cat flea and deer fly.

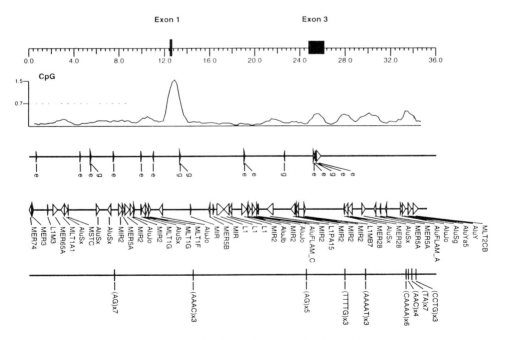

Fig 6. Structure of the human PrP gene. (see legend, figure 2)

and one clone from a human brain cDNA library, the human PrP gene was thought to have a simple structure, analogous to that of the Syrian hamster, namely one small exon encoding most of the 5'UTR, a 12 kb intron, and a second coding exon which includes the 3'UTR(Puckett et al. 1991)(Kretzschmar et al. 1986). Thus an unexpected finding was the presence of intron sequences closely analogous to those that encode the second, 5' untranslated region exons in the murine and ovine PrP genes (Fig.7). Like the authentic mouse and sheep sequences, this exon 2-like sequence is flanked by consensus splice donor and acceptor sites, and is preceded by a polypyrimidine tract. These data indicate that either the environment of the human exon 2-like sequence precludes splicing into mature mRNA, or that this is a functional exon expressed in extraneuronal tissues.

Search for novel protein coding sequences

Three approaches were used to locate novel coding regions within human, mouse, and sheep PrP genomic clones. First, similarity analysis of the PrP region against the GenBank 91 release of DNA sequences identified PrP coding exons in each DNA sequence, and the aforementioned intracisternal A particle (IAP) and mariner DNA transposable elements in the mouse *Prn-pa* intron 2 and sheep PrP 3' UTR, respectively. Second, we searched for potential exons using the gene-finding program, GRAIL, which uses a "neural-network" based algorithm(Uberbacher and

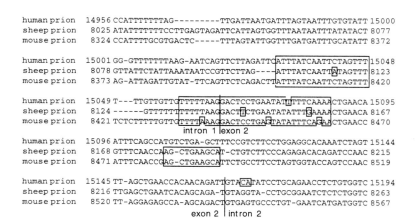

Sequences outside the coding region
conserved between human, sheep and mouse
1

A)   In intron 1

```
human prion   14408 GATGCATTAAGAAGCTGGAAGCTGTGACCCAGAAACCCCACTCCTGAGAA 14457
sheep prion    7607 GAAAAAT--GGATCCTTTAAGCCATGACCCTGAAACCCCACTCCTGGGAA 7654
mouse prion    7773 GAGGTATTATTA-GTCGTGTGCTGTGACCCAGAAACCCCACTCCTGGCAA 7821

human prion   14458 CTTACCTGCAATGGAAGAAACAAACAAA---CAAAAACAGGCATGTATTC 14504
sheep prion    7655 CTTACCTGCAATGGAAGAAATTCGGAAAGAAGAAAAGCTG-CATTCACCC 7703
mouse prion    7822 TTTTAC-TG----GGAAGGAACAAACAAAGGGCTAGGGGAGCCATATGGCC 7865
```

B)   Around exon2

```
human prion   14956 CCATTTTTTTAG---------TTGATTAATGATTTAGTAATTTGTGTATT 15000
sheep prion    8025 ATATTTTTTTCCTTGAGTAGATTCATTAGTGGTTTAATAATTTATATACT 8077
mouse prion    8324 CCATTTTGCGTGACTC-----TTTAGTATTGGTTTGATGATTTGCATATT 8372

human prion   15001 GG-GTTTTTTTAAG-AATCAGTTCTTAGATTCATTTATCAATTCTAGTTT 15048
sheep prion    8078 GTTATTCTATTAAATAATCCGTTCTTAG----ATTTATCAATTATAGTTT 8123
mouse prion    8373 AG-ATTAGATTGTAT-TTCAGTTCTCAGACTTATTTATCAATTCTAGTTT 8420

human prion   15049 T---TTGTTGTTGTTTTTAAGGACTCCTGAATATTTTTTCAAAACTGAACA 15095
sheep prion    8124 ------GTTTTTTTTTTTAAGGACTTCTGAATATATTTGAAAACTGAACA 8167
mouse prion    8421 TCTCTTTTTGTTGTTTTAAAGGACTCCTGAGTATATTTCAGAACTGAACC 8470
                                              intron 1 exon 2
human prion   15096 ATTTCAGCCATGTCTGA-GCTTTCCGTCTTCCTGGAGGCACAAATCTAGT 15144
sheep prion    8168 GTTTCAACCAAG-CTGAAGCAT-CTGTCTTCCCAGAGACACAGATCCAAC 8215
mouse prion    8471 ATTTCAACCGAG-CTGAAGCATTCTGCCTTCCTAGTGGTACCAGTCCAAC 8519

human prion   15145 TT-AGCTGAACCACAACAGATTGTACATATCCTGCAGAACCTCTGTGGTC 15194
sheep prion    8216 TTGAGCTGAATCACAGCAGA-TGTAGGTA-CCTGCGGAATCTCTCTGGTC 8263
mouse prion    8520 TT-AGGAGAGCCA-AGCAGACTGTGAGTGCCCTGT-GAATCATGATGGTC 8567
                              exon 2  intron 2
```

Fig. 7. Conserved non-coding sequences in intron 1 (Panel A), around exon 2 (Panel B), at the polyadenylation site (Panel C), and 3' flanking region(Panel D). Note that the *PRNP* gene contains an "exon 2-like" sequence within intron 1 (panel B).

Mural 1991).   The GRAIL program identified the prion coding regions (but not the anticoding strand ORF, present in three PrP alleles analyzed here), and the mutationally "drifted" versions of the gag, pol, and env genes present within the *Prn-pa* IAP element. The third type of analysis was for large open reading frames (ORFs). Excluding PrP and IAP protein-coding regions there are 7 ORFs greater than 300bp in length in the human PrP gene, 10 ORFs in the mouse *Prn-pa*, and 5 ORFs in the sheep gene. With one exception, i.e. PrP, these ORFs are not conserved between species, making their functional significance unclear. Similarly, with the exception of PrP, coding regions predicted by the GRAIL program do not coincide with these ORFs. Taken together, data from these analyses indicate that additional genes are not present in the introns and flanking sequences common to *Prn-pa*, sheep, and human PrP molecular clones examined here.

Sequences outside the coding region
conserved between human, sheep and mouse

2

C) Polyadenylation region

```
human prion   27129 ATCGTTTCATGTAAGAATCCAAAGTGGACACCATTAACAGGTCTTTGAAA 27178
sheep prion   26167 ATAGTTTTGTATAAGAATCCAGAGTGA-------------TATTTGAAA 26202
mouse prion   21521 ---GTTCC--GTAGGATTCCAAAGCAGACCCC---TAGCTGGTCTTTGAA- 21563
mink prion     2260 ATAGTTT--TGTATGA--------TAGAGCCCAT---GTGGTCTTCGAAA  2296

human prion   27179 TATGCATGTACTTTATATTTTCTATATTTGTAACTTTGCATGTTCCTTGT 27228
sheep prion   26202 TACGCATGTGCTTAATATTTTTTATATTTGTAACTTTGCATGTAC-TTGT 26250
mouse prion   21564 TCTGCATGTACTTGACGTTTTCTATATTTGTAACTTTGCATGTATTTTGT 21612
mink prion     2297 TATGCATGTACTTTATATTTTCTATATTTGTAACTGGGCATGTAC----- 2341

human prion   27229 TTTGTTATATAAAAAAAATTGTAAATGTTTAATATCT-GACTGAAATTAAA 27276
sheep prion   26249 TTTGT---GTTAAAAGTTTATAAATATTTAATATCT-GACTAAAATTAAA 26296
mouse prion   21613 TTTGTCATATAAAAAGTTTATAAATGTTTGCTATCA-GACTGACATTAAA 21661
mink prion     2342 ----TTGTATAAAAAATGTATAAACATTCGAACTCTTGACTAGAATTAAA 2386

human prion   27277 ACG--AGCGAAGATGAGCACC ACCTCCC--GTGTCTG 27509       polyA
sheep prion   26297 ACAGGAGCTAAAAGGAGTATC TTCCACGGAGTGTCTG 26332       signal
mouse prion   21661 ATAGAAGCTATGATGAACACC T--GGCGGGGTTTGTT 21695
mink prion     2387 ACAGGAACTGAG-TGTGTCCC AT-------GTGTTTG 2416
                                    |                ‾‾‾‾‾‾‾‾
                                polyA site      GT-rich stretch
                                                usually following polyA site
```

D) 3' flanking region

```
mer5A cons       122 TGCTGGGCCCCACCCCCAGAGTTTCTGATTCAGTAGGTCTGGGG-TGGGG 79
human prion    30553 GGCTGAACCCCATCCCCAGAGTTTCTGATTCCTAGTACCAGGG-TGGGG 30601
sheep prion    30424 TGCTGCATCCCA-CCCCAGAGTTTCTGATTCCAGAGTGC-AGGG-TAGGA 30472
mouse prion    24467 GA-TGGGTCCCACCCATAGAGTTTCTGATTCCTGAG-ACCAGGGATGGGG 24514

mer5A cons        78 CCCGAGAATCTGCATTTCTAACAAGTTCCCAGGTGATGCTGATGCTGCTG
human prion    30602 TCTGAGAATTTGCATTTCTAACAAATTCCCAGGTGATACTGATGCTGCTG 30651
sheep prion    30471 CCAGAGAATTTACATTTCTAACAGAGTCCCCGGCAAT---GATGCTGTTG 30517
mouse prion    24515 CCTGGAAATGTACGTTTCTAACAAATTCCCAAATAATAAC-ATGTTGCTG 24563
```

Fig. 7. (*Continued*).

69

DISCUSSION

Insertion events and the genesis of natural prion diseases

Sequencing of the coding sequence of the human PrP gene, *PRN-P*, from patients with familial prion diseases often referred to as Creutzfeldt-Jakob disease (CJD), Gerstmann-Straussler-Scheinker (GSS) disease and fatal familial insomnia (FFI) identified mutations 19 non-conservative missense substitutions, as well as expansions of the tandemly repeated octapeptide motifs(Hsiao et al. 1989)(Goldfarb et al. 1991). Genetic linkage has been established for 5 of the 19 known mutations that segregate with the inherited prion diseases. Recent studies suggest that mutations destabilize PrPC which leads to its conversion into PrPSc: once PrPSc is formed, then it acts as a ligand which binds to PrPC resulting in its conversion into a second molecule of PrPSc. In addition to the documented role of point mutations, data presented here reveals that transposition events have contributed to the evolution of PrP genes and should also be considered among the "forward" mutational events in mammalian PrP loci. Insertion events might initiate prion disease by causing mis-regulation of PrP mRNA expression(Westaway et al. 1994), or indirectly, by creating chromosomal rearrangements .

In the case of the mouse PrP gene, the IAP insertion defines the structural basis for an intron polymorphism which distinguishes (in addition to missense substitutions at codons 108 and 189) the *a* and *b* alleles. DNA sequences reveal that the IAP insertion site lies 565 bp 5' to a BstEII polymorphism that also differs between a and b alleles (*Prn-pa* = GGTGACC, *Prn-pb* = GGTGGCC). Extrapolating from published restriction digests with BamH1, RIIIS/J mice (*Prn-pc* haplotype) and Molf/Ei mice (*Prn-pf* haplotype) may also contain an IAP genome within intron 2. Cast/Ei (*Prn-pe* haplotype) mice bear a constellation restriction sites similar to I/LnJ and inconsistent with an IAP insertion, but instead suggesting a crossover event between a Taq1 site in intron 2 and the aforementioned BstEII site. Finally, MaMy/J mice (*Prn-pd* haplotype), may contain a more extensively deleted or rearranged IAP genome since they exhibit a diagnostic SacI site (perhaps corresponding to nucleotides 21248-21253 of the *Prn-pa* IAP sequence) yet exhibit an IAP-intron 2 BamH1 fragment smaller than that of *Prn-pa* mice(Westaway et al. 1994). In sum, since 4/6 *Prn-p* haplotypes carry IAP elements (close to full-length in the case of the most prevalent *Prn-pa* haplotype, and partial and/or rearranged in the case of haplotypes c, d, and f), insertionally mutated alleles can be said to represent the "wild-type" PrP gene in laboratory strains of mice.

At this time there is insufficient data to estimate the antiquity of the IAP-insertion. However, the identical LTRs present in *Prn-pa* genomic clones, as well as full-length "gag" and "env" genes argue for a comparatively recent origin or, perhaps less likely, selection pressure to maintain the functionality of these IAP genes and control elements. These considerations beg the question of transcriptional status, both for the IAP element and *Prn-pa* itself. Regarding the former, as the LTRs are intact, there is no evidence to exclude that the IAP element in intron 2 is actively transcribed: however, direct demonstration of transcription of an individual member of a multigene family is difficult (when the transcriptional start- and stop-sites lie within the boundaries of the transposable element, as is the case for IAPs). Regarding the latter, steady-state levels of *Prn-pa* mRNA are known to approximate those of *Prn-pb* mice (Westaway et al. 1987). This result indicates that if the IAP element is transcribed, it has little effect upon transcription on the other DNA strand, with the caveat that *Prn-pa* and *Prn-pb* alleles differ in other regions (aside from presence/absence of the IAP) that might influence PrP mRNA levels: for example in the putative promoter (Westaway et al. 1994), and the 3' UTR (two substitutions ansd one one base-pair deletion). Furthermore, since the IAP element is present in common wild-type *Prn-p* alleles in mice, and since mice are not thought to exhibit spontaneous prion disease, it is highly unlikely that the intron 2 IAP is pathogenic. On the other hand, the connection between the mariner-like element present in the PrP gene of sheep (and presumably cows) and prion disease pathogenesis is more intriguing.

In humans, a mariner-like element most closely related to the African malaria mosquito and the green lacewing (but also revealing homology in alignments to the sub-family which includes mellifera elements) is located within the boundaries of a chromosome 17 hot-spot for homologous recombination that gives rise to Charcot-Marie tooth disease type 1A (CMT1A: ~1.5 Mb duplication) or to hereditary neuropathy with pressure palsies (HNPP: deletion)(Rieter et al. 1996). Rieter et. al. speculate that these sequences, in concert with a mariner transposase, perhaps encoded elsewhere in the human genome, result in double-stranded breaks and ultimately chromosomal rearrangments. Though this specific hypothesis can be challenged on the grounds that human mariner-like elements appear ancient, with "decayed" terminal inverted repeats that are most likely a poor substrate for transposase, the phenomenology is nonetheless provocative and suggests an investigation of chromosomal instability adjacent to the sheep PrP gene.

The incentive for undertaking such a study is that the origin of natural scrapie is obscure. One obvious possibility is that overexpression of PrPC, arising through such hypothetical chromosomal rearrangements might give rise to "sporadic" natural scrapie (by de novo formation of PrPSc): this could then be disseminated horizontally within flocks by an infectious mechanism. With regard to this hypothesis it is known that overexpression of sheep PrPC is pathogenic, in transgenic mice, and is associated with the synthesis of prion titer in the brains of such animals (Westaway et al. 1994) and unpublished data of D. Westaway, G. Carlson, S. DeArmond, and S. Prusiner). Since the bovine PrP gene also contains an extended 3'UTR(Goldmann et al. 1990) it is very likely that it too harbors a mariner element. Further, as some subscribe to the view that BSE did not arise as a result of infection with scrapie-contaminated feed, but instead from a prions arising from a natural disease of cattle, it is possible that the hypothesis of mis-regulated PrPC expression caused by chromosomal rearrangement might be germane to the origin of the BSE epidemic.

A cryptic untranslated exon in the human PrP gene?

Three-way alignments of PrP gene sequences revealed several blocks of conserved nucleotides clustered around exon 2. Previous PCR analyses of cDNA have indicated that this exon is present in the majority of mature PrP mRNAs in brain RNA of sheep and mice. Though the function of this 98 bp 5' UTR is obscure, it is notable that this exon is conserved, most markedly in the 5' half, and indeed is far more conserved than the exon 1 sequences adjacent to the PrP promoter (e.g. human and mouse exon 1 sequences show negligible homology). Surprisingly, data presented here reveal that highly related sequences are also present in an analogous position within the human PrP gene, bracketed by consensus splice donor and acceptor sites (Fig.7). One possibility is that the "conservation" of these sequences in humans is illusory and merely reflects the so-called hominid "slow-down" in evolution. The second possibility is that these sequences are expressed.

Though 3 human PrP cDNAs isolated to date demonstrate exon 1+3 splice events (using the exon numbering system of the mouse and sheep PrP genes), human exon 2 may be actively spliced and included within a subset of mRNAs in the brain, or may be present in mRNAs from other tissues. Recent reanalyses of the structure of the Syrian hamster (SHa) PrP gene are consistent with the former scenario. Li and Bolton have demonstrated that while 90% of SHa PrP mRNAs in the brain are of the exon 1+3 variety, 10% include exon 2 sequences closely-related those of sheep and mouse. Increased abundace of exon 2 containing mRNAs is observed in scrapie-infected hamsters, which may reflect preferential expression in astrocytes (which become activated during the course of prion infections)(Li and Bolton 1996).

Since exon 2-containing cDNAs have now been described in cattle(Yoshimoto et al. 1992), this accumulation of data forces us to consider the notion that exon 2 sequences are a ubiquitous feature of mammalian PrP genes: this in turn raises the issue of function. Reading frames in-phase with the PrP ORF are not conserved(Westaway et al. 1994)(Westaway et al. 1994) so a role for exon 2 in encoding amino-terminally extended PrPs can be excluded. At this time, a role in the regulation of gene expression is perhaps more plausible.

Prions and retroviruses

Lastly, it is important to note that although our data define an IAP element within intron 2 of the mouse *Prn-p*a gene, these findings cannot be used to support the contention that CJD is caused by a retrovirus, or retrovirus-like particle, as suggested by some(Akowitz et al. 1993)(Akowitz et al. 1994). The following points should be noted in this regard: (i) though humans can be considered the natural host for CJD, retroviral insertions are absent from the human PrP genes (Fig. 1), as well as the sheep PrP gene (Fig.4) and the *Prn-p*b gene (unpublished data of I.L.), (ii) insofar as IAP transpositions contribute significantly to the burden of new mutations in rodent cells, the insertion event described here may have no special significance, and (iii) at this time we have no data to indicate that the *Prn-p*a-IAP element is transcribed, as integrity of LTR sequences is necessary but not sufficient for transcription. Instead a wealth of evidence drawn from the disciplines of cell-biology, biochemistry, and genetics indicates that prions are proteinaceous infectious pathogens, fundamentally distinct from retroviruses and other RNA and DNA viruses(Baldwin et al. 1995). The complete gene sequences described here will form the basis for future experiments into the biology of these fascinating infectious agents.

ACKNOWLEDGMENTS

We thank Dr. Bruce Chesebro and Dr. Sylvia Perryman for communicating the sequence of the IAP-related cDNA up-regulated in scrapie infected cells. We also thank technicians at the University of Washington sequencing core facility managed by J. Seto for performing DNA sequencing, and Dr. Lee Rowen for discussions. This work was supported by grants from the NIH, the American Health Assistance Foundation, and by gifts from the Sherman Fairchild Foundation and the Osher Foundation (SBP and LEH), the Alzheimer Association of Ontario and the Bayer/Canadian Red Cross Research Fund (DW).

REFERENCES

Akowitz, A., E. E. Manuelidis and L. Manuelidis (1993). Protected endogenous retroviral sequences copurify with infectivity in experimental Creutzfeldt-Jakob disease. *Arch. Virol.* **130**: 301-316.

Akowitz, A., T. Sklaviadis and L. Manuelidis (1994). Endogenous viral complexes with long RNA cosediment with the agent of Creutzfeldt-Jakob disease. *Nucleic Acids Res.* **22**: 1101-1107.

Amouyel, P., Vidal, O., Launay, J.-M., and Laplanche J.-L. (1995). The apoliloprotein E alleles as major susceptibility factors for Creutzfeldt-Jakob Disease. *Lancet* **344**: 1315-1318.

Auge-Gouillou, C., Y. Bigot, N. Pollet, M. H. Hamelin, M. Meunier-Rotival and G. Periquet (1995). Human and other mammalian genomes contain transposons of the mariner family. *FEBS letters* **368**: 541-546.

Baldwin, M. A., F. E. Cohen and S. B. Prusiner (1995). Prion protein isoforms, a convergence of biological and structural investigations. *J. Biol. Chem.* **270**: 19197-19200.

Bessen, R. A., Kocisko, D.A., Raymond, G.J., Nandan, S., Lansbury, P.T., and Caughey, B. (1995). Non-genetic propagation of strain-specific properties of scrapie prion protein. *Nature* **375**: 698-700.

Bessen, R. A. and R. F. Marsh (1992). Identification of two biologically distinct strains of transmissible mink encephalopathy in hamsters. *J. Gen. Virol.* **73**: 329-334.

Bruce, M. E. (1993). Scrapie strain variation and mutation. *Brit. Med. Bul* **49**(4): 822-838.

Bueler, H., M. Fischer, Y. Lang, H. Bluethmann, H.-P. Lipp, S. J. DeArmand, S. B. Prusiner, M. Aguet and C. Weissmann (1992). Normal development and behaviour of mice lacking the neuronal cell-surface PrP protein. *Nature* **356**: 577-582.

Carlson, G. A., C. Ebeling, S.-L. Yang, G. Telling, M. Torchia, D. Groth, D. Westaway, S. J. DeArmond and S. B. Prusiner (1994). Prion isolate specified allotypic interactions between the cellular and scrapie prion proteins in congenic and transgenic mice. *Proc. Natl. Acad. Sci. USA* **91**: 5690-5694.

Carlson, G. A., P. A. Goodman, M. Lovett, B. A. Taylor, S. T. Marshall, M. Peterson-Torchia, D. Westaway and S. B. Prusiner (1988). Genetics and polymorphism of the mouse prion gene complex: the control of scrapie incubation time. *Mol. Cell. Biol.* **8**: 5528-5540.

Carlson, G. A., D. T. Kingsbury, P. A. Goodman, S. Coleman, S. T. Marshall, S. J. DeArmond, D. Westaway and S. B. Prusiner (1986). Linkage of prion protein and scrapie incubation time genes. *Cell* **46**: 503-511.

Cohen, F. E., K.-M. Pan, Z. Huang, M. Baldwin, R. J. Fletterick and S. B. Prusiner (1994). Structural clues to prion replication. *Science* **264**: 530-531.

Doh-ura, K., S. Perryman, R. Race and B. Chesebro (1995). Identification of differentially expressed genes in scrapie-infected mouse neuroblastoma cells. *Microbial Pathogenesis* **18**: 1-9.

Gabriel, J.-M., B. Oesch, H. Kretzschmar, M. Scott and S. B. Prusiner (1992). Molecular cloning of a candidate chicken prion protein. *Proc. Natl. Acad. Sci. USA* **89**: 9097-9101.

Gabriel, J.-M., B. Oesch, M. Scott and S. B. Prusiner (1991). Molecular cloning and evolutionary analysis of a candidate chicken prion protein. *Prion Diseases in Humans and Animals Symposium, London, Sept. 2-4, 1991.*

Goldfarb, L. G., P. Brown, W. R. McCombie, D. Goldgaber, G. D. Swergold, P. R. Wills, L. Cervenakova, H. Baron, C. J. J. Gibbs and D. C. Gajdusek (1991). Transmissible familial Creutzfeldt-Jakob disease associated with five, seven, and eight extra octapeptide coding repeats in the *PRNP* gene. *Proc. Natl. Acad. Sci. USA* **88**: 10926-10930.

Goldmann, W., N. Hunter, J. D. Foster, J. M. Salbaum, K. Beyreuther and J. Hope (1990). Two alleles of a neural protein gene linked to scrapie in sheep. *Proc. Natl. Acad. Sci. USA* **87**: 2476-2480.

Hsiao, K., H. F. Baker and T. J. Crow (1989). Linkage of a prion protein missense variant to Gerstmann-Straussler Syndrome. *Nature* **338**: 342-345.

Hsiao, K. K., D. Groth, M. Scott, S.-L. Yang, A. Serban, D. Rapp, D. Foster, M. Torchia, S. J. DeArmond and S. B. Prusiner (1992). Genetic and transgenic studies of prion proteins in Gerstmann-Straussler–Scheinker disease. *Prion Diseases of Humans and Animals*. S. B. Prusiner, J. Collinge, J. Powell and B. Anderton. London, Ellis Horwood: 120-128.

Kretzschmar, H. A., L. E. Stowring, D. Westaway, W. H. Stubblebine, S. B. Prusiner and S. J. DeArmond (1986). Molecular cloning of a human prion protein cDNA. *DNA* **5**: 315-324.

Kuff, E. L. and K. K. Lueders (1988). The intracisternal A-Particle gene family: structure and functional aspects. *Advances in Cancer Research* **51**(183-276).

Li, G. and D. C. Bolton (1996). A novel prion protein mRNA contains an extra exon:increased expression in scrapie. *Brain Research*. in press

Manson, J. C., A. R. Clarke, M. L. Hooper, L. Aitchison, I. McConnel and J. Hope (1994). 129/Ola mice carrying a null mutation in PrP that abolishes mRNA production are developmentally normal. *Molecular Neurobiology* **8**: 121-127.

Medori, R., H.-J. Tritschler, A. LeBlanc, F. Villare, V. Manetto, H. Y. Chen, R. Xue, S. Leal, P. Montagna, P. Cortelli, P. Tinuper, P. Avoni, M. Mochi, A. Baruzzi, J. J. Hauw, J. Ott, E. Lugaresi, L. Autilio-Gambetti and P. Gambetti (1992). Fatal familial insomnia, a prion disease with a mutation at codon 178 of the prion protein gene. *N. Engl. J. Med.* **326**: 444-449.

Oesch, B., D. Westaway and S. B. Prusiner (1991). Prion protein genes: evolutionary and functional aspects. *Curr. Top. Microbiol. Immunol.* **172**: 109-124.

Oosumi, T., W. R. Belknap and B. Garlick (1995). Mariner transposons in humans. *Nature* **378**: 672.

Prusiner, S. B., M. Scott, D. Foster, K.-M. Pan, D. Groth, C. Mirenda, M. Torchia, S.-L. Yang, D. Serban, G. A. Carlson, P. C. Hoppe, D. Westaway and S. J. DeArmond (1990). Transgenetic studies implicate interactions between homologous PrP isoforms in scrapie prion replication. *Cell* **63**: 673-686.

Puckett, C., P. Concannon, C. Casey and L. Hood (1991). Genomic structure of the human prion protein gene. *Am. J. Hum. Genet.* **49**: 320-329.

Rieter, L. T., T. Murakami, T. Koeuth, L. Pentao, D. M. Muzny, R. A. Gibbs and J. R. Lupski (1996). A recombination hotspot responsible for two inherited peripheral neuropathies is located near a mariner transposon-like element. *Nature genetics* **12**: 288-297.

Robertson, H. M. (1995). *J. Insect Physiology* **41**: 99-105.

Sakaguchi, S., S. Katamine, N. Nishida, R. Moriuchi, K. Shigematsu, T. Sugimoto, A. Nakatani, Y. Kataoka, T. Houtani, S. Shirabe, H. Okada, S. Hasegawa, T.

Miyamoto and T. Noda (1996). Loss of cerebellar purkinje cells in aged mice homozygous for a disrupted PrP gene. *Nature* **380**: 528-531.

Saunders, A. M., K. Schmader, J. C. S. Breitner and e. al. (1993). Apolipoprotein E epsilon 4 allele distributions in late-onset Alzheimer's Disease and in other amyloid-forming diseases. *Lancet* **342**: 710-711.

Schätzl, H. M., M. Da Costa, L. Taylor, F. E. Cohen and S. B. Prusiner (1994). Prion protein gene variation among primates. *J. Mol. Biol.* **245**: 362-374.

Sparkes, R. S., M. Simon, V. H. Cohn, R. E. K. Fournier, J. Lem, I. Klisak, C. Heinzmann, C. Blatt, M. Lucero, T. Mohandas, S. J. DeArmond, D. Westaway, S. B. Prusiner and L. P. Weiner (1986). Assignment of the human and mouse prion protein genes to homologous chromosomes. *Proc. Natl. Acad. Sci. USA* **83**: 7358-7362.

Stahl, N., D. R. Borchelt, K. Hsiao and S. B. Prusiner (1987). Scrapie prion protein contains a phosphatidylinositol glycolipid. *Cell* **51**: 229-240.

Telling, G. C., M. Scott, J. Mastrianni, R. Gabizon, M. Torchia, F. E. Cohen, S. J. DeArmond and S. B. Prusiner (1995). Prion Propagation in Mice Expressing Human and Chimeric PrP transgenes implicates the Interaction of Cellular PrP with Another Protein. *Cell* **83**: 79-90.

Uberbacher, E. C. and R. J. Mural (1991). Locating protein-coding regions in human DNA sequences by a multiple sensor-neural network approach. *Proc. Natl. Acad. Sci. U.S.A.* **88**: 11261-11265.

Westaway, D., C. Cooper, S. Turner, M. Da Costa, G. A. Carlson and S. B. Prusiner (1994). Structure and polymorphism of the mouse prion protein gene. *Proc. Natl. Acad. Sci. USA* **91**: 6418-6422.

Westaway, D., S. J. DeArmond, J. Cayetano-Canlas, D. Groth, D. Foster, S.-L. Yang, M. Torchia, G. A. Carlson and S. B. Prusiner (1994). Degeneration of skeletal N. muscle, peripheral nerves, and the central nervous system in transgenic mice overexpressing wild-type prion proteins. *Cell* **76**: 117-129.

Westaway, D., P. A. Goodman, C. A. Mirenda, M. P. McKinley, G. A. Carlson and S. B. Prusiner (1987). Distinct prion proteins in short and long scrapie incubation period mice. *Cell* **51**: 651-662.

Westaway, D., V. Zuliani, C. M. Cooper, M. Da Costa, S. Neuman, A. L. Jenny, L. Detwiler and S. B. Prusiner (1994). Homozygosity for prion protein alleles-encoding glutamine-171 renders sheep susceptible to natural scrapie. *Genes Dev.* **8**: 959-969.

Windl, O., M. Dempster, P. Estibeiro and R. Lathe (1995). A candidate marsupial PrP gene reveals two domains conserved in mammalian PrP proteins. *Gene* **159**: 181-186.

Yoshimoto, J., T. Iinuma, N. Ishiguro, M. Horiuchi, M. Imamura and M. Shinagawa (1992). Comparative sequence analysis and expression of bovine PrP gene in mouse L-929 cells. *Virus Genes* **6**: 343-356.

Zerr, I., M. Helmhold and T. Weber (1994). Apolipoprotein E in Creutzfeldt-Jakob Disease [letter]. *Lancet* **345**: 68-69.

ELECTRON MICROSCOPY IN PRION RESEARCH: TUBULOVESICULAR STRUCTURES ARE NOT COMPOSED OF PRION PROTEIN (PrP) BUT THEY MAY BE INTIMATELY ASSOCIATED WITH PrP AMYLOID FIBRILS

P.P. Liberski,[1,2] M. Jeffrey,[3] and C. Goodsir[3]

[1]Laboratory of Electron Microscopy and Neuropathology, Laboratories of Tumor Biology, Chair of Oncology, Medical Academy Lodz, Poland and [2]Laboratory of Electron Microscopy, Department of Pathomorphology, Polish Mother memorial Hospital, Lodz and [3]Central Veterinary Laboratory, Lasswade Veterinary Laboratory, Bush   Estate, Penicuik, Edinburgh, Scotland

INTRODUCTION

In a recent monograph on electron microscopy of prion diseases one of us wrote „Attempts, using thin-section transmission electron microscopy, to define the structure of the infectious virus of scrapie (and analogously, Creutzfeldt-Jakob disease, kuru and, recently, bovine spongiform encephalopathy) is an repeatable example of failures, falstarts and misinterpretations."[1] The presence of tubulovesicular structures is, however, a consistent finding.[2]

Tubulovesicular structures (TVS) are disease-specific particles found by thin-section electron microscopy in all transmissible spongiform encephalopathies (TSE) or prion diseases, including Creutzfeldt-Jakob disease and Gerstmann-Straussler-Scheinker disease.[3-7] These virus-like particles, discovered more than a quarter of century ago by David-Ferreira and collaborators,[8] are visualized by thin-section (as opposed to the negative staining) electron microscopy as approximately regular spheres of less than 30 nm in diameter or as a short tubules (Fig. 1). TVS have been observed only in the brain (but not in the spleen) at both sides of synaptic cleft, either in dendrites or in axonal preterminals and terminals. The number of affected processes correlate roughly with the titer of infectivity in the brain.[5] For more than a decade TVS have been ignored because they could not be found in hamsters infected with the 263K strain of scrapie which, incidently, produced the highest titre of infectivity in the brain.[9-10] Subsequently, TVS were found also in this scrapie model which rejuvenate an interest in these peculiar particles.[5-6]

TSE are caused by a still incompletely characterized infectious agent variously referred to as a slow, unconventional virus, prion or virino.[11-15] It has also been suggested that this agent is a single stranded (ss)DNA virus of unusual structure, called "nemavirus", which is visualized as TVS (designated "tubulofilamentous particles") by thin-section electron microscopy or as "thick tubules" when the touch preparation method was applied.[16-18] Furthermore, it was proposed that TVS are composed of an abnormal isoform of prion protein (PrP).[16]

Using thin-section electron microscopy and immunogold techniques we report here that TVS are not stained with anti-PrP antisera and thus are not composed of PrP. We further report here the immunogold-electron microscopic studies on primitive plaques in scrapie-affected hamster brains. While such plaques were first observed a decade ago,[19-20] this is the first detailed ultrastructural study of such plaques which differ from compact "kuru" plaques encountered in kuru, CJD and GSS and in scrapie-affected mouse brain.[21-22] It is thus plausible, that morphology of PrP deposits is influenced entirely by the local brain environment. For instance, if PrP is released from dendrites, a classical plaque will form. If, however, PrP is released early in the incubation period, over a large neuroanatomical

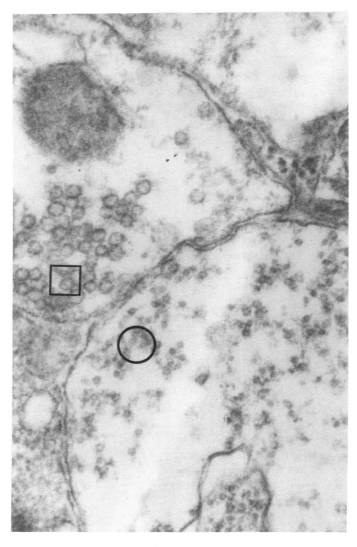

Fig. 1. Typical TVS (circle) in a process in close contact with axonal terminal containing synaptic vesicles (square). Note that TVS are smaller and of higher electron density than synaptic vesicles. Lead citrate and uranyl acetate. Original magnification, x 50 000.

area, a diffuse plaque will form. Finally, if PrP is released late in the incubation period , from only selected neurons, perineuronal pattern will form. Of note, TVS were found in intimate contact with amyloid fibrils within "loose" plaques.

MATERIAL AND METHODS

Animals

Outbred, 6-week-old golden Syrian hamsters were inoculated intracerebrally with 0.05 ml of a 10% brain suspension of the 263K and 22C-H strains of scrapie (kindly supplied by Dr. Richard Kimberlin, SARDAS, Edinburgh, Scotland and Dr. Richard Carp, IBR, Staten Island, new York, USA, respectively), perfused with 100 ml of 1.25% glutaraldehyde and 1% paraformaldehyde prepared in cacodylate buffer (pH 7.4) followed by 50 ml of 5% glutaraldehyde and 4% paraformaldehyde. Mice infected with the 87V or ME7 murine scrapie strains were also available for this study. These mice were perfused with a 2.5% glutaraldehyde and 2.5% paraformaldehyde when terminally ill.

Immunogold procedures

The immunogold methods employed for ultrastructural localization of PrP are as previously reported.[23-24] Both 10 nm immunogold and the 1 nm immunogold silver enhanced (IGSS) methods were applied. Briefly, these methods are as follows. 65-80 nm sections were taken from blocks previously identified as containing, on immunostained 1 mm thick sections, accumulations of PrP either in the form of amyloid plaques or as other pathological accumulations of PrP. Most sections originated from cerebral cortex, thalamus or hippocampus, but occasional samples taken from the cerebellum, midbrain or medulla were also examined. Sections placed on 400 mesh nickel grids were etched with sodium periodate for 60 minutes or in potassium methoxide 18 crown DMSO for 15 minutes. Endogenous peroxidase was blocked and sections de-osmicated with 3% hydrogen peroxide for 3 minutes. Primary antibodies (1B3 and 1A8 kindly supplied by Dr. James Hope, MRC & AFRC Neuropathogenesis Unit, Edinburgh, Scotland) was then applied 1:100 dilution or 1:400 dilution, respectively in incubation buffer for 1 hour. After rinsing, sections were incubated with Extravidin 10 nm colloidal gold diluted 1:10 in

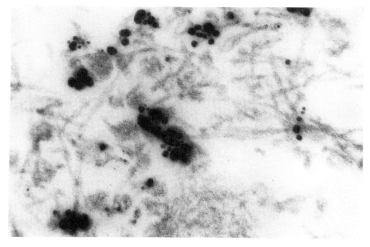

Fig. 2. Amyloid fibrils decorated by anti PrP antibody. IGSS method.

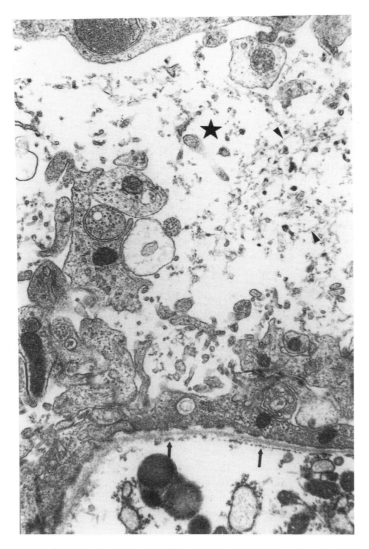

Fig. 3. A typical „loose" plaques in hamster brain infected with the 263K strain of scrapie. Note amyloid fibrils (arrowheads) floating in distended extracellular space (star). Note a blood vessel lined by a collar of glial cells supported by a basal membrane(arrows). Lead citrate and uranyl acetate. Original magnification, x 7000.

incubation buffer for 1 hour. Sections prepared for the IGSS method were stained with 1nm gold and silver enhanced. Grids were postixed with 2.5% glutaraldehyde in PBS and counterstained with uranyl acetate and lead citrate.

RESULTS

In the 87V murine model, PrP-conjugated gold particles decorated typical stellate amyloid plaques and the cell surface of numerous dendrites.[23-24] By light microscopy and semi-thin (1 µm) sections, discrete PrP-immunopositive plaques were observed in both the 263K and 22C-H models (not shown) in the subependymal region but not in the deep brain neuroparenchyma. These plaques were not discernible by routine H & E staining.

Ultrastructurally, plaques were recognized as areas of low electron density containing haphazardly-oriented fibrils heavily decorated with PrP-conjugated gold particles (Fig. 2) but not as stellate compact structures. These plaques were located beneath the basal border of the ependymal cells, the fine structure of the latter was also clearly recognizable (Fig. 3). When dystrophic neurites (Fig. 4) containing electron-dense inclusion bodies were seen within the plaque perimeter, they always remained PrP-negative. Numerous microglial cells were observed in a close contact with PrP-positive plaques and secondary lysosomes (Fig. 5) within these cells were heavily decorated with gold particles. In these two scrapie models neither stellate plaques nor PrP-immunodecorated dendrites were observed.

In all models TVS-containing processes were readily detected and neither these processes nor TVS themselves were decorated with gold particles. Even if amyloid plaques were observed in close contact with a TVS-containing neuronal processes, the plaques were decorated with gold particles while the processes remained unstained (Fig. 6).

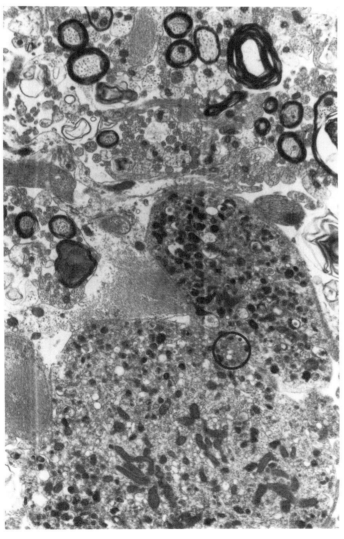

Fig. 4. Large branching dystrophic neurite (circle) in area of „loose" plaque in hamster brain infected with the 263K strain of scrapie. Lead citrate and uranyl acetate. Original magnification, x 7000.

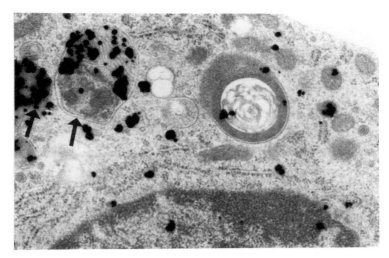

Fig. 5. A fragment of a Kohler cell containing a lysosome heavily decorated with silver deposits (arrows). IGSS method

At higher magnification, amyloid fibrils were clearly visible with numerous round or short tubular particles attached to them (Fig. 7). At such a magnification, these particles were membrane-bound and their diameter was approximately twice that of amyloid fibrils; they were virtually indistinguishable from TVS. No such particles were visible within amyloid plaques of six cases of Alzheimer's disease which serve as a negative control. TVS located in areas adjacent to plaques in the 87V model and in areas of diffuse PrP immunolabelling in ME7 model were also unlabelled with anti PrP antisera.

DISCUSSION

We demonstrated here that PrP-immunoreactive amyloid plaques in scrapie-affected hamster brains are frequently located beneath the ependymal border. They do not exhibit compact structure of the classical kuru plaque but rather randomly-oriented amyloid fibrils accumulated within expanded extracellular space with an admixture of dystrophic neurites and microglial cells. Thus, using alternative approach, we could confirmed that plaque pattern first described by DeArmond et al.[19-20] In contrast to the observations of the latter investigators, microglial cells contained not merely cellular debris but were heavily labelled within secondary lysosomes. Taken together, PrP plaques in scrapie-affected hamster brains are different from classical kuru plaques predominate.[2, 23-24]

The sequence of events which leads to formation of the amyloid plaque is not yet clear. In the 87V mouse model, immunogold electron microscopy demonstrated PrP on the cell membrane before fibrilization which suggests that PrP is initially released into extracellular space where the amyloid fibrils are formed.[23-24] The assembly of fibrils may be subsequently an entirely spontaneous process of nucleation and indeed, recent *cell-free* experiments suggest that PrP^{SC} which, in predominantly β-sheet pleated conformation, is a constituent of a fibril, is generated from its precursor (PrP^C) as a result of a conversion of PrP^C into PrP^{SC} (probably through "seeded" nucleation process).[25-28] How these *in vitro* experiments reflect the *in vivo* situation of the amyloid plaque formation is yet to be established. The presence of PrP within the lysosomal-endosomal system was clearly demonstrated by us in these and previous studies.[23-24] These findings probably reflect the fibrocytosis of excess of abnormal PrP. However, the exact role of the microglial cell is unclear.[29-31] It is possible that interleukines produced by microglia may help contribute towards an environment in which fibrilisation of PrP may occur.

82

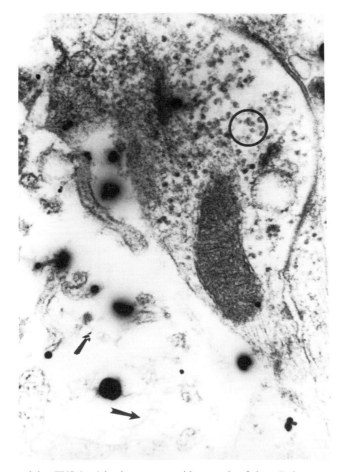

Fig. 6. A process containing TVS (star) in close contact with a margin of „loose" plaque containing numerous amyloid fibrils (arrows). Note that silver deposits decorate only a plaque. IGSS method.

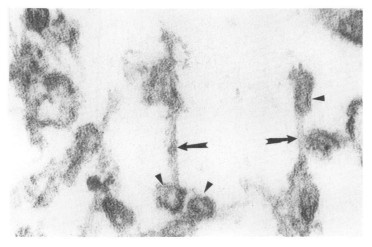

Fig. 7. High power electron micrograph showing amyloid fibrils (arrows) with TVS-like structures (arrowheads) attached to them. Lead citrate and uranyl acetate. Original magnification, x 140 000.

The concept of scrapie as a brain amyloidosis[15] has evolved for the last two decades since the description of various forms of brain amyloids in kuru[11] and a demonstration of amyloid fibrils (designated scrapie-associated fibrils, SAF[33-34] or prion rods[35]) in all TSE. Like all other amyloidoses in general and brain amyloidoses in particular, accumulation of PrP follows the same characteristic pattern.[11-12,15] Mutations in a gene (*Prn-p* or *Sinc* in mice, *PRNP* in humans) which encodes for amyloid precursor cause amyloidosis in familial forms of a disease (GSS or CJD) and different mutations are linked to different phenotypic expression of these diseases.

Using immunogold techniques we were unable to label TVS with anti-PrP antibodies. As these techniques proved to be sensitive enough to immunolabel not only amyloid plaques but also pre-amyloid accumulations of PrP,[23-24] we believe that the absence of staining reflects the structure of TVS which are not composed of PrP. That TVS are PrP-negative may have several important implications for hypotheses trying to explain their nature. Principally, it does not support the suggestion put forward by Narang that TVS are cross-sections of "thick tubules" or "tubulofilamentous particles" visualized by touch-preparations of scrapie-affected mouse and hamster brains.[16-18] These "thick tubules" were claimed to represent an ultrastructural correlate of the "nemavirus" which in turn was proposed to be elusive scrapie agent. Indeed, one of us had previously suggested that "thick tubules" represent merely swollen microtubules as they are observed in both scrapie-infected and sham-inoculated animals.[36] Furthermore, Chasey[37] in a recent comment to the Narang papers[16-18] suggested that "thick tubules" may represent microtubular doublets, again these are entirely normal subcellular component.

The number of neuritic processes containing TVS increases through the incubation period and it correlates with the titre of infectivity in two longitudinal disease studies of scrapie and Creutzfeldt-Jakob disease.[5] It has therefore been suggested that TVS may represent the infectious agent or aggregates of it rather than a pathological product of disease. The latter hypothesis was advocated, however, by Gibson and Doughty[7] who interpreted TVS as breakdown products of microtubules.

The predominant theory of the scrapie agent is now the "prion hypothesis"[13] and its derivatives which implies that a conformationally altered abnormal isoform (PrP^{Sc}) of a normal cellular membrane glycoprotein (PrP^C) is the agent and its accumulation merely mimicks replication. As already mentioned, recent *cell-free* studies elegantly showed the protein-protein (PrP^{Sc}-PrP^C) interactions[25-28] which may underly *de novo* formation of the agent. However, *de novo* generation of the infectivity was neither addressed nor proven in these experiments, and thus, they did not discriminate between two possibilities: PrP^{Sc} as the scrapie agent versus PrP being merely the amyloidogenic protein. If PrP is the infectious agent, as suggested by the prion hypothesis, the absence of stainable PrP in TVS would indicate that TVS are not the ultrastructural correlate of the agent. If, however, TVS turn out to be more than merely useful ultrastructural marker for the whole group of TSE, it may suggest that PrP and the agent are two separate entities. The analogies to molecular pathogenesis of another brain amyloidosis, Alzheimer's disease,[11] are so striking, that it is tempting to suggest that PrP plays the role of the amyloidogenic protein in TSE, the processing of which is merely triggered by still elusive agent of which TVS may be an ultrastructural correlate.

As with other disease-specific particles seen in TSE, we do not know the pathogenetic significance, if any, of these virus-like structures, TVS. We have not seen such structures associated with plaques of Creutzfeldt-Jakob disease or Gerstmann-Straussler-Scheinker disease or 87V murine scrapie possibly because of the compact nature of stellate plaques in these diseases prevents visualization of small structures attached to isolated amyloid fibrils.[21] If these PrP negative, virus-like TVS structures associated with amyloid fibrils are an essential determinant of infectivity of the scrapie agent, then their close relationship with amyloid would offer an explanation for the well known linkage of PrP and infectivity noticed under majority[38] but not all[39-40]experimental conditions and it would also explain why only a fraction of abnormal PrP generated in diseased brains appears to correlate with infectivity.[41]

ACKNOWLEDGEMENTS

This paper is a part of European Community Concerted Action "Prion Diseases: from neuropathology to pathobiology and molecular biology: (Project leader: Prof. Herbert Budka, Vienna, Austria). Prof. Pawel P. Liberski is supported by grants from the KBN and the Maria Sklodowska-Curie Foundation while in Poland and by the British Council fellowship while in the United Kingdom. Ms. Jolanta Bratosiewicz, Mr. Ryszard Kurczewski, Ms. Elzbieta Naganska, Ms. Leokadia Romanska and Mr. Kazimerz Smoktunowicz are kindly acknowledged for skilfull technical assistance.

REFERENCES

1. P.P. Liberski, H. Budka, R. Yanagihara, C.J. Gibbs, and D.C. Gajdusek, Tubulovesicular structures, in: Light and Electron Microscopic Neuropathology of Slow Virus Disorders, P.P. Liberski, CRC Press, Boca Raton, (1992).
2. M. Jeffrey, I.A. Goodbrand, and M. Goodsir, Pathology of the transmissible spongiform encephalopathies with special emphasis on ultrastructure, *Micron* 26: 277 (1995).
3. P.P. Liberski, and H. Budka, Tubulovesicular structures in Gerstmann-Straussler-Scheinker disease. *Acta Neuropathol. (Berl)* 88: 491 (1994).
4. P.P. Liberski, H. Budka, E. Sluga, M. Barcikowska, and H. Kwiecinski, Tubulovesicular structures in Creutzfeldt-Jakob disease. *Acta Neuropathol. (Berl)* 84: 238 (1992).
5. P.P. Liberski, R. Yanagihara, C.J. Gibbs, Jr, and D.C. Gajdusek, Appearance of tubulovesicular structures in Creutzfeldt-Jakob disease and scrapie precedes the onset of clinical disease. *Acta Neuropathol. (Berlin)* 79: 349 (1989).
6. P.P. Liberski, R. Yanagihara, C.J. Gibbs, and D.C. Gajdusek, Tubulovesicular structures in experimental Creutzfeldt-Jakob disease and scrapie. *Intervirology* 29: 115 (1988).
7. P.H. Gibson, and L.A. Doughty, An electron microscopic study of inclusion bodies in synaptic terminals of scrapie-infected animals. *Acta Neuropathol. (Berl)* 77: 420 (1989)
8. J.F. David-Ferreira, K.L. David-Ferreira, C.J. Gibbs, J.A. and Morris, Scrapie in mice: ultrastructural observations in the cerebral cortex. *Proc Soc Exp Biol Med* 127: 313 (1968).
9. J.R. Baringer, and S.B. Prusiner, Experimental scrapie in mice: ultrastructural observations. *Ann Neurol* 4: 205 (1978).
10. J.R. Baringer, S.B. Prusiner, and J.S. Wong, Scrapie-associated particles in postsynaptic processes. Further ultrastructural studies. *J Neuropathol Exp Neurol* 40: 281 (1981).
11. D.C. Gajdusek DC, Infectious amyloids: subacute spongiform encephalopathies as transmissible cerebral amyloidoses, in: „Fields Virology", 3rd ed, B. Fields, D.M. Knipe, P.M. Howley, eds, Lippincott - Raven Publishers, Philadelphia-New York, (1995).
12. P.P. Liberski, Prions, β-sheets and transmissible dementias: is ther still something missing ? *Acta Neuropathol. (Berl)* 90: 113 (1995).
13. S.B. Prusiner, Prions, in: „Fields virology", 3rd ed, B. Fields, D.M. Knipe, P.M. Howley, eds, Lippincott - Raven Publishers, Philadelphia-New York, (1995).
14. R.H. Kimberlin, Scrapie and possible relationships with viroids. *Sem Virol* 1: 153 (1990)
15. H. Diringer, Hidden amyloidoses. *Exp Clin Immunogenet* 9: 212 (1993).
16. H.K. Narang, Evidence that scrapie-associated tubulofilamentous particles contain a single stranded DNA. *Intervirology* 36: 1 (1994).
17. H.K. Narang, D.M. Asher, and D.C. Gajdusek, Tubulofilaments in negatively stained scrapie-infecetd brains: relationships to scrapie-associated fibrils. *Proc. Natl. Acad. Sci. USA.* 84: 7730 (1987).
18. H.K. Narang, D.M. Asher, and D.C. Gajdusek, Evidence that DNA is present in abnormal tubulofilamentous structures found in scrapie. *Proc. Natl. Acad. Sci. USA* 85: 3375 (1988)
19. S.J. DeArmond, M.P. McKinley , R.A. Barry, M.B. Braunfeld, J.R. McCulloch, and S.B. Prusiner, Identification of prion filaments in scrapie-infected brain, *Cell* 41: 221 (1985).
20. C.A. Wiley, P.C. Burrola, M.J. Buchmeier, M.K. Wooddell, R.A. Barry, S.B. Prusiner, and P. Lampert, Immuno-gold localization of prion filaments in scrapie-infected hamster brain, *Lab Invest* 57: 646 (1987).

21. P.P. Liberski, D.C. Guiroy, E.S. Williams, R. Yanagihara, P. Brown, and D.C. Gajdusek, The amyloid plaque, in: „Light and Electron Microscopy Neuropathology of Slow Virus Disorders", P.P. Liberski, ed, CRC Press, Boca Raton, (1993).
22. P.P. Liberski, and H. Budka, Ultrastructural pathology of Gerstmann-Straussler-Scheinker disease. *Ultrastr Pathol* 19: 23 (1995).
23. M. Jeffrey, C. Goodsir, M.E. Bruce, P.A. McBride, and C. Farquhar, Morhogenesis of amyloid plaque in 87V murine scrapie. *Neuropathol Appl Neurobiol.* 20: 535 (1994).
24. M. Jeffrey, C. Goodsir, M.E. Bruce, P.A. McBride, and W.G. Halliday, Correlative light and electron microscopy studies of PrP localization in 87V scrapie. *Brain Res.* 656: 329 (1994).
25. D.A. Kocisko, J.H. Come, S.A. Priola, B. Chesebro, G.J. Raymond, P.T. Lansbury, and B. Caughey, Cell free formation of protease-resistant prion protein. *Nature* 370: 471 (1994).
26. R.A. Bessen, D.A. Kocisko, G.J. Raymond, S. Nandan, P.T. Lansbury, and B. Caughey, Non genetic propagation of strain-specific properties of scrapie prion protein, *Nature* 375: 698 (1995).
27. B. Caughey, D.A. Kocisko, G.J. Raymond, P.T. and Lansbury, Aggregates of scrapie-associated prion protein induce the cell-free conversion of protease-sensitive prion protein to the protease-resistant state, *Curr Biol.* 2: 807 (1995).
28. P.T. Lansbury, B. and Caughey, The chemistry of scrapie infection: implications of the „ice 9" metaphor, *Curr Biol* 2: 1(1995).
29. M. Barcikowska, P.P. Liberski, J. Boellaard, P. Brown, D.C. Gajdusek, and H.Budka, Microglia is a component of the prion protein amyloid plaque in the Gerstmann-Straussler-Scheinker syndrome. *Acta Neuropathol. (Berl.)* 85: 623 (1993).
30. D.C. Guiroy, I. Wakayama, P.P. Liberski, and D.C. Gajdusek, Relationship of microglia and scrapie amyloid-immunoreactive plaques in kuru, Creutzfeldt-Jakob disease and Gerstmann-Strausler syndrome. *Acta Neuropathol. (Berl)* 87: 526 (1994).
31. D.R. Brown, B. Schmidt, H. Kretzschmar, Role of microglia and host prion protein in neurotoxicity of a prion protein fragment. *Nature* 380: 345 (1996).
32. M.P. McKinley, A. Taraboulos, A. Kenaga, L. Serban, A. Steiber, S.J. DeArmond, and S.B. Prusiner, Ultrastructural localization of scrapie prion proteins of cytoplasmic vesicles of infected cultured cells, *Lab. Invest.* 65: 622 (1991).
33. P.A. Merz, R.A. Somerville, H.M. Wisniewski, and J. Iqbal, Abnormal fibrils from scrapie-infected brain. *Acta Neuropathol. (Berl.)* 60: 63 (1981).
34. H. Diringer, H. Gerdelblom, H. Hilmert, M. Ozel M, C..Edelbluth, and R.H. Kimberlin, Scrapie infectivity, fibrils and low molecular weight protein, *Nature* 306: 476 (1983).
35. M.P. McKinley, R.K. Meyer, L. Kenaga, F. Rahbar, R. Cotter, A. Serban, and S.B. Prusiner, Scrapie prion rod formation in vitro requires both detergent extraction and limited proteolysis. *J. Virol.* 65: 1340 (1991).
36. Liberski, "Tubulofilamentous particles" are not scrapie-specific and are unrelated to tubulovesicular structures. *Acta Neurobiologiae Exp.* 55: 149 (1995).
37. D. Chasey, Comment on the paper of H.K. Narang "Evidence that scrapie-associated tubulofilamentous particles contain a single-stranded DNA". *Intervirology* 37: 106 (1994).
38. M.P. McKinley, D.C. Bolton, and S.B. Prusiner, A protease resistant protein is a structural component of the scrapie prion, *Cell* 35: 57 (1994).
39. L. Manuelidis, T. Sklaviadis, and E.E. Manuelidis, Evidence suggesting that PrP is not the infectious agent in Creutzfeldt-Jakob disease, *EMBO J* 6: 341 (1987).
40. Y.G. Xi, L. Ingrosso, and C. Massullo, Amphoterricin B treatment dissociates in vivo replication of the scrapie agent from PrP accumulation, *Nature* 356: 598 (1992).
41. R.G. Rohwer, The scrapie agent: "a virus by any other name", in: Transmissible Spongiform Encephalopathies, Current Topics in Microbiology and Immunology, Vol. 172, B. Chesebro, ed, Springer Verlag, Berlin (1991).

FAMILIAL PRION DISEASES MODELED IN CELL CULTURE

David A. Harris, Sylvain Lehmann, and Nathalie Daude

Department of Cell Biology and Physiology
Washington University School of Medicine
St. Louis, MO 63110 USA

CELL BIOLOGY OF PRION DISEASES

The central pathogenic event in prion diseases is the posttranslational conversion of PrP^C, a normal cell-surface glycoprotein, into PrP^{Sc}, the principal constituent of infectious prion particles (reviewed by Prusiner, 1996). This transformation, which is hypothesized result from a change in the conformation of the polypeptide chain, has been studied both in vivo using transgenic mice (Scott et al., 1996) and in vitro using purified proteins and synthetic peptides (Bessen et al., 1995; Kaneko et al., 1995). The approach we have chosen to take is a cell biological one, focusing on the biosynthesis, posttranslational processing, and cellular trafficking of PrP^C and PrP^{Sc} in cultured cells (Harris et al., 1996). The techniques we employ include metabolic labeling, immunochemical analysis, microscopy, and subcellular fractionation. The primary advantage of a cell biological approach is the ability to study the formation of PrP^{Sc} in the context of the membranes, organelles, and cellular cofactors that are likely to be important in mediating the conversion process.

Cell culture models of prion infection have proven to be invaluable in understanding the molecular mechanisms underlying these diseases. Several different transformed lines, including N2a mouse neuroblastoma, HaB hamster brain, and PC12 rat pheochromocytoma cells can be infected in culture with prions derived from scrapie-infected brain (Butler et al., 1988; Race et al., 1988; Taraboulos et al., 1990; Rubinstein et al., 1991). Remarkably, the infected cells show no obvious cytopathology, and can be cloned and maintained in culture for many passages. They continuously convert about 1% of their PrP^C into PrP^{Sc}, which can be recognized by its protease-resistance and infectivity in animal bioassays. Using such cell culture models, it has been possible to establish that generation of PrP^{Sc} occurs in part along an endocytic pathway, and that the kinetics of this process is quite slow, requiring many hours for maximal accumulation of the scrapie isoform (Caughey and Raymond, 1991; Borchelt et al., 1992).

These cell culture systems can be considered to be models of the infectious manifestation of prion diseases. However, prion disorders also exist in familial forms, and until now there has been no cell culture model of these cases. The work to be described here is our attempt to develop such a model.

FAMILIAL PRION DISEASES

About 10% of the cases of Creutzfeldt-Jakob disease (CJD), and all cases of Gerstmann-Sträussler syndrome (GSS) and fatal familial insomnia (FFI) are inherited in an autosomal dominant pattern with nearly 100% penetrance (reviewed in Prusiner and Hsiao, 1994). These familial cases all result from one of several insertional or point mutations in the human gene that encodes PrP. The point mutations all occur in the C-terminal half of the PrP molecule, and are associated with either CJD, GSS or FFI. The insertion mutations, which are all associated with CJD, occur in the N-terminal half of the protein, which normally contains four copies of the octapeptide repeat PQGGGWGQ, and one copy of the nonapeptide PHGGGWGQ. The insertions consist of up to 9 additional copies of the octapeptide repeat. One of the most well characterized is a 6-octapeptide insertion that has been described in an English family that spans seven generations, all of whose members are descended from a single founder born in the late 18th century (Poulter et al., 1992). The precise phenotype of a given disease is strongly influenced by the mutation that causes it. For example, P102L causes an ataxic form of GSS (Hsiao et al., 1989), while F198S and Q217R produce a dementing variant of GSS with prominent neurofibrillary tangles (Hsiao et al., 1992). However, even patients with the same PrP mutation can show considerable clinical variability.

The PrP gene also encodes several amino acid polymorphisms which are found in the normal population, but which significantly influence prion disease susceptibility and symptomatology. The most well characterized is a met/val polymorphism at codon 129 (Owen et al., 1990). Homozygosity at this locus enhances susceptibility to sporadic and iatrogenic CJD in Caucasians (Palmer et al., 1991; Deslys et al., 1994), and decreases the age of onset and/or duration of illness in some forms of familial prion diseases (Goldfarb et al., 1992). In addition, the type of disease produced by the D178N mutation depends on

TABLE 1. Mutant moPrPs expressed in CHO cells, their human homologues, and associated phenotypes. Unless otherwise indicated, moPrPs contain Met at position 128. (Reprinted from Harris, 1997)

Mouse PrP	Human PrP	Phenotype	Reference
PG11	6 octapeptide insertion	CJD	Poulter et al., 1992
P101L	P102L	GSS	Hsiao et al., 1989
M128V	M129V	Normal	Owen et al., 1990
D177N/Met128	D178N/Met129	FFI	Goldfarb et al., 1992
D177N/Val128	D178N/Val129	CJD	Goldfarb et al., 1992
F197S/Val128	F198S/Val129	GSS	Dlouhy et al., 1992
E199K	E200K	CJD	Hsiao et al., 1991; Goldfarb et al., 1991

the amino acid at codon 129: met is associated with FFI, and val with CJD (Goldfarb et al., 1992).

TRANSFECTED CHO CELLS EXPRESSING MUTANT PrPs

In order to develop a cell culture model of inherited prion diseases, we have constructed stably transfected lines of Chinese hamster ovary (CHO) cells that express mouse PrP (moPrP) molecules carrying mutations homologous to six different pathogenic mutations of humans (Table 1). As a negative control, we have also analyzed cells expressing moPrP with a substitution of valine for methionine at codon 128, homologous to the non-pathogenic polymorphism at codon 129 in human PrP. We have carried out a detailed analysis of the cellular and biochemical properties of these mutant PrPs (Lehmann and Harris, 1995; 1996a; 1996b).

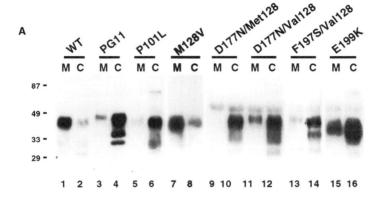

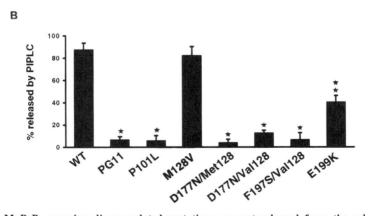

FIGURE 1. **MoPrPs carrying disease-related mutations are not released from the cell surface by PIPLC**. *Panel A*: CHO cells expressing each protein were biotinylated with the membrane-impermeant reagent sulfo-biotin-X-NHS, and were then incubated with PIPLC prior to lysis. MoPrP in the PIPLC incubation media (**M** lanes) and cell lysates (**C** lanes) was immunoprecipitated, separated by SDS-PAGE, and visualized by developing blots of the gel with HRP-streptavidin and enhanced chemiluminescence. *Panel B*: The amount of PrP released by PIPLC was plotted as a percentage of the total amount of PrP (medium + cell lysate). Each bar represents the mean ± SD. Values that are significantly different from wild-type (WT) moPrP by t-test are indicated by single (p<0.001), and double (p<0.01) asterisks. (Reprinted from Harris, 1997).

MUTANT PrPs DISPLAY AN ABNORMAL MEMBRANE ATTACHMENT

Wild-type PrP^C is attached to the plasma membrane by a glycosyl-phosphatidylinositol (GPI) anchor (Stahl et al., 1987). The core structure of this anchor, which is attached posttranslationally in the endoplasmic reticulum (ER) after cleavage of a short C-terminal hydrophobic segment of the polypeptide chain, consists of a phosphoethanolamine residue, three mannose residues, a glucosamine residue, and finally a phosphatidylinositol molecule which is part of the outer leaflet of the lipid bilayer (Fig. 2A) (Englund, 1993).   This anchor can be cleaved by the bacterial enzyme phosphatidylinositol-specific phospholipase C (PIPLC), which removes the diacylglycerol moiety and releases polypeptide chain with the rest of the anchor into the extracellular medium.

We have found that all mutant moPrPs transit the ER and Golgi, and reach the cell surface in a glycosylated form with a time-course that is not markedly different than that for wild-type PrP (Lehmann and Harris, 1995). However, a remarkable result is seen when cells are treated with PIPLC (Lehmann and Harris, 1995; 1996a). Each of the moPrPs carrying a disease-related mutation is much less efficiently released into the medium than is wild-type moPrP (Fig. 1). Of note, M128V moPrP behaves like the wild-type protein, implying that the effect is specific for pathogenic mutations. Failure of PIPLC to release the mutant PrPs is not due to the absence of the GPI anchor, or to a modification of the anchor (acylation of the inositol ring, for example) that renders its intrinsically PIPLC-resistant.   This conclusion is supported by the metabolic incorporation of the anchor precursors [3H]ethanolamine, [3H]stearate, and [3H]palmitate into the mutant proteins, and removal of the [3H]fatty acid label by incubation of intact cells with PIPLC.

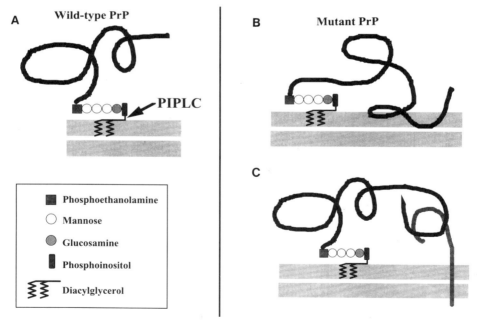

FIGURE 2. Models for the membrane topology of wild-type and mutant PrPs. *Panel A*: Wild-type PrP is anchored to the membrane exclusively by its GPI anchor, the core structure of which is illustrated, along with the site cleaved by PIPLC. The polypeptide chain of mutant PrP may be integrated into the lipid bilayer (*Panel B*), or bind tightly to another membrane-associated molecule (*Panel C*); in both cases, the mutant protein would be retained on the cell surface after cleavage of the GPI anchor by PIPLC. (Reprinted from Harris, 1997).

These results have led us to hypothesize that PrPs carrying disease-associated mutations have a second mechanism of membrane association in addition to the GPI anchor (Lehmann and Harris, 1995). The nature of this accessory attachment is under active investigation, but could involve integration of the PrP polypeptide chain into the lipid bilayer (Fig. 2B), tight binding to other membrane components (Fig. 2C), or formation of hydrophobic aggregates on the cell surface. We have found that PrP[Sc] synthesized by scrapie-infected neuroblastoma cells is also retained on the cell surface after PIPLC cleavage of its GPI anchor, an observation consistent with numerous previous reports that prion infectivity is tightly associated with cellular membranes (Lehmann and Harris, 1996a). These results make it likely that alterations in membrane topology are an important feature of prion biosynthesis.

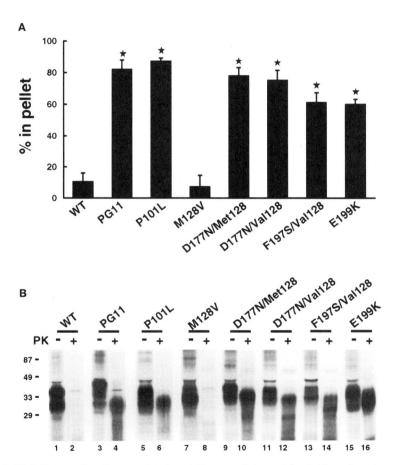

FIGURE 3. **MoPrPs carrying disease-related mutations are detergent-insoluble and protease-resistant when expressed in cultured CHO cells**. *Panel A*: Lysates of metabolically labeled CHO cells expressing each moPrP were centrifuged first at 16,000 x g for 5 min, and then at 265,000 x g for 40 min. MoPrP in the supernatants and pellets from the second centrifugation was quantitated by immunoprecipitation, and the percentage of PrP in the pellet was calculated. Each bar represents the mean ± SD of values from three experiments. Values that are significantly different from wild-type (WT) moPrP by t-test (p<0.001) are indicated by an asterisk. *Panel B*: Proteins in lysates of metabolically labeled cells were either digested at 37°C for 10 minutes with 3.3 µg/ml of proteinase K (+ lanes), or were untreated (- lanes), prior to recovery of moPrP by immunoprecipitation and analysis by SDS-PAGE. Five times as many cell-equivalents were loaded in the + lanes as in the - lanes. (Reprinted from Harris, 1997).

MUTANT PrPs DISPLAY BIOCHEMICAL PROPERTIES OF PrP[Sc]

We were interested in determining whether mutant PrPs synthesized in CHO cells were converted to PrP[Sc] (Lehmann and Harris, 1996a; 1996b). PrP[Sc] can be distinguished operationally from PrP[C] by several biochemical properties, including insolubility in non-denaturing detergents and resistance to protease digestion (Prusiner, 1996). To test the detergent-insolubility of the PrPs, we centrifuged detergent lysates of metabolically labeled cells at 265,000 x g for 40 minutes. We observed that while most the wild-type PrP remained in the supernatant under these conditions, the majority of each of the PrPs carrying a pathogenic mutation was found in the pellet (Fig. 3A). Again, M128V moPrP behaved like the wild-type protein. In addition, treatment of the mutant PrPs with 3.3 μg/ml proteinase K for 10 minutes resulted in formation of a protease-resistant fragment that migrated between 27 and 30 kDa, the same size as PrP 27-30 from scrapie-infected brain (Fig. 3B). Under the same conditions, wild-type and M128V moPrPs were completely digested. These results indicate that moPrPs carrying pathogenic mutations acquire two of the key biochemical properties of PrP[Sc]: detergent-insolubility and protease-resistance.

We note that mutant PrPs expressed in CHO cells are less protease-resistant than authentic PrP[Sc] from infected cells and brain, which can withstand digestion with higher concentrations of proteinase K for longer periods of time (Borchelt et al., 1992). However, the most feature of our data is not the absolute level of protease resistance of the mutant PrPs, but the fact that all of the proteins carrying pathogenic mutations are significantly more resistant than wild-type and M128V moPrP. Moreover, it is known that the degree of protease resistance of PrP[Sc] can vary considerably, depending on the prion strain and cell type from which it is derived (Bessen and Marsh, 1992; Meiner et al., 1992).

KINETICS OF ACQUISITION OF PrP[Sc]-LIKE PROPERTIES

We used pulse-chase metabolic labeling to determine the time course with which mutant PrPs became detergent-insoluble and protease-resistant (Lehmann and Harris, 1996b). We observed that both of these properties were acquired primarily during the chase period, with maximal detergent-resistance peaking at 1 hour (Fig. 4B) and maximal protease-resistance peaking at 6 hours (Fig. 4C). We also noted that once the mutant PrP acquired these properties, it was quite metabolically stable, decaying very little during a 16-hour chase (Figs. 4B and 4C). We conclude from these results that mutant PrPs undergo conversion to a PrP[Sc]-like state in a slow posttranslational process that confers metabolic stability on the molecule. These kinetic properties are very reminiscent of those of authentic PrP[Sc] synthesized in scrapie-infected neuroblastoma and hamster brain cells, which is synthesized with a $t_{1/2}$ of several hours, and is then stable for over 24 hours (Caughey and Raymond 1991; Borchelt et al., 1992).

STEPS IN THE GENERATION OF PrP[Sc]

Our kinetic studies suggested that conversion of mutant PrPs to a PrP[Sc]-like state may be a step-wise process, with acquisition of detergent-insolubility preceding attainment of protease-resistance. Additional experiments suggested the existence of an even earlier step, one related to an alteration in membrane topology. The first hint of this earlier step was the observation that that mutant PrP molecules were already resistant to PIPLC release by the time they reached the cell surface, within 30 min after pulse-labeling (Fig. 4A).

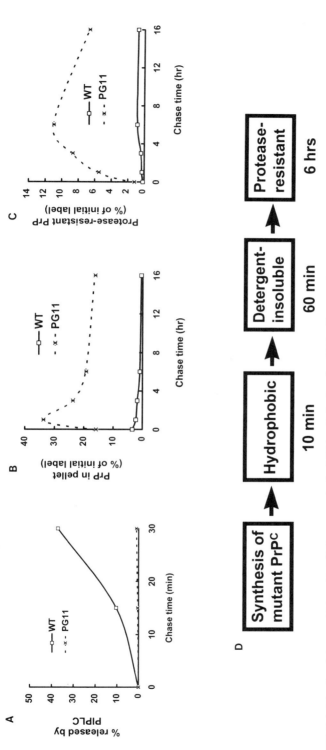

FIGURE 4. Kinetics of the acquisition of PrP^Sc^-like properties by PG11 moPrP. *Panel A:* CHO cells expressing wild-type (WT) or PG11 moPrP were labeled with [^{35}S]methionine for 20 minutes, and then chased for the indicated times. At the end of the chase period, cells were treated with PIPLC before lysis, and moPrP in cell lysates and PIPLC incubation medium was immunoprecipitated. The amount of PrP released by PIPLC was plotted as a percentage of the total amount of PrP present at each time point (medium + cells). *Panel B:* CHO cells were labeled for 30 minutes with [^{35}S]methionine and chased for the indicated times. At the end of each chase period, cell lysates were centrifuged as described in Fig. 3A, and moPrP immunoprecipitated from supernatants and pellets of the 265,000 x g spin. The amount of PrP in the pellet at each time point was plotted as a percentage of the total amount of radioactive PrP present at the end of the labeling period. *Panel C:* CHO cells were labeled for 45 min with [S^{35}]methionine, chased for the indicated times and then lysed. Proteins in cell lysates were digested at 37°C for 10 minutes with proteinase K (3.3 μg/ml), and moPrP recovered by immunoprecipitation. The amount of protease-resistant PrP (M$_r$ 27-30 kDa) at each time point was plotted as a percentage of the total amount of radioactive PrP present at the end of the labeling period. *Panel D:* A scheme for generation of PrP^Sc^ from mutant PrPs. (Modified from Harris, 1997).

93

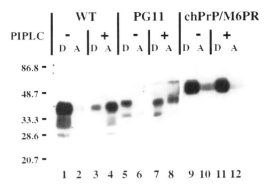

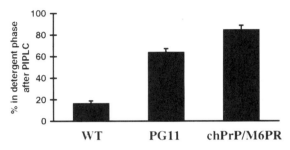

FIGURE 5. PIPLC-treated PG11 moPrP is retained in the detergent phase after Triton X-114 phase partitioning. *Panel A*: CHO cells expressing wild-type (WT) or PG11 moPrP, and N2a cells expressing a chicken PrP/mannose 6-phosphate receptor (chPrP/M6PR) transmembrane chimera were surface-biotinylated, and then lysed at 4°C in a buffer containing 1% Triton X-114. After the temperature was raised to 37°C, the detergent phase was recovered, diluted to the original volume, and split in half. One half was incubated with PIPLC at 4°C for 2 h (+ lanes), and the other half was left untreated (- lanes). Phase separation was then repeated, and moPrP (lanes 1-8) and chPrP/M6PR (lanes 9-12) in the detergent (**D**) and aqueous (**A**) phases was immunoprecipitated. SDS polyacrylamide gels of the immunoprecipitates were electroblotted, and the blots developed with HRP-streptavidin and ECL to visualize biotinylated proteins. *Panel B*: PrP bands from the experiment shown in panel A, and from 2 additional experiments, were quantitated by densitometry, and the amount of PrP in the detergent phase after PIPLC treatment (lanes 3, 7, and 11) was plotted as a percentage of the total amount of PrP present (detergent + aqueous phases). Each bar represents the mean ± SD. (Reprinted from Lehmann and Harris, 1995).

This result suggested that whatever molecular change was responsible for the abnormal membrane attachment of mutant PrPs, it occurred soon after synthesis of the proteins.

To study the kinetics of this early step in more detail, it would be desirable to assess the PIPLC-releasability of molecules before they reach the cell surface. Since this is technically difficult, we chose instead to assay a property that is likely to be correlated with abnormal membrane attachment: increased hydrophobicity, as measured by Triton X-114 phase partitioning (Lehmann and Harris, 1995). When cell lysates containing this detergent are warmed to 37°C, two phases are formed: integral membrane proteins partition into the detergent phase, and hydrophilic proteins into the aqueous phase. We observed that surface-biotinylated molecules of both wild-type and PG11 moPrP partitioned into the detergent phase before treatment with PIPLC, as expected because of the presence of the hydrophobic glycolipid anchor (Fig. 5). After removal of the diacylglycerol moiety of the anchor with PIPLC, wild-type moPrP shifted almost completely into the aqueous phase. In contrast, PG11 moPrP remained predominantly in the detergent phase after PIPLC treatment, as did a <u>bona fide</u> transmembrane protein (chPrP/mannose 6-phosphate receptor chimera) used as a control. These results indicated that the PG11 polypeptide chain itself was hydrophobic, possibly because it was integrated into the lipid bilayer (Fig. 2B).

To determine the time course with which this hydrophobic character was acquired, we performed pulse-labeling experiments. We observed that after metabolic labeling for only 10 minutes, PG11 moPrP partitioned predominantly into the detergent phase after PIPLC treatment, suggesting that the protein becomes hydrophobic during or very soon after synthesis (Daude, Lehmann, and Harris, manuscript in preparation). Similar results were obtained when hydrophobicity was measured by binding to phenylsepharose (Davitz, 1988).

Overall, our results are consistent with the idea that mutant PrPs are initially synthesized as PrP^C, and then are converted to the PrP^{Sc} state in a stepwise manner with discrete biochemical intermediates (Fig. 4D). The three steps we have identified thus far are acquisition, first, of hydrophobicity (which is probably correlated with abnormal membrane attachment as assayed by PIPLC-releasability), then detergent-insolubility, and finally protease-resistance. How these operationally defined biochemical changes relate to the postulated conformational alteration that underlies formation of PrP^{Sc} remains to be determined; one possibility is that mutant PrP molecules undergo a concerted conformational switch early after synthesis, which only subsequently manifests itself by gradual changes in measurable biochemical properties. In any case, one implication of our data is that the first observable change, increased hydrophobicity, may occur along the secretory pathway (ER or Golgi), while the two later steps may occur on the cell surface or in endocytic organelles. We have used procedures which block the secretory pathway at various points, including treatment with the fungal metabolite brefeldin A and incubation at 18°C, to help identify the cellular compartments where the three steps occur (Daude, Lehmann, and Harris, manuscript in preparation).

STRAIN-LIKE PROPERTIES OF MUTANT PrPs

We have observed that mutant PrPs expressed in CHO cells display subtle variations in their biochemical properties that may relate to the distinctive phenotype with which each of homologous human mutations is associated (Lehmann and Harris, 1996a). Differences among the mutant PrPs can be seen in the PIPLC releasability (note E199K in Fig. 1B), glycosylation pattern (Fig. 1A), and size of the protease-resistant fragment (Fig. 3B). Differences in the last two properties have been documented in PrP^{Sc} from the brains of patients with inherited, sporadic and iatrogenic prion diseases, and these have been used to define several different strains of the infectious agent (Monari et al., 1994; Parchi et al., 1996; Collinge et al., 1996).

INTERACTION OF MUTANT AND WILD-TYPE PrPs IN THE SAME CELL

Almost all patients with inherited prion diseases are heterozygous for the mutant allele (Prusiner and Hsiao, 1994), implying that their cells express both mutant and wild-type PrPs. To model this situation, we have prepared stably transfected lines of CHO cells that express approximately equal levels of wild-type and PG11 moPrPs. The PG11 protein was chosen for these experiments, because it can be readily distinguished from wild-type PrP on SDS-PAGE because of its lower electrophoretic mobility due to the insertion of 48 additional amino acids. The cells we have utilized have been cloned at least once to ensure that each cell in the population expresses both proteins.

We find that wild-type moPrP does not acquire any of the biochemical properties of PrP^{Sc} when co-expressed with PG11 moPrP (Lehmann, Daude, and Harris, manuscript in preparation). The properties we have examined include detergent-insolubility, protease-resistance, PIPLC-releasability, and hydrophobicity measured by Triton X-114 phase partitioning. This result is consistent with studies of PrP^{Sc}-containing amyloid fibrils extracted from patients carrying the F198S and Q217R mutations, which appear to contain only protein encoded by the mutant allele (Tagliavini et al., 1994). In contrast, patients with the E200K mutation have been reported to contain a form of wild-type PrP that is partially converted to PrP^{Sc}, in that it is detergent-insoluble but not protease-resistant (Gabizon et al., 1996). It may therefore be that differences exist among mutant PrPs with respect to their interaction with wild-type PrP in the same cell.

CONCLUSIONS

Taken together, our results document the usefulness of cultured CHO cells expressing mutant PrPs as a model system for studying familial prion diseases of humans. To our knowledge, these results provide the first evidence that a PrP^{Sc}-like molecule can be produced in cultured cells not exposed to exogenous infectious agent. The biochemical similarities between mutant moPrPs synthesized in CHO cells and authentic infectious PrP^{Sc} include detergent-insolubility, protease-resistance, slow metabolic generation and turnover, abnormal membrane attachment, and strain-like biochemical variations. It remains to be determined whether mutant PrP molecules derived from CHO cells are infectious in animal bioassays.

This cell culture system can now be used to address a number of important issues, including the cellular trafficking of mutant PrPs, the identification of additional intermediates in the generation of PrP^{Sc}, the role of cellular chaperones and cofactors in the conversion process, neurotoxicity of mutant PrPs, and the development of permeabilized cell systems for PrP^{Sc} formation. The results of these studies, in conjunction with those obtained using scrapie-infected cell models, will greatly enrich our understanding of this unusual group of neurodegenerative disorders.

ACKNOWLEDGMENTS

This work was supported by grants to D.A.H. from the National Institutes of Health and the Alzheimer's Association. S.L. is the recipient of a Postdoctoral Fellowship for Physicians from the Howard Hughes Medical Institute.

REFERENCES

Bessen, R.A., and Marsh, R.F.,1992, Biochemical and physical properties of the prion protein from two strains of the transmissible mink encephalopathy agent, *J. Virol.* 66:2096.

Bessen, R.A., Kocisko, D.A., Raymond, G.J., Nandan, S., Landsbury, P.T., and Caughey, B., 1995, Non-genetic propagation of strain-specific properties of scrapie prion protein, *Nature* 375:698.

Borchelt, D.R., Taraboulos, A., and Prusiner, S.B., 1992, Evidence for synthesis of scrapie prion proteins in the endocytic pathway, *J. Biol. Chem.* 267:16188.

Butler D.A., Scott, M.R.D., Bockman, J.M., Borchelt, D.R., Taraboulos, A., Hsiao, K.K., Kingsbury, D.T., and Prusiner, S.B., 1988, Scrapie-infected murine neuroblastoma cells produce protease-resistant prion proteins, *J. Virol.* 62:1558.

Caughey, B., and Raymond, G.J., 1991, The scrapie-associated form of PrP is made from a cell surface precursor that is both protease- and phospholipase-sensitive, *J. Biol. Chem.* 266:18217.

Collinge, J., Sidle, K.C., Meads, J., Ironside, J., and Hill, A.F., 1996, Molecular analysis of prion strain variation and the aetiology of "new variant" CJD, *Nature* 383:685.

Davitz, M.A., 1988, Use of phenylsepharose to discriminate between hydrophilic and hydrophobic forms of decay accelerating factor, in: *Posttranslational Modification of Proteins by Lipids*, U. Brodbeck and C. Bordier, eds., Springer-Verlag, Berlin, p. 40.

Deslys, J.-P., MarcJ, D., and Dormont, D.,1994, Similar genetic susceptibility in iatrogenic and sporadic Creutzfeldt-Jakob disease, *J. Gen. Virol.* 75:23.

Dlouhy, S.R., Hsiao, K., Farlow, M.R., Foroud, T., Conneally, P.M., Johnson, P., Prusiner, S.B., Hodes, M.E., and Ghetti, B., 1992, Linkage of the Indiana kindred of Gerstmann-Sträussler-Scheinker disease to the prion protein gene, *Nature Genet.* 1:64.

Englund, P. T.,1993, The structure and biosynthesis of glycosyl phosphatidylinositol protein anchors, *Ann. Rev. Biochem.* 62:121.

Gabizon, R., Telling, G., Meiner, Z., Halimi, M., Kahana, I., and Prusiner, S.B., 1996, Insoluble wild-type and protease-resistant mutant prion protein in brains of patients with inherited prion disease, *Nature Med.* 2:59.

Goldfarb, L.G., Brown, P., Mitrova, E., Cervenakova, L., Goldin, L., Korczyn, A.D., Chapman, J., Galvez, S., Cartier, L., Rubenstein, R., Gajdusek, D.C., 1991, Creutzfeldt-Jakob disease associated with the PRNP codon 200^{LYS} mutation: an analysis of 45 families, *Eur. J. Epidem.* 7:477.

Goldfarb, L.G., Petersen, R.B., Tabaton, M., Brown, P., LeBlanc, A.C., Montagna, P., Cortelli, P., Julien, J., Vital, C., Pendelbury, W.W., Haltia, M., Wills, P.R., Hauw, J.J., McKeever, P.E., Monari, L., Schrank, B., Swergold, G.D., Autilio-Gambetti, L., Gajdusek, D.C., Lugaresi, E., and Gambetti, P.,1992, Fatal familial insomnia and familial Creutzfeldt-Jakob disease: disease phenotype determined by a DNA polymorphism, *Science* 258:806.

Harris, D.A.., 1997, Cell biological insights into prion diseases, in: Prion Diseases: Current State and Perspectives in Prion Research, H. Kretzschmar, and D. Riesner, eds., Chapman and Hall, New York, In Press.

Harris, D.A., Gorodinsky, A., Lehmann, S., Moulder, K., and Shyng, S.-L., 1996, Cell biology of the prion protein, *Curr. Topics in Microbiol. and Immunol.* 207:79.

Hsiao, K., Baker H.F., Crow, T.J., Poulter, M., Owen, F., Terwilliger, J.D., Westaway, D., Ott, J., and Prusiner, S.B., 1989, Linkage of a prion protein missense variant to Gerstmann-Sträussler syndrome, *Nature* 338:342.

Hsiao, K., Meiner, Z., Kahana, E., Cass, C., Kahana, I., Avrahami, D., Scarlato, G., Abramsky, O., Prusiner, S.B., Gabizon, R., 1991, Mutation of the prion protein in Libyan Jews with Creutzfeldt-Jakob disease, *N. Engl. J. Med.* 324:1091.

Hsiao, K., Dlouhy, S.R., Farlow, M.R., Cass, C., Da Costa, M., Conneally, P.M., Hodes, M.E., Ghetti, B., and Prusiner, S.B., 1992, Mutant prion proteins in Gerstmann-Sträussler-Scheinker disease with neurofibrillary tangles, *Nature Genet.* 1:68.

Kaneko, K., Peretz, D., Pan, K.-M., Blochberger, T.C., Wille, H., Gabizon, R., Griffith, O.H., Cohen, F.E., Baldwin, M.A., and Prusiner, S.B., 1995, Prion protein (PrP) synthetic peptides induce cellular PrP to acquire properties of the scrapie isoform, *Proc. Natl. Acad. Sci. USA* 92:11160.

Lehmann, S., and Harris, D.A., 1995, A mutant prion protein displays an aberrant membrane association when expressed in cultured cells, *J. Biol. Chem.* 270:24589.

Lehmann, S., and Harris, D.A., 1996a, Mutant and infectious prion proteins display common biochemical properties in cultured cells. *J. Biol .Chem.* 271:1633.

Lehmann, S., and Harris, D.A., 1996b, Two mutant prion proteins expressed in cultured cells acquire biochemical properties reminiscent of the scrapie isoform, *Proc. Natl. Acad. Sci. USA* 93:5610.

Meiner, Z., Halimi, M., Polakiewicz, R.D., Prusiner, S.B., and Gabizon R., 1992, Presence of prion protein in peripheral tissues of Libyan Jews with Creutzfeldt-Jakob disease, *Neurol.* 42:1355.

Monari, L., Chen, S.G., Brown, P., Parchi, P., Petersen, R.B., Mikol, J., Gray, F., Cortelli, P., Montagna, P., Ghetti, B., Goldfarb, L., Gajdusek, D.C., Lugaresi, E., Gambetti, P., Autilio-Gambetti, L., 1994, Fatal familial insomnia and familial Creutzfeldt-Jakob disease: different prion proteins determined by a DNA polymorphism, *Proc. Natl. Acad. Sci. USA* 91:2839.

Owen, F., Poulter, M., Collinge, J., and Crow, T.J., 1990, Codon 129 changes in the prion protein gene in Caucasians, *Am. J. Hum. Genet.* 46:1215.

Palmer, M.S., Dryden, A.J., Hughes, J.T., and Collinge, J.,1991, Homozygous prion protein genotype predisposes to sporadic Creutzfeldt-Jakob disease, *Nature* 352:340.

Parchi, P., Castellani, R., Capellari, S., Ghetti, B., Young, K., Chen, S.G., Farlow, M., Dickson, D.W., Sima, A.A.F., Trojanowski, J.Q., Petersen, R.B., and Gambetti, P., 1996, Molecular basis of phenotype variability in sporadic Creutzfeldt-Jakob disease, *Ann. Neurol.* 39:767.

Poulter, M., Baker, H.F., Frith, C.D., Leach, M., Lofthouse, R., Ridley, R.M., Shah, T., Owen, F., Collinge, J., Brown, J., Hardy, J., Mullan, M.J., Harding, A.E., Bennett, C., Doshi, R., Crow, T.J., 1992,

Inherited prion disease with 144 base pair gene insertion. 1. Genealogical and molecular studies. *Brain* 115:675.

Prusiner, S.B., 1996, Prions, in: *Virology, 3rd Edition*, B.N. Fields, D.M. Knipe, and P.M. Howley, eds., Lipincott-Raven Press, Philadelphia, p. 2901.

Prusiner, S.B., and Hsiao, K.K., 1994, Human prion diseases, *Ann. Neurol.* 35:385.

Race, R.E., Caughey, B., Graham, K., Ernst, D., and Chesebro, B., 1988, Analyses of frequency of infection, specific infectivity, and prion protein biosynthesis in scrapie-infected neuroblastoma cell clones, *J. Virol.* 62:2845.

Rubinstein, R., Deng, H., Scalici, C.L., and Papini, C., 1991, Alterations in neurotransmitter-related enzyme activity in scrapie-infected PC12 cells, *J. Gen. Virol.* 72:1279.

Scott, M.R.D., Telling, G.C., and Prusiner, S.B., 1996, Transgenetics and gene targeting in studies of prion diseases, *Curr. Top. Microbiol. Immunol.* 207:95.

Stahl, N., Borchelt, D.R., Hsiao, K., and Prusiner, S.B., 1987, Scrapie prion protein contains a phosphatidylinositol glycolipid anchor, *Cell* 51:229.

Tagliavini, F., Prelli, F., Porro, M., Rossi, G., Fiaccone, G., Farlow, M.R., Dlouhy, S.R., Ghetti, B., Bugiani, O., and Fragione, B., 1994, Amyloid fibrils in Gerstmann-Sträussler-Scheinker disease (Indiana kindreds) express only PrP peptides encoded by the mutant allele, *Cell* 79:695.

Taraboulos, A., Serban, D., and Prusiner, S.B., 1990, Scrapie prion proteins accumulate in the cytoplasm of persistently infected cultured cells, *J. Cell Biol.* 110:2117.

YEAST APPROACH TO PROTEIN "PRIONIZATION": SUP35 - [PSI] SYSTEM

S.G.Inge-Vechtomov , E.A.Ilmov, L.N.Mironova,
V.L.Tikchomirova, K.V.Volkov, and S.P.Zadorsky

Dept. of Genetics and Breeding
St.Petersburg University
199034 St.Petersburg, Russia

INTRODUCTION

Some human neurodegenerative diseases, such as Kuru, Creutzfeld-Jacob, Gerstmann-Streussler-Scheinker, as well as the animal ones (scrapie in sheeps, mad cow desease, etc.), are transmitted by prion, an infectious protein PrP^{Sc}. This protein is an oligomer form of the normal cellular neuropeptide PrP^C. Both proteins are encoded by the same gene PrP, which is conserved at least among mammals. Mutations within the gene cause familial forms of the same prion diseases in men. In the latter case duplications of characteristic aminoacid repeats had been discovered in the N-terminal part of PrP protein. Deletion of PrP gene in mice confers resistance to prion infection in the animal. Overexpression of the same gene induces PrP^{Sc} and prion disease in mice (see a review by Wickner et al[1]).

Two systems comparable to PrP gene - prion protein system of mammals had been decifered recently in *Saccharomyces* yeast[1]. Cytoplasmic genetic factors, [PSI], a dominant omnipotent nonsense-(allo)suppressor[2], and [URE3], a factor, regulating nitrogene metabolism[3], are known in this object. It was established that the nuclear gene SUP35, encoding eukaryotic translation termination factor eRF3[4], is the structural gene for the [PSI]-factor[5,6]. Mutations of this gene, as well as mutations of SUP45, coding for another translational release factor, eRF1[7], were described as recessive omnipotent nonsense-suppressors sup2 and sup1, respectively[8,9]. Mutations in either of these genes show identical spectra of pleiotropic effects: sensitivity to high and low temperature, osmosensivity, sensitivity to paromomycin and benomyle, etc.[10,11]. Sup35p and Sup45p function as a single complex in translation termination[4,12].

The facts proving that SUP35 is the structural gene of [PSI] as an yeast prion are as follows:
1) Amplification of SUP35 (but not of SUP45) on moderate copy number plasmid induces [PSI] factor[5], the same way as amplification of URE2 induces [URE3] factor[13]. Basing upon these results R.Wickner[13] proposed a hypothesis of prion nature for both [PSI] and [URE3] factors.

2) Overexpression of SUP35 on centromeric plasmid under inducible Gal-promoter also induces [PSI]. It was shown that it is overproduction of Sup35 protein, but neither of SUP35 DNA nor mRNA responsible for [PSI] induction[6].

3) The PNM2 mutation preventing [PSI] propagation, described by B.S. Cox, was mapped to the left part of SUP35, coding for aminoacid repeats, comparable to the N-terminal repeats of PrP protein[14].

4) Deletion analysis of SUP35 confirmed that the N-terminal 1/3 of Sup35p contains a peptide responsible for propagation of the [PSI] factor. Deletion of this part of SUP35 prevents [PSI] propagation in haploids and expresses itself as a semidominant antisuppressor against [PSI][15].

5) The same part of the Sup35p is necessary for [PSI] induction when SUP35 is overexpressed. Deletion, partially covering aminoacid repeat coding region, prevented [PSI] induction[6].

6) Mutations of SUP35 (and of SUP45) cause the same omnipotent nonsense-suppressor phenotype as [PSI] factor does, although the mutations are recessive and have many pleiotropic effects in contrast to [PSI] that is dominant and has no pleiotropy.

7) [PSI] is transferred by cytoduction - transfer of cytoplasm without transfer of nuclei. The process is considered as an yeast analogue of prion infection in mammals.

8) It was shown recently, that [PSI+], but not [PSI-] strains, contain oligomers of eRF3 most probably inactive in translation termination[16].

9) Conversion of [PSI-] to [PSI+] cells depends on an optimal expression level of HSP104, coding for one of the yeast chaperone[17]. Deletion or high overexpression of this gene prevents [PSI] propagation, meanwhile its moderate overexpression only partially neutralizes [PSI] mediated suppression effect and does not cure the cells of [PSI].

10) [PSI] factor may be cured by 5 mM Guanidine HCl (GuHCl)[18]. Elimination of suppressor effect after GuHCl treatment is utilized for identification of [PSI] containing yeast strains. It is necessary to emphasize that new [PSI+] clones may be repeatedly isolated from the cured cells[19].

As was shown recently, there exist several forms of [PSI] factors estimated by their suppressor efficiency and stability after GuHCl treatment. They may be found in different [PSI+] strains or selected in the same [PSI-] strain. It meens that the "vigour" of [PSI] factor is its own characteristic, but not of some additional nuclear modifiers, which might be induced simultaneously with [PSI][6].

The two systems, PrP gene - prion of mammals and SUP35 - [PSI] of yeast, have much in common[1], but polypeptides encoded in these systems have different functions, and Sup35p is known as a part of a complex protein aggregate[4,12]. It is important to note also that SUP35 function is indispensable one[10], whereas the complete inactivation of PrP gene (its deletion) is not lethal.

It is more or less commonly accepted, that process of "prionization" - transition from PrP^C to PrP^{Sc} or from eRF3 to [PSI] - depends somehow upon the N-terminal repeats of corresponding proteins (this is not exactly applicable to the system of URE2-[URE3] as Ure2p has no these repeats) and includes stable self-reproducing conformational changes of the protein[1]. The most intriguing questions concerning the prionization phenomenon are: What is its mechanism? and How common is this phenomenon in regulation of protein activity and in epigenetic inheritance/variability?

Yeast as an object has many advantages for the study of these problems: well developed genetics and molecular biology, simplicity in the obtaining of mutants, the possibility to use highly specific selective methods for isolation of sup35 and sup45 mutants[9], curability of [PSI] with GuHCl[18]. In addition yeast prion (namely [PSI]) is not dangerous for humans, as we believe. Though sequencing data showed the presence of a peptide comparable to the prionisation peptide in eRF3 of another yeast species, Pichia methanolica (see Fig.1), there was no such peptide found in human

```
                            *
              -----→        ─────────→    ------→          ──
S.cerevisiae   ...55 YQQGGYQQYNPDAGYQQQ---YNP--QGGYQQYNP------QGGY
P.methanolica  ...71 YNQGGYNNYNNRGGYSNNRGGYNNSNRGGYSNYNSYNTNSNQGGY
                     ······→      ······→  ······→    ······→        ·······→


S.cerevisiae   104 ───QQQFNPQGG--R...
P.methanolica  116 SNYNNNYANNSYNNN...
```

Figure 1. N-terminal aminoacid repeats (indicated by arrows) in Sup35p of *S.cerevisiae* and of *P.methanolica*[20]. Position of PNM2 mutant substitution in *S. cerevisiae* SUP35 is indicated by *.

and Xenopus eRF3[4,10]. One more advantage of yeast - clear phenotype change after [PSI⁻] ⇔ [PSI⁺] transition.

This work was undertaken for the study of variability of [PSI], depending on sup35 and sup45 mutations and on interspecific variations in SUP35 structure.

MATERIALS AND METHODS

Strains and Media

The Saccharomyces cerevisiae strains, used in this study, belong to the Peterhoff Breeding Stocks (Table 1).

We used standard yeast growth media described previously[21]. Yeast cells were grown at 30⁰ or 20⁰C. Non-fermentable media contained the ethanol (20 ml/l) as a sole carbon source.

E.coli strains HB101 and DH5α[22] were used for cloning experiments. Bacteria were grown at 37⁰.

Table 1. Strains used in this study

Strain	Genotype
33G-D373	MATα pheA10 ade2-144,791 his7-1(UAA) lys9-A21(UAA) ura3-52 leu2-3,112 trp1-289(UAG)
11-P5213	MATa ade1-14(UGA) his7-1 met13-A1 leu2-1 [PSI⁺]
du8-132-L28-2V-P3982	MATα ade1-14 his7-1 lys2-87 Δura3 thr4- B15(UGA) leu2-1
42B-P3990	MATa ade1-14 his7-1 lys2-A12(UAG)
39-D796	MATα ade1-14 his7-1 lys9-A21 ura3-52 trp1-289 leu2-3,112
31V-D543	MATa ade1-14 his7-1 ura3-52 leu2-3,112 trp1-289
16A-P5154	MATa his7-1 lys2-87 met13-A1 thr4-B15 leu2-1

Plasmids and DNA Manipulation

All DNA manipulations, plasmid construction techniques and Southern blot hybridization were carried out according to standard protocols[22]. The plasmid construction for replacement of SUP35S for SUP35P in Saccharomyces chromosome was obtained as follows. 3.5 kb HindIII fragment of pTR30-1 plasmid, containing SUP35P gene[20], and 2.1 kb PstI fragment of pLL12 plasmid, containing LEU2 gene[24], were cloned subsequently into pBluescript vector. XbaI - BssHII fragment of the resulting plasmid containing SUP35P and LEU2 genes was ligated to XbaI - MluI fragment of pSTR7 plasmid, containing S.cerevisiae SUP35 gene The resulting constract was designated as pSPZ1. 7.2 kb fragment of this plasmid containing regions flanking SUP35 in S. cerevisiae chromosome and SUP35P and LEU2 genes in between was used to transform 33G-D373.

Genetic Methods

We used standard yeast genetic procedures[21]. The sup35 and sup45 mutants were induced with UV light (300J/m[2]) and isolated as described previously[9,25]. Yeast strains were cured of the [PSI] by incubation on YPD medium supplemented with 5mM GuHCl[18] if not indicated otherwise.

RESULTS

Recessive [psi]'s with Allo- and Anti-suppressor Effects

It is known that in vivo cytoplasmic [PSI]-factor behaves as a dominant omnipotent nonsense-suppressor or dominant allosuppressor, interacting with low effective recessive sup35 and sup45 mutations. These recessive mutations become dominant in suppression of his7-1, when crossed with suppressorless strains carrying [PSI⁺] factor. Now we have shown that induction of both sup35 and sup45 mutations in some [PSI⁺] strains can convert the dominant [PSI] factor into unusual recessive form.

In two strains, 42B-P3990 and 39-D796, bearing the dominant [PSI], recessive suppressors were selected at 20°C on ethanol containing medium. 19 mutants, His⁺ on glucose at 20°C, but His⁻ at 30°, were obtained. So [PSI] factor of the original strains expressed allosuppressor effect toward these suppressor mutations. Complementation test showed that there were 13 sup35 mutations and 6 sup45 mutations among them. After GuHCl treatment these 19 conditional mutants became His⁻, it means that suppression effect was caused by interaction of sup35 and sup45 cryptic mutations with [PSI] factor. The ability of these mutants to pass through the complementation test means that originally dominant allosuppressor form of [PSI] factor had been converted into the recessive one as a result of induction of both sup35 and sup45 mutations.

The unusual forms of [PSI] factor were discovered also among sup35 and sup45 mutants selected in the originally [PSI⁻] strain du8-132-L28-2V-P3982. After GuHCl treatment 4 sup35 mutants from 61 investigated and 1 sup45 mutant from 42 investigated changed their phenotypes (Table 2). In addition to the effect of a decrease of suppression efficiency after GuHCl treatment, revealed in three mutants, the elevation of suppression efficiency was discovered in two sup35 mutants. Thus, some mutant sup35 alleles generate probably the antisuppressor form of [PSI] factor, decreasing the suppression efficiency of these alleles.

Two sup35 mutants bearing the antisuppressor [PSI] factor were crossed with [PSI⁻] strain 16A-P5154. Diploids obtained were homozygous for his7-1 and lys2-87. No segregation for suppression of his7-1 and lys2-87 was found in the meiotic progeny of these diploids. All tetrads analysed (16 for one diploid and 18 for the other one) showed 4:0 for both characters. After three subsequent replica platings of these tetrads on GuHCl containing medium in most of them the segregation 2(His Lys)⁺: 2(His Lys)⁻ appeared (fig. 2). When the same sup35 mutants were treated by GuHCl before to be crossed with 16A-P5154, 2:2 segregation for suppression of both nonsense mutations was obtained (data not shown). It means that recessive antisuppressor [psi] is inherited cytopolasmically and its characteristics are not dependent on any other genetic factors besides sup35 mutation.

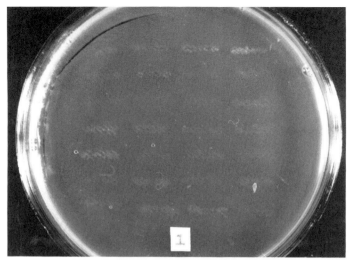

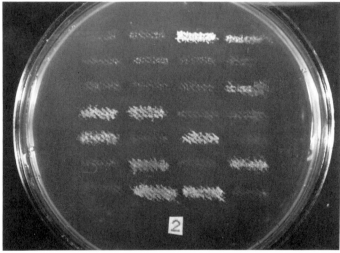

Figure 2. Tetrads of one of the diploids, bearing antisuppressor [psi], immediately after dissection (1) and after three replica platings on 5mM GuHCl (2).

SUP35 Substitution in S. cerevisiae by Its Homologue from P. methanolica

For further study of specificity of prionization phenomenon in yeast we substituted SUP35 gene of S.cerevisiae for its homologue from P. methanolica. The latter one had been cloned and sequenced previously[20]. A high degree of homology between the C-terminal 2/3 of Sup35 proteins from S. cerevisiae and from P. methanolica was shown. At the same time there is no considerable homology of primary structure among the N- terminal parts (about 1/3) of Sup35 proteins in these species. Nevertheless some resemblance in secondary structure (β-sheet enriched in Asn, Gln, Tyr, Gly) of these parts is very probable. There are oligopeptide repeats in this part of Pichia Sup35p, though different from the ones in Saccharomyces Sup35p (see fig.1). We decided to study the SUP35P expression and possibility of its protein product prionization in S.cerevisiae cells.

Table 2. The change of phenotypes of sup35 and sup45 mutants of strain du8-132-L28-2V-P3982 after GuHCl treatment

Mutants	Suppression of		
	his7-1 (UAA)	lys2-87 (UGA)	thr4-B15 (UGA)
48 (sup45)	+/-	-/+	-
After GuHCl	-	-	-
12 (sup35)	+	+	-
After GuHCl	-	-	-
28 (sup35)	+	+	+
After GuHCl	+	+	-
10 (sup35)	-	-	-
AfterGuHCl	+/-	+	-
25 (sup35)	-	-	-
AfterGuHCl	-/+	+	-

+, +/-, -/+, - : normal growth, poor growth, traces of growth, the absence of growth, respectively, on omission media.

For this purpose Leu⁻ strain 33G-D373 was transformed by the fragment of pSPZ1 plasmid, containing SUP35 of P. methanolica (SUP35P) under its own promoter and LEU2 gene of S. cerevisiae (see Materials and Methods for details). 8 stable Leu⁺ transformants were obtained, in two of them SUP35S gene was substituted by SUP35P. This was proved by the following facts.
1) Absense of LEU2 - MAT linkage in diploids obtained from the crosses of transformants with the tester strains 11-P5213 and 31V-D543 (Table 3). LEU2 and MAT are closely linked markers of the yeast chromosome III, so the absense of their linkage revealed in diploids under investigation indicates the integration of LEU2 into another location.

Table 3. Absence of linkage between MAT and LEU2 in diploids obtained from the crosses of 33G-D3731 transformants with tester strains

Parent strains	Number of tetrads:			
	P	N	T	
1-SUP35P x 11- P5213	5	2	12	$\chi^2 = 0,11$
6-SUP35P x 31V-543	2	3	8	$\chi^2 = 0,15$

2) Southern blot hybridisation of radioactive DNA probe, containing SUP35P gene, with chromosomal DNA of these two Leu⁺ transformants (data not shown).
3) The analysis of the transformants phenotype revealed the suppression of UAA mutations (his7-1 and lys9-A21) absent in the original strain. In complementation test performed by the cross of these transformants with sup35 and sup45 mutants this suppressor effect behaved as a recessive one, caused by sup35 mutation. In the random ascospore sample of diploids obtained in cross of one of the transformants with the [PSI⁻] derivative of 11-P5213 the linkage of suppressor effect with LEU2 was observed (Table 4, left column). Thus the suppressor effect is a consequense of substitution for the wild type allele of SUP35P. Suppression had been registered not in all 39 LEU⁺ segregants analysed, possibly because of a genetic background variation.

Table 4. Suppression effect segregation in diploids heterozygous for SUP35S ⟹ substitution SUP35P

Phenotypes		Number of segregants obtained in crosses of:	
		1-SUP35P-33G-D373 x 11-P5213 [PSI⁻]	1-SUP35P-33G-D373 x 11-P5213 [PSI⁺]
Ade±	Leu⁺	4	8
	Leu⁻	0	2
His±	Leu⁺	10	21
	Leu⁻	0	31
Lys±	Leu⁺	3	17
	Leu⁻	0	11

Suppression effect was registered as poor growth of Ade± (ade1-14), Lys± (lys9-21), His± (his 7-1) segregants on the omission media up to 21 day.

At the same time there were no Sup⁺ segregants among 38 Leu⁻ ones obtained in the progeny of this diploid. It is evident from the same cross that the SUP35 substitution suppresses not only UAA (his7-1 and lys9-A21) but also UGA (ade1-14) nonsenses. Suppression of all this mutations was shown in both Leu⁺ and Leu⁻ segregants of diploid obtained in the cross with 11-P5213 [PSI⁺] (Table 4, right column).

It is important to mention that suppressor efficiency was different in two strains bearing substitutions. This difference was especially clear on the medium without lysine (Table 5). We designated the stronger suppressor efficiency phenotype of 1-SUP35P-33G-D373 strain as Sup⁺⁺⁺, and weaker suppressor efficiency phenotype of 6-SUP35P-33G-D373 strain as Sup⁺. Later some Sup⁺⁺⁺ mitotic segregants of originally Sup⁺ 6-SUP35P-33G-D373 strain were obtained. It may reflect [PSI⁺] induction in these segregants.

We tested the existence of [PSI]-like factors in both original strains with SUP35 substitution and Sup⁺⁺⁺ segregant of 6-SUP35P-33G-D373. After treatment with 7 mM or 8,5 mM GuHCL two types of suppression efficiency were converted to each other with a frequency of 18 - 50% (see Table 5).

DISCUSSION

Data obtained so far allow one to consider [PSI] factor of Saccharomyces yeast as a prion, or a conformational variant of translation termination factor eRF3 (Sup35p)[1]. Like the mammalian prion PrP^Sc, Sup35p forms in [PSI⁺] cells high molecular weight aggregates accumulating most of Sup35p molecules[16]. However, mechanisms of the prionization process and its molecular specificity are not known yet. Here we studied a process of Sup35p prionization in S. cerevisiae depending on the state of SUP35 gene and of SUP45 gene, closely interacting with it[4,12]. Thus, we studied a process of [PSI⁻] to [PSI⁺] conversion and of [PSI] modification at the background of sup35 and sup45 mutations or SUP35 substitution for its homologue from P. methanolica. We found that some sup35 (or sup45) mutations can change dominant [PSI] factor into recessive form, as was shown in crosses with sup35 and sup45 [PSI⁻] testers. Among recessive [psi] factors there are either allosuppressors to sup35 and sup45 mutations or antisuppressors to sup35. Further study of sup35 mutant alleles to which recessive [psi] factors serve as antisuppressors may provide information about the reason of the antisuppressor effect.

Table 5. Influence of GuHCl on suppression efficiency in strains of S. cerevisiae with substitution of SUP35S $\Rightarrow$ SUP35P

Strain	Original phenotype	Phenotypes and number of clones tested after 3 platings on the media:						% of clones that changed suppression efficiency after 3 platings on the media:		
		YPD		7 mM GuHCl		8,5 mM GuHCl		YPD	7 mM GuHCl	8,5 mM GuHCl
		Sup^{+++}	Sup^{+}	Sup^{+++}	Sup^{+}	Sup^{+++}	Sup^{+}			
1-SUP35P	Sup^{+++}	80	0	25	7	18	14	0	21,9%	43,8%
6-SUP35P	Sup^{+}	0	80	16	32	18	30	0	33,3%	37,5%
6-SUP35P*	Sup^{+++}	80	0	26	6	16	16	0	18,8%	50%

Sup^{+++} - clear growth on glucose medium without lysine at 30° C after 4- 5 days; Sup^{+} - the weak growth on ethanole medium without lysine and no growth or residual growth on glucose medium without lysine at 30°C after 4-5 days; * - obtained as spontaneous mitotic segregant with elevated suppression efficiency from the original transformant 6-SUP35P-33G-D373

We have shown previously that in some recessive sup35 and sup45 mutants dominant [PSI] is present. The identification of these mutants by complementation test is possible only after elimination of [PSI] factor by GuHCl[19]. Moreover the efficient sup35 and sup45 suppressors are usually incompatible with [PSI][26].

Now it is evident, that [PSI] factor is compatible with and may be changed by certain sup35 and sup45 mutations, which convert it into recessive form. It looks logic for sup35 mutants, because SUP35 is the structural gene for [PSI]. At the same time the effect of sup45 mutations upon [PSI] expression shows that interaction of heterologous proteins in the cell may influence dominant-recessive characteristics of [PSI] and the process of Sup35p prionization. The recessivity of [psi] may have at least two alternative explanations:

1) The ability of normal eRF3 to correct abnormal conformation of prion eRF3[psi]. Than the so-called here recessive [psi] should not persist in cytoplasm of heterozygous SUP35/sup35 diploids.

2) The inability of prion conformation of mutant eRF3 to convert normal eRF3 into prion form eRF3[psi]. In that instance [psi] may be preserved in heterozygotes mentioned above.

Second explanation is preferable in the case of two antisuppressor [psi] factors at the background of sup35 mutations, because the recessive [psi] segregates out of the diploids, but cannot reappear in segregants if the [psi] factor had been cured off before the cross.

For further estimation of the specificity of prionization process we studied a hybrid translation termination system, including homologue of SUP35 gene from P. methanolica (SUP35P) instead of authentic SUP35S. The SUP35P gene can compensate for the absence of SUP35 gene in Saccharomyces chromosome, though incompletely, as follows from nonsense suppressor effect of such substitution. It is interesting that two substitutions obtained had different suppressor phenotypes, estimated by the efficiency of the lys9-A21(UAA) suppression. Spontaneous $Sup^+ \Rightarrow Sup^{+++}$ conversion and GuHCl induced $Sup^+ \Leftrightarrow Sup^{+++}$ interconversion of these two forms had been demonstrated (see table 5). We assumed that two types of suppressor phenotypes may reflect two different conformations of Sup35P protein, one of those may be considered as a prion form of another (or, alternatively, both of those are prion forms of each other).

Elimination of nonsense-suppressor effect after treatment with GuHCl is one of the main criteria of the [PSI] factor presence in the yeast strains. Thus our results could be considered as a preliminary indication of the existence of [PSI]-like factors in Saccharomyces cells with SUP35 substituted by its homologue from P.methanolica.

However, we observed nontypical effect of GuHCl on strains with substitution of $SUP35S \Rightarrow SUP35P$. In addition to the previously known effect of elimination of [PSI+] mediated suppression, GuHCl treatment both lowered and elevated suppression efficiency in our experiments. $Sup^+ \Rightarrow Sup^{+++}$ conversion may be considered as elimination of antisuppressor [PSI], similar to [PSI] factor revealed in sup35 mutants described in this paper. Furthermore, the effective change of suppressor phenotype was registered only at increased concentrations of GuHCl in the media containing 7 mM or 8.5 mM instead of 5 mM, sufficient for elimination of the typical [PSI] factor. At 5mM GuHCl we observed only rare Sup^+ to Sup^{+++} transition with a frequency of 1-3% (data not shown). Probably, no one of Sup35P conformations dominate strongly to another, therefore GuHCl cannot convert all Sup35P molecules to one of the conformations, unlike in the case of authentic SUP35. At the same time it increases strongly a frequency of some transient state, which may be converted to either one or another conformation with a comparable frequency.

Basing on our results different forms of [PSI] may be considered as different conformational variants of the same protein with local minimization of the free energy state. We shall admit that these different conformational variants may be obtained either without[6] or with the changes of the protein primary structure. Yet we cannot exclude that our results could be interpreted as an indication of negative (in the case of dominant [PSI]) or positive (in the case of recessive [psi]) complementation between identical Sup35p subunits whose activity may either be lowered or elevated as a result of prionization. The latter case may become evident only at the background of deficient function, like in sup35 mutants or in the strains with substitution of SUP35. The influence of nonhomologous subunits interaction upon prionization phenomenon is evident from the effect of sup45 mutations, also described in this work. The exact notion of quarternary structure of translation termination complex in yeast and of the eRF3, in particular, is necessary for the understanding of the mechanism under discussion.

This work was supported by Russian Foundation for Basic Research (RFBR grant 96-04-48236) and by the Russian State Program Frontiers in Genetics.

REFERENCES

1. R.B.Wickner, D.C.Masison, H.K.Edskes, [PSI] and [URE3] as yeast prions. *Yeast.* 11:1671 (1995)
2. B.S.Cox, Psi, a cytoplasmic suppressor of super-suppressor in yeast. *Heredity.* 20:505 (1965)
3. F.J.Lacroute, Nonmendelian mutation allowing ureidosuccinic acid uptake in yeast. *J.Bacteriol.*106:519 (1971)
4. G.Zhouravleva, L.Frolova, X. Le Goff, R.Le Guilles, S.Inge-Vechtomov, L.Kisselev, M.Philippe, Termination of translation in eukaryotes is governed by two interacting polypeptide chain release factors eRF1 and eRF3. *The EMBO J.* 14: 4065 (1995)
5. Y.O.Chernoff, I.L.Derkatch, S.G.Inge-Vechtomov, Multicopy SUP35 gene induces de-novo appearance of psi-like factors in yeast S.cerevisiae. *Curr. Genet.* V:268 (1993)
6. I.L.Derkatch, Yu.O.Chernoff, V.V.Kushnirov, S.G.Inge-Vechtomov, S.W.Liebman, Genesis and variability of prion-like [PSI] factors in S.cerevisiae. *Genetics,* In press. (1996)
7. L.I.Frolova, X.Le Goff, H.H.Rasmussen, S.Cheperegin, G. Drugeon, M.Kress, I.Arman, A-L.Haenni, J.E.Celis, M.Philippe, J.Justesen, L.Kisselev. A highly conserved eukaryotic protein family possessing properties of polypeptide chain release factor. *Nature.* 372: 701 (1994)
8. S.G.Inge-Vechtomov, Reversions to prototrophy in yeast, requiring adenine. *Vestnik LGU.* Ser. Biol. N2 (9):112 (1964) (In Russian)
9. S.G.Inge-Vechtomov, V.M.Andrianova, Recessive supersuppressors in yeast. *Genetika.* 6:103 (1970) (In Russian)
10. S.G.Inge-Vechtomov, L.N.Mironova and M.D.Ter-Avanesyan, Ambiguity of translation : An eukaryotic version? *Russian J.Genet.* 30:890 (1994)
11. V.L.Tikhomirova, S.G.Inge-Vechtomov, Sensitivity of sup35 and sup45 suppressor mutants in S.cerevisiae to the anti-microtubule drug benomyle. *Current Genet.* 30:44 (1996)
12. I.Stansfield, K.M.Jones, V.V.Kushnirov, A.R.Dagkesamanskaya, A.I.Poznyakovski, S.V.Paushkin, C.R.Nierras, B.S.Cox, M.D.Ter-Avanesyan, M.F.Tuite, The products of the SUP45 (eRF1) and SUP35 genes interact to mediate translation termination in S.cerevisiae. *The EMBO J.* 14:4365 (1995)
13. R.Wickner, [URE3] as an altered URE2 protein: evidence for a prion analog in Saccharomyces cerevisiae. *Science.* 264:566 (1994)
14. S.M.Doel, S.J.McCready, C.R.Nierras and B.S.Cox, The dominant PNM2 mutation which eliminates the Ψ factor of S.cerevisiae is the result of a missense mutation in the SUP35 gene. *Genetics.* 137:659 (1994)

15. M.D.Ter-Avanesyan, A.R.Dagkesamanskaja, V.V.Kushnirov and V.N.Smirnov, The SUP35 omnipotent suppressor gene is involved in the maintenance of the non-mendelian determinant [psi+] in the yeast S.cerevisiae. *Genetics.* 137:671 (1994)

16. S.V.Paushkin, V.V.Kushnirov, V.N.Smirnov and M.D.Ter-Avanesyan, Propagation of the yeast prion-like [psi+] determinant is mediated by oligomerization of the SUP35-encoded polypeptide chain release factor. *The EMBO J.* 15:3127 (1996)

17. Y.O.Chernoff, S.L Lindquist., B.-I.Ono, S.G.Inge-Vechtomov and S.W.Liebman, Role of the chaperone-like protein Hsp104 in propagation of the yeast prion-like factor psi. *Science.* 268:880 (1995)

18. M.F.Tuite, C.R.Mundy, B.S.Cox, Agents that cause a high frequency of genetic change from [PSI+] to [PSI-] in S.cerevisiae. *Genetics.* 98:691 (1981)

19. O.N.Tikhodeev, E.V.Getmanova, V.L.Tikhomirova, S.G.Inge-Vechtomov, Ambiguity of translation in yeasts: Genetic control and modifications, in: "Molecular Mechanisms of Genetic Processes", Yu.P.Altukhov, ed, "Nauka", Moscow (1990) (In Russian)

20. V.V.Kushnirov, M.D.Ter-Avanesyan, S.A.Didichenko, V.N.Smirnov, Y.O.Chernoff, I.L.Derkach, O.N.Novikova, S.G.Inge-Vechtomov, M.A.Neistat, I.I.Tolstorukov, Divergence and conservation of SUP2 (SUP35) gene of yeast Pichia pinus and S.cerevisiae. *Yeast.* 6:461 (1990)

21. F.Sherman, G.R.Fink, J.B.Hicks, "Methods in Yeast Genetics", Cold Spring Harbor Laboratory Press, NY, (1986)

22. J.Sambrook, E.E.Fritsch and T.Maniatis, "Molecular Cloning: A Laboratory Manual", Cold Spring Harbor Laboratory Press, NY. (1989)

23. D.A.Gordenin, M.V.Trofimova, O.N.Shaburova, Yu.I.Pavlov, Yu.O.Chernoff, Yu.O.Chekuolene, Y.Y.Proscyavichus, K.V.Sasnauskas, A.A.Yanulaitis, Precise exision of bacterial transposon Tn5 in yeast. *Mol.Gen.Genet.* 213:388 (1988)

24. M.V.Telckov, A.P.Surguchev, A.R.Dagkesamanskaja, M.D.Ter-Avanesyan, Isolation of a chromosomal DNA fragment containing SUP2 gene of the yeast S.cerevisiae. *Genetika.* 22:17 (1986) (In Russian)

25. S.G.Inge-Vechtomov, O.N.Tikhodeev, T.S.Karpova, The selective systems for obtaining recessive ribosomal suppressors in the yeast Saccharomyces. *Genetika.* 24:1159 (1988) (in Russian).

26. S.G.Inge-Vechtomov, Yu.O.Chernoff, I.L.Derkatch, G.Richter, V.V.Kushnirov, M.D.Ter-Avanesyan, SUP2 (SUP35) - PSI factor interaction in S.cerevisiae: incompatibility and interdependency. Yeast, 6 (Spec. issue: Abstracts of XV International Conference on Yeast Genetics and Molecular Biology, The Hague, Netherlands, July 21-26, 1990: 08-21B. (1990)

PRIONS OF YEAST: GENETIC EVIDENCE THAT THE NON-MENDELIAN ELEMENTS, [PSI] AND [URE3] ARE ALTERED SELF-REPLICATING FORMS OF Sup35p AND Ure2p, RESPECTIVELY

Reed B. Wickner, Daniel C. Masison,
Herman Edskes and Marie-Lise Maddelein

Laboratory of Biochemistry and Genetics,
National Institute of Diabetes,
Digestive and Kidney Diseases,
National Institutes of Health, Bldg.8, Room 225,
8 Center Dr. MSC 0830, Bethesda, MD 20892-0830, U.S.A.

SUMMARY

The word "prion" means **infectious protein**, an altered form of a cellular protein which may have lost its normal function, but has acquired the ability to convert the normal form of the protein into the same altered (prion) form. The yeast non-Mendelian genetic elements, [PSI] and [URE3], can be understood if they are viewed as prions arising from the normal proteins, Sup35p and Ure2p, respectively. Both [PSI] and [URE3] are a) reversibly curable, b) induced to arise by overproduction of the corresponding normal protein and c) depend upon their normal form chromosomal gene (*SUP35* and *URE2*) for propagation and yet confer on the cell the same dominant phenotype as does a recessive mutation in this corresponding normal form chromosomal gene. Our 'yeast prion hypotheses' have now been supported by biochemical data indicating that Sup35p is altered in [PSI] strains and Ure2p is altered in [URE3] strains. We have defined a prion-inducing domain of Ure2p, an asparagine - rich N-terminal region which mediates the prion alteration in *cis* and in *trans*.

INTRODUCTION

The prion concept

As is well documented elsewhere in this volume, the prion concept arose from studies of scrapie of sheep, Creutzfeldt-Jakob disease and Kuru of man, and the corresponding disease transmitted to mice, goats, etc. The unusual resistance of the scrapie agent to UV light suggested to Alper that it might lack an essential nucleic acid (Alper et al. 1967). This led Griffith (Griffith 1967) to suggest the prion hypothesis (without the word 'prion') in what is essentially its current form. Griffith's abstract idea was given substance with Prusiner's discovery of the PrP protein (Bolton et al. 1982) as a protease - resistant protein found only in partially purified preparations of the scrapie agent. The cloning of the PrP gene by Weissmann (Oesch et al. 1985) and Chesebro (Chesebro et al. 1985) showed that it was not propagated by 'reverse translation' or 'protein-dependent protein synthesis' but was encoded by a cellular gene. The PrP gene proved to be identical to the mouse *Sinc* gene identified in the 1960's by Dickinson as critical to determining scrapie **inc**ubation periods (Dickinson et al. 1968; Carlson et al. 1986). Further genetic studies proved the importance of PrP in this disease (Prusiner et al. 1990; Bueler et al. 1993).

Yeast non-Mendelian genetic elements

Chromosomal and non-chromosomal genetic elements (synonyms include 'cytoplasmic genetic elements' and 'non-Mendelian genetic elements') are easily distinguished by their inheritance properties (Table 1). For example, the non-chromosomal inheritance of resistance to the antibiotics erythromycin and chloramphenicol for growth on non-fermentable carbon sources led to the discovery of mitochondrial DNA (reviewed by (Dujon 1981). Similarly, the non-chromosomal inheritance of ability to secrete a protein toxin led to discovery of the dsRNA viruses of yeast (reveiwed by (Wickner 1996).

Table 1. Genetic properties of non-Mendelian elements

Property	Description	Chromosomal mutations	Non-chromosomal elements
Meiotic segregation	Mate wild-type (+) and mutant (-). Induce meiosis, analyze phenotypes of the four meiotic progeny of each diploid cell.	2+ : 2-	4+ : 0 or irregular
Mitotic segregation	Mate + and -. Allow diploid to grow mitotically.	Heterozygosity is maintained	+ and - traits segregate during mitotic growth if '-' is due to a altered genome, not if '-' is due to absence of the genome.
Curing	Grow cells in presence of special agent (e.g., ethidium bromide 100% cures mitDNA)	no effect or rare mutations	efficient elimination of the element
Cytoduction	Use *kar1* mutant defective in nuclear fusion. Cytoplasmic mixing occurs in heterokaryon without diploid formation.	cytoductants do not acquire chromosomal traits	cytoductants acquire non-chromosomal traits, usually with high efficiency

Two non-Mendelian genetic elements which have long been known to yeast genetics, [PSI] (Cox 1965; reviewed by Cox et al. 1988; Cox 1993; Wickner 1995) and [URE3] (Lacroute 1971), reviewed by (Wickner 1995), have resisted explanation until recently. [URE3] is a non-chromosomal genetic element which allows cells to take up **ure**idosuccinate on medium containing a rich nitrogen source, such as ammonia. Ureidosuccinate is an intermediate in uracil biosynthesis, the product of aspartate transcarbamylase, the first step in the pathway. The accidental chemical similarity of ureidosuccinate to allantoate, a poor but usable source of nitrogen for yeast, results in ureidosuccinate being taken up by Dal5p, the allantoate transporter (Turoscy and Cooper 1987), Fig. 1). Since *DAL5* is repressed in the presence of a good nitrogen source, like ammonia, ureidosuccinate uptake is too (Rai et al. 1987). The repression system involves Ure2p (Lacroute 1971; Rai et al. 1987; Courchesne and Magasanik 1988), so mutation of Ure2p leads to uptake of ureidosuccinate, even in the presence of ammonia, the same phenotype as for [URE3] (Fig.1).

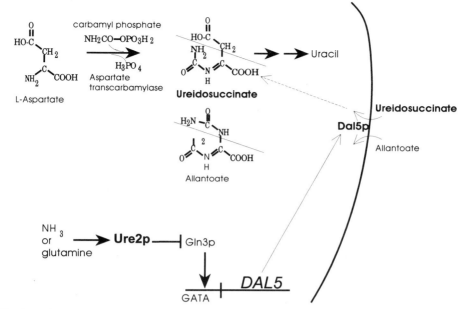

Fig. 1. Relation of *URE2* and ureidosuccinate uptake. Ureidosuccinate is an intermediate in the uracil biosynthesis pathway, the product of aspartate transcarbamylase. However, ureidosuccinate closely resembles allantoate, a poor but usable nitrogen source for yeast. Because of this resemblance, Dal5p, the allantoate permease, can take up ureidosuccinate allowing growth of a Ura⁻ mutant defective in aspartate transcarbamylase (Turoscy and Cooper 1987). However, Dal5p is normally repressed when the cell has a rich nitrogen source, like ammonia or glutamine. This nitrogen regulation or 'nitrogen catabolite repression' system involves Ure2p, which is thought to sense the nitrogen supply, blocking the activation by Gln3p of many genes involved in nitrogen metabolism (reviewed by Magasanik 1992). Thus, *ure2* mutants, or strains in which Ure2p is in the prion (inactive) form, do not repress ureidosuccinate uptake on media containing ammonia. Although they lack asparate transcarbamylase, they grow on ureidosuccinate without added uracil.

[PSI] is a non-Mendelian genetic element that increases the efficiency of suppressor tRNAs working at translation termination codons. Attempts to correlate [PSI] with the presence of the various known non-chromosomal DNA or RNA molecules of yeast were unsuccessful (reviewed by Cox et al. 1988; Cox 1993).

RESULTS

The genetic properties expected of yeast prions

A prion is an infectious protein, so a prion of yeast should resemble other infectious entities of yeast, such as the dsRNA viruses of yeast or the yeast retroviruses. No viruses of yeast or other fungi are known to leave one cell and enter another via the external environment (reviewed by (Wickner 1996). Mating and hyphal fusion of yeasts and fungi are so frequent in nature that these viruses are nonetheless very widespread in their respective species. Most yeast strains carry several different viruses, which are passed from cell to cell at high efficiency in the course of mating. They are, in effect, sexually transmitted diseases. The dsRNA viruses and (+) ssRNA viruses appear as non-Mendelian genetic elements. In the same way, a prion of yeast should appear as non-Mendelian genetic elements.

We have proposed three genetic characteristics or criteria for deciding if a non-Mendelian genetic element is a yeast prion (Wickner 1994; Table 2). The first is **reversible curability**. If the prion can be cured by some treatment, it should be possible for it to arise again, at some low frequency, in the progeny of the cured strain. This is because the stochastic event that gave rise to the prion in the first case could happen again, without introducing anything from the outside. The normal protein is present in the cell and can, with some low probability, change into the prion form. This is not true of a nucleic acid replicon. Once the mitochondrial DNA or one of the yeast viruses is lost from the cell, it will not return to any of the progeny without its introduction from another strain.

The second genetic criterion for a yeast prion is that **overproduction of the normal form should increase the frequency with which the prion arises.** If there is more of the normal form in the cell, there are more molecules that have the potential to undergo the spontaneous prion change. Once the change has occured in one molecule (or whatever the minimum number is), the change will propagate through the population of molecules. This is again in contrast to a nucleic acid replicon. Overproduction of a DNA polymerase or RNA polymerase (for example) will not promote the de novo generation of the mitochondrial DNA or the genome of an RNA virus.

Table 2. Genetic Properties of a Prion

Property expected	[URE3]	[PSI]	Scrapie
Reversible curability	√	√	not yet
Overproduction of the normal form increases the frequency with which the prion arises	√	√	not yet
Mutation of chromosomal gene for the protein results in			
a)loss of prion	√	√	√
b) same phenotype as presence of the prion	√	√	no

The third genetic criterion for a yeast prion has to do with the phenotypes of 1) the presence of the prion and 2) a defect in the chromosomal gene for the protein. These **phenotypes are the same.** In the first case the normal form is made but quickly converted to the prion form, so the normal function of the protein is not carried out. In the second case, the normal protein is not made in the first place because the gene is mutant. The gene for the protein is, of course, necessary for the propagation of the prion. This relationship is different from what is observed in the case of a nucleic acid replicon and one of the chromosomal genes on which it depends (see Table 3). For example, the mitochondrial genome requires a chromosomally - encoded DNA polymerase for its replication. A mutant in the chromosomal polymerase gene loses mitochondrial DNA and so cannot grow on a non-fermentable carbon source, like glycerol, that needs mitochondrial oxidation for its metabolism. But this is the **opposite** of the phenotype confered by the presence of the mitochondrial DNA, namely, ability to use glycerol.

[PSI] and [URE3] have the genetic properties expected of yeast prions

[PSI] can be cured by either growth in hyperosmotic media (Singh et al. 1979) or by growth in the presence of 5 mM guanidine HCl (Tuite et al. 1981). However, in each case, from the cured purified cells can again be isolated, at some low frequency, subclones which have again acquired the [PSI] genetic element (Lund and Cox 1981; Chernoff et al. 1993). Thus [PSI] shows reversible curability.

Table 3. Relation of phenotypes of a prion and mutants in the gene for the protein

Non-Mendelian Element	Presence of non-Mendelian element	Chromosomal mutant that loses the element	Relation	Does replacing the chrom. mutant gene restore the phenotype
	Phenotypes			
M dsRNA	killer +	killer —	opposite	No
mitDNA	glycerol +	glycerol —	opposite	No
mitDNA-DI	glycerol —	glycerol —	same	No
Theoretical Prio	defective	defective	same	Yes
[URE3]	USA uptake +	USA uptake +	same	Yes
[PSI]	suppressor ⇑	suppressor ⇑	same	Yes

The effects of the presence of [PSI] are similar to those of the recessive 'omnipotent suppressor' chromosomal mutants in the *SUP35* gene (Hawthorne and Mortimer 1968). Moreover, the intact *SUP35* gene is necessary for the propagation of [PSI] (Doel et al. 1994; TerAvanesyan et al. 1994). This is the relationship expected for a yeast prion, not that seen in the known nucleic acid replicons.

Overproduction of Sup35p results in the *de novo* generation of [PSI] in strains which had not carried it before (Chernoff et al. 1993).

These properties of [PSI] led us to propose that [PSI] is a prion form of Sup35p, an altered form of the protein that has lost its normal function, but has acquired the ability to convert the normal form into the same inactive (prion) form (Wickner 1994).

[URE3], like [PSI] can be cured by growth on medium containing 5 mM guanidine HCl (Wickner 1994). It is important to emphasize that it is not the curing of [URE3] (or [PSI]) by guanidine that makes one suspect it is a prion. 5 mM is a very low concentration of guanidine, unlikely to denature any protein. Rather, it is the **reversibility** of this curing, the fact that from cured strains can be again derived strains again carrying the element (Wickner 1994).

The chromosomal *ure2* mutants have essentially the same phenotype as do the non-chromosomal [URE3] strains (Lacroute 1971; Aigle and Lacroute 1975), and yet the normal *URE2* gene is necessary for the propagation of [URE3] (Aigle and Lacroute 1975; Wickner 1994). This is the expected relation of a prion and the chromosomal gene encoding its normal form (Wickner 1994).

Overproduction of Ure2p results in a 20 to 100 fold increase in the frequency with which [URE3] arises (Wickner 1994). It would be of interest to precisely determine the relation of Ure2p concentration and frequency of [URE3] since this would have implications concerning the mechanism of [URE3] generation.

We thus have suggested that [URE3] is a prion form of Ure2p and [PSI] is a prion form of Sup35p (Wickner 1994). The way in which [URE3] and [PSI] are propagated as altered forms of Ure2p and Sup35p respectively and the relation of the genes and prions to the phenotypes are shown in Fig. 2.

Ure2p is protease-resistant in [URE3] strains

Since protease-resistance was the means by which Prusiner and colleagues discovered the altered PrP in scrapie brain (Bolton et al. 1982), we examined the relative proteinase K-sensitivity of Ure2p in extracts of isogenic wild-type and [URE3] strains (Masison et al. 1995). We found that Ure2p was relatively proteinase K - resistant in [URE3] strains compared to either the parent strain or to a strain which had been cured of [URE3]. The intact Ure2p is 40 kDa and digestion of an extract of a [URE3] strain leaves relatively stable fragments of 30 and 32 kDa. These are, however, eventually degraded.

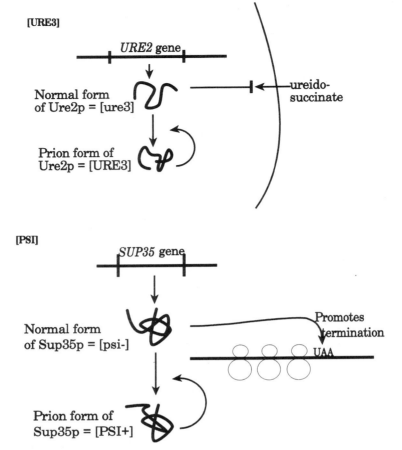

Fig. 2. [URE3] and [PSI] as prion forms of Ure2p and Sup35p, respectively (Wickner 1994).

In contrast, Ure2p, in a wild-type extract, is completely degraded at the earliest time points examined (Masison et al. 1995).

The prion notion predicts that Ure2p will be structurally altered in a [URE3] strain compared to a wild-type strain, but it is not necessary that the protein be protease - resistant. It might have been (and may be for some other prion) that the prion form is **more** sensistive than the normal form, or the difference between prion form and normal form may not be detectable by protease sensitivity. What is demanded is only that there be a difference.

Sup35p is aggregated in [PSI] strains

Sup35p is one subunit of the translation release factor that recognizes the translation termination codons and cleaves the completed peptide from the tRNA. Normally, Sup35p is a soluble protein that is highly sensitive to protease digestion, but in strains carrying [PSI], Sup35p sediments rapidly and is protease-resistant (Paushkin et al. 1996). Fluorescent microscopy of cells expressing a Sup35p-Green fluorescent protein fusion shows even distribution of the fusion protein in normal strains, but one or a few large aggregates in [PSI] strains (Patino et al. 1996). This evidence of alteration of Sup35p in [PSI] strains supports the prion explanation of [PSI]. It also explains how the

[PSI] phenotype may be related to changes in Sup35p. If Sup35p is aggregated, it cannot efficiently carry out its normal function of peptide chain termination, allowing weak nonsense suppressors to be more efficient.

Prion-inducing and nitrogen regulatory domains of Ure2p

Various fragments of the *URE2* gene were overexpressed from a high copy plasmid in yeast strains to examine 1) their ability to induce the formation of [URE3] and 2) their ability to complement a total deletion of the chromosomal copy of *URE2* (Masison and Wickner 1995), Fig. 3). Deletions at any of several places in the C-terminal part of the molecule resulted in loss of ability to complement the chromosomal deletion of *URE2*. Deletion of the N-terminal 65 residues was, however, without effect on the nitrogen regulatory activity of Ure2p.

As described above, overexpression of Ure2p increased the frequency with which [URE3]-carrying colonies were detected (Wickner 1994). We found that over expression of mutants with deletions in the C-terminal domain were not only still able to induce

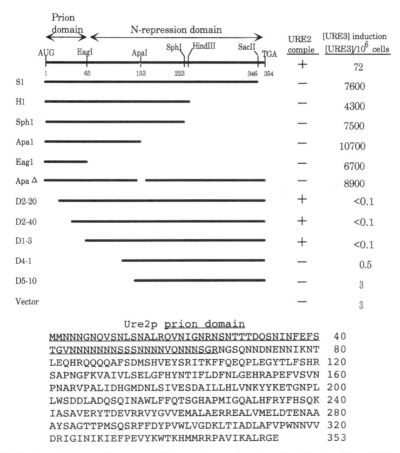

Fig. 3. Definition of the Prion Domain and Nitrogen-Repression Domain of Ure2p. The solid lines indicate the regions overexpressed. Each fragment of Ure2p was tested for its ability to complement a chromosomal *ure2Δ* and for its ability to induce [URE3] in a strain with an intact *URE2* gene. These studies define a prion - inducing domain, the 65 N-terminal residues, that are sufficient to induce [URE3] and are necessary for a molecule to be subject to the [URE3] change (from Masison and Wickner 1995). The prion-inducing domain is underlined in the sequence of Ure2p.

117

formation of [URE3], they were 100-fold better at doing so than was the intact protein expressed at similar levels (Masison and Wickner 1995). In fact, overexpression of just the N-terminal 65 amino acid residues of Ure2p were sufficient to give this same high level of [URE3] induction. We call this the prion-inducing domain of Ure2p to describe this fact.

Not only did this N-terminal 65 residue region suffice to induce prion formation when overexpressed, but we also found that only Ure2p molecules which included this domain were subject to inactivation by the [URE3] prion change (Masison and Wickner 1995). The deletion mutant D1-3 ((Masison and Wickner 1995), Fig. 3) lacks the first 65 residues of Ure2p, and complements a deletion mutation of the chromosomal copy of *URE2*. However, the overexpression of this C-terminal part of Ure2p does not induce the generation of [URE3] (Masison and Wickner 1995, Fig. 3). In fact, the spontaneous [URE3] events are not seen in a strain expressing this fragment, indicating that it is not subject to the [URE3] change. Thus we have suggested that the prion change of Ure2p is mediated via the N-terminal part of the molecules.

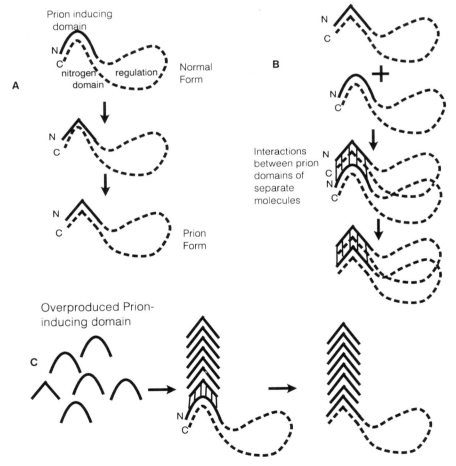

Fig. 4. Proposed mechanism of propagation of the [URE3] prion. A spontaneous (stochastic) change in the N-terminal prion domain of a Ure2p molecule (A) induces a change in the C-terminal nitrogen regulation domain that inactivates it producing the phenotype. The change is propagated to other molecules by an interaction between the N-terminal domains of separate molecules (B). Whether this happens as a heterodimer between normal and prion forms or in a crystal seed is not yet clear. Overproduction of the N-terminal prion domain dramatically increases the frequency of [URE3] (C).

Our working model for the propagation of [URE3] is shown in Fig. 4. We believe that the prion change in Ure2p is propagated by interaction between the N-terminal prion domains of separate molecules. Ure2p molecules with their prion domain in the prion form then alter the C-terminal nitrogen regulation domain of their own molecule. Molecules lacking the N-terminal prion domain cannot participate in the prion change, remain active, and prevent a phenotype from being apparent even if the full length Ure2p molecules in the same cell have undergone the prion change.

Comparison of prion systems

The prion domain of Ure2p is very rich in asparagine residues, these constituting 40% of total residues. The part of of Sup35p necessary for the propagation of [PSI] has been determined by TerAvanesyan et al. (TerAvanesyan et al. 1994). Although those authors interpreted Sup35p as a trans-acting factor needed for [PSI], they now agree with our interpretation that this is the prion domain of the molecule, analogous to that we have determined for Ure2p. It too is rich in asparagine and glutamine residues, with 45% of the 114 residues of this domain consisting of these two amino acids (Fig. 5).

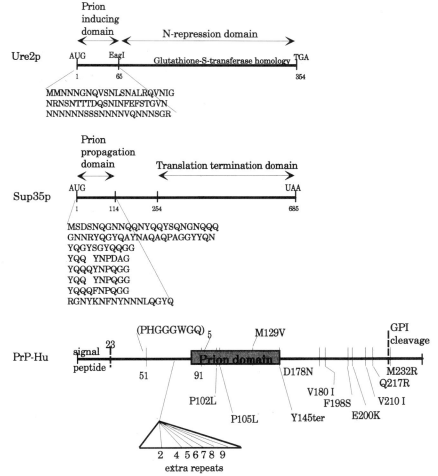

Fig. 5. Comparison of three likely prion systems, Ure2p - [URE3], Sup35p - [PSI], and PrP - scrapie (CJD).

PrP is not particularly rich in asparagines or glutamines, but the basis of prion propagation in these different systems need not be the same.

DISCUSSION

While it has often been speculated that other human degenerative diseases may be due to prions, this notion has yet to bear fruit. Our findings are the first concrete evidence that there may be more than one prion system. Indeed, the evidence that [URE3] and [PSI] represent prion systems is in some respects, better than that for the scrapie system in mammals. Specifically, the genetic evidence that [URE3] and [PSI] are prions is probably better than that for PrP and scrapie. Questions remain about whether scrapie is caused by a cryptic (or not so cryptic) virus. There are many such entities lurking in the mammalian genome. Yeast is a more well defined system, whose viral systems have probably now all been discovered (reviewed in Wickner 1996).

The yeast system is ideally suited to discovering trans-acting factors that influence prion propagation. Chernoff et al. searched for a high copy plasmid which eliminated [PSI] from cells (Chernoff et al. 1995). They discovered that the heat-shock protein, Hsp104, when overexpressed, eliminated [PSI]. They also found that deletion of *HSP104* had the same result. Since Hsp104 is known to disaggregate abnormally aggregated proteins (Parsell et al. 1994), and Sup35p is aggregated in [PSI] strains (Paushkin et al. 1996), it is likely that overproduced Hsp104 disaggregates Sup35p. That Hsp104 cures [PSI] is, in fact, evidence that the mechanism of prion formation in this case is aggregation, and not some covalent change of the Sup35 molecule. Paradoxically, heat does not significantly cure [PSI], although it does induce overproduction of Hsp104. However, artificially overproduced of Hsp104 in a strain with [PSI] finds little but Sup35p to work on, whereas perhaps in a heat-shocked cell, many proteins are aggregated and it is too busy with those to concentrate on Sup35p.

There is as yet no cure for scrapie, but there are cures for [URE3] and [PSI], namely, growth of cells in 5 mM guanidine HCl (Tuite et al. 1981; Wickner 1994). [PSI] is cured by high osmotic strength (Singh et al. 1979). We are developing drug-screening systems using yeast to look for potentially useful anti-scrapie (anti CJD) compounds. The same methods will be used to screen for agents that are prion-inducing, in analogy with the Ames tests for mutagens.

Finally, our work is the first evidence in any system that inheritance can be mediated by a prion. It is likely that other yeast or fungal non-Mendelian elements will be found to be prion-based, since in these organisms mating pools all the cytoplasm of both parents. In contrast, mammalian sperm is mostly a compact DNA packet, and bacteria exchange only a single strand of DNA in their mating.

REFERENCES

Aigle, M. and F. Lacroute (1975). "Genetical aspects of [URE3], a non-Mendelian, cytoplasmically inherited mutation in yeast." Molec. Gen. Genet. **136**: 327 - 335.

Alper, T., W. A. Cramp, et al. (1967). "Does the agent of scrapie replicate without nucleic acid?" Nature **214**: 764 - 766.

Bolton, D. C., M. P. McKinley, et al. (1982). "Identification of a protein that purifies with the scrapie prion." Science **218**: 1309 - 1311.

Bueler, H., A. Aguzzi, et al. (1993). "Mice devoid of PrP are resistant to Scrapie." Cell **73**: 1339 - 1347.

Carlson, G. A., D. T. Kingsbury, et al. (1986). "Linkagae of prion protein and scrapie incubation time genes." Cell **46**: 503 - 511.

Chernoff, Y. O., I. L. Derkach, et al. (1993). "Multicopy SUP35 gene induces de-novo appearance of psi-like factors in the yeast *Saccharomyces cerevisiae*." Curr. Genet. **24**: 268 - 270.

Chernoff, Y. O., S. L. Lindquist, et al. (1995). "Role of the chaperone protein Hsp104 in propagation of the yeast prion-like factor [psi$^+$]." Science **268**: 880 - 884.

Chesebro, B., R. Race, et al. (1985). "Identification of scrapie prion protein-specific mRNA in scrapie-infected brain." Nature **315**: 331 - 333.

Courchesne, W. E. and B. Magasanik (1988). "Regulation of nitrogen assimilation in *Saccharomyces cerevisiae:* roles of the *URE2* and *GLN3* genes." J. Bacteriol. **170**: 708 - 713.

Cox, B. S. (1965). "PSI, a cytoplasmic suppressor of super-suppressor in yeast." Heredity **20**: 505 - 521.

Cox, B. S. (1993). Psi phenomena in yeast. The early days of yeast genetics. M. N. Hall and P. Linder. Cold Spring Harbor, Cold Spring Harbor Laboratory Press: 219 - 239.

Cox, B. S., M. F. Tuite, et al. (1988). "The Psi factor of yeast: a problem in inheritance." Yeast **4**: 159 - 179.

Dickinson, A. G., V. M. H. Meikle, et al. (1968). "Identification of a gene which controls the incubation period of some strains of scrapie in mice." J. Comp. Path. **78**: 293 - 299.

Doel, S. M., S. J. McCready, et al. (1994). "The dominant *PNM2$^-$* mutation which eliminates the [PSI] factor of *Saccharomyces cerevisiae* is the result of a missense mutation in the *SUP35* gene." Genetics **137**: 659 - 670.

Dujon, B. (1981). Mitochondrial genetics and functions. The molecular biology of the yeast *Saccharomyces*: Life cycle and inheritance. J. N. Strathern, E. W. Jones and J. R. Broach. Cold Spring Harbor, N.Y., Cold Spring Harbor Laboratory. **1**: 505 - 635.

Griffith, J. S. (1967). "Self-replication and scrapie." Nature **215**: 1043 - 1044.

Hawthorne, D. C. and R. K. Mortimer (1968). "Genetic mapping of nonsense suppressors in yeast." Genetics **60**: 735 - 742.

Lacroute, F. (1971). "Non-Mendelian mutation allowing ureidosuccinic acid uptake in yeast." J. Bacteriol. **106**: 519 - 522.

Lund, P. M. and B. S. Cox (1981). "Reversion analysis of [psi-] mutations in *Saccharomyces cerevisiae*." Genet. Res. **37**: 173 - 182.

Magasanik, B. (1992). Regulation of nitrogen utilization. The molecular and cellular biology of the yeast *Saccharomyces*. E. W. Jones, J. R. Pringle and J. R. Broach. Cold Spring Harbor, N.Y., Cold Spring Harbor Laboratory Press. **2**: 283 - 317.

Masison, D. C., A. Blanc, et al. (1995). "Decoying the cap- mRNA degradation system by a dsRNA virus and poly(A)- mRNA surveillance by a yeast antiviral system." Mol. Cell. Biol. **15**: 2763 - 2771.

Masison, D. C. and R. B. Wickner (1995). "Prion-inducing domain of yeast Ure2p and protease resistance of Ure2p in prion-containing cells." Science **270**: 93 - 95.

Oesch, B., D. Westaway, et al. (1985). "A cellular gene encodes scrapie PrP 27-30 protein." Cell **40**: 735 - 746.

Parsell, D. A., A. S. Kowal, et al. (1994). "Protein disaggregation mediated by heat-shock protein Hsp104." Nature **372**: 475 - 478.

Patino, M. M., J.-J. Liu, et al. (1996). "Support for the prion hypothesis for inheritance of a phenotypic trait in yeast." Science **273**: 622 - 626.

Paushkin, S. V., V. V. Kushnirov, et al. (1996). "Propagation of the yeast prion-like [*psi$^+$*] determinant is mediated by oligomerization of the *SUP35*-encoded polypeptide chain release factor." EMBO J. **15**: 3127 - 3134.

Prusiner, S. B., M. Scott, et al. (1990). "Transgenic studies implicate interactions between homologous PrP isoforms in scrapie prion replication." Cell **63**: 673 - 686.

Rai, R., F. Genbauffe, et al. (1987). "Transcriptional regulation of the *DAL5* gene in *Saccharomyces cerevisiae*." J. Bacteriol. **169**: 3521 - 3524.

Singh, A. C., C. Helms, et al. (1979). "Mutation of the non-Mendelian suppressor [PSI] in yeast by hypertonic media." Proc. Natl. Acad. Sci. USA **76**: 1952 - 1956.

TerAvanesyan, A., A. R. Dagkesamanskaya, et al. (1994). "The *SUP35* omnipotent suppressor gene is involved in the maintenance of the non-Mendelian determinant [psi+] in the yeast *Saccharomyces cerevisiae*." Genetics **137**: 671 - 676.

Tuite, M. F., C. R. Mundy, et al. (1981). "Agents that cause a high frequency of genetic change from [*psi+*] to [*psi-*] in *Saccharomyces cerevisiae*." Genetics **98**: 691 - 711.

Turoscy, V. and T. G. Cooper (1987). "Ureidosuccinate is transported by the allantoate transport system in *Saccharomyces cerevisiae*." J. Bacteriol. **169**: 2598 - 2600.

Wickner, R. B. (1994). "Evidence for a prion analog in *S. cerevisiae*: the [URE3] non-Mendelian genetic element as an altered *URE2* protein." Science **264**(22 April 1994): 566 - 569.

Wickner, R. B. (1995). Non-Mendelian Genetic Elements in S. cerevisiae: RNA viruses, 2 micron DNA, [Psi], [URE3], 20S RNA and other wonders of nature. The Yeasts. A. H. R. a. A. E. Wheals. London, Academic Press. **6**: 309 - 356.

Wickner, R. B. (1996). Viruses of yeasts, fungi and parasitic microorganisms. Fields Virology. B. N. Fields, D. M. Knipe and P. M. Howley. New York, Raven Press. **1**: 557 - 585.

EARLY CLINICAL DETECTION OF BRAIN DISEASES IN ANIMALS

B.F. Semenov[1], S.V. Ozherelkov[2], and V.V. Vargin[2]

[1] Mechnikov Research Institute for Vaccines and Sera
[2] Chumakov Institute of Poliomyelitis and Viral Encephalitides
Moscow, Russia

According to recent data of literature, alterations in behavior can serve as an index of functional CNS impairment in the absence of, or prior to, more readily apparent symptoms. A number of reports have been concerned with mice infected with conventional viruses of prions.[7,9]

The present work summarizes the results obtained in the study of the maze behavior of mice, infected by viruses (members of the Togaviridae family) causing, under certain conditions, latent CNS infections.

MATERIALS AND METHODS

Viruses. In this work yellow fever virus, vaccine strain 17D, tick-borne encephalitis (TBE) virus (high-virulent strain Sofyin and low-virulent strain Scalica) and Langat virus, strain TP-21, were used.

Strain 17D was inoculated intracerebrally (i.c.), TBE and Langat viruses were inoculated intraperitoneally (i.p.) in doses causing mainly or only latent infection (Table 1).

Animals. All experiments were made on male BALB/c mice weighing 18–20 g.

Mazes. The infected animals were studies in a linear maze[4] or a complicated ├──┴──│─ shaped maze.[13,14]

Only the results of the experiments obtained on those mice which exhibited no clinical signs of CNS pathology during the whole period of observation (days 30–110) were considered. Those animals in which such signs appeared were eliminated from the experiments. In addition, at the end of behavior experiments the sera of all mice were studied for the presence of antibodies and the brain of some of the animals was studied for the presence of the virus and pathomorphological changes.

The linear maze was a plexiglas box sized 1000 x 400 x 120 mm (Figure 1). It consisted of the open field, 9 sections divided by partitions with openings and food cells. The walls of the maze and partitions were painted black.

The linear maze was used for the evaluation of search activity.

After 24–hour food deprivation a mouse was placed in the open field and the time spent by the animal for the search of food was measured.

Table 1. Viruses, doses of viruses, their biological effect

Virus	Strain	LD_{50}	Route	Day[1]	Survival (%)	Seroconversion (%)	CNS pathology (%)
YF[2]	17D	1,000	i.p.	35	80	23	12 –14
TBE[3]	Sofyin	100	i.p.	40	70	37	19
	Scalica	100,000	i.p.	30	100	66	40
Langat	TP–21	10,000	i.p.	30	85	60	46

[1] Postinoculation day.
[2] YF – yellow fever vurus.
[3] TBE – tick-borne encephalitis virus.

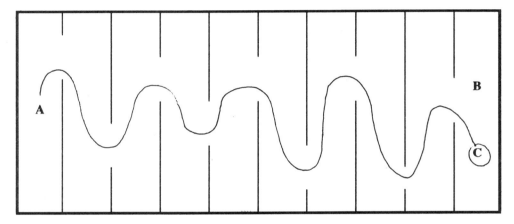

Figure 1. Scheme of the linear maze. A, B – open fields. AB – the path of a mouse to the refreshed cell. C – the refreshed cell.

The ⊢—⊣ – shaped maze consisted of two parts: the open field and the multialternative maze, connected by 3 doors which could be easily opened by the animals (Figure 2). The central door was the entrance to the maze and two other doors were the entrances to the open field. There were 4 food cells in the maze. Of these, two were refreshed and two were false. The acquisition of ⊢—⊣ – shaped maze learning consisted of several stages: pressing on the shelf, entering the maze, finding one of the two refreshed food cells and leaving the maze to the open field. Only after that the animal could repeat this way to get a new reinforcement.

The sessions lasted 10 min. The subsequent food deprivation lasted 24 hours. Before and after sessions mice were kept in cages in groups of 10 animals.

Two experimental protocols were used:

(a) the study of the effect of latent virus infection on maze learning;

(b) the evaluation of the effect of latent virus infection on maze–acquired experience (first mice underwent a course of 10–20 sessions of maze learning, then the animals were infected with the virus and afrer that tested in the maze).

The following measures were recorded during each 10–min. session:

(a) the latent period of response;

124

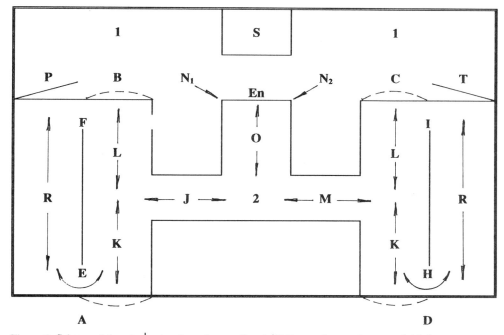

Figure 2. Scheme of the ⊢⊥─| – shaped maze. S – shelf. En – entrance. 1 – open field. 2 – maze area. Letters denoting the possible movements of a mice in the maze: A,D – looking into a refreshed food cell; B,S – looking into a false food cell; O,I,M,K,L,R – passage from one section of the maze to another section; P,T – exit to the open field; N_1, N_2 – approaching the entrance from the left or from the right.

(b) activity level (the number of sets of correct movements: entrance to the maze – a refreshed cell – exit – entrance, etc.):

(c) the number of erroneous and false movements:

(d) the number and duration of freezing and grooming.

Movements first towards a false food cell, and then towards a refreshed one were regarded as erroneous. After finding the food cell, movements not towards the exit, but in a different direction, e.g. towards the entrance, or circular movements, were regarded as false. After making false movements a mouse found the entrance. The combination of correct and false moviments was denoted as correct–false.

Denoting the movements of a mouse in the maze by letters (Figure 2), we can describe one of erroneous movements as OILB, and correct–false movements as OIKA + ERFLMKD + HRT. In the ⊢⊥─| – shaped maze a mouse can make 16 correct movements, 64 erroneous movements and an unlimited number of correct–false movements.

To determine the frequency of one concrete movement (Pm) or a combination of movements (Pmc), made during a 10–min. session, the following formulae are used:

$$Pm_1 = \frac{nm_1}{\Sigma\,(m_1 \ldots m_n)}\;;\qquad\qquad Pm_1 = \frac{n\,(mc_1 \ldots mc_k)}{\Sigma\,(mc_1 \ldots mc_n)}\;;$$

where

nm_1 and $n\,(mc_1 \ldots mc_k)$ – the number of performances of one movement or one combination of movements during the session;

Σ (m$_1$ .... m$_n$) and Σ (mc$_1$ .... mc$_n$) — the sum of all movements or combination of movements, registered within 10 min.

Following the completion of the maze behavior testing, the mice were deeply anesthetized by ether.

The brain of the animals was studied for the presence of the virus and/or morphological pathology.

To isolate the virus, the brain was homogenized in Hanks' balanced salt solution, cell debris was removed by centrifugation and the resulting sepernatant was inoculated intracerebrally to 1– or 2-day old suckling mice.

For histopathological examination the brains were fixed overnight in buffered aqueous formalin solution, then dehydrated through graded alcohols to xylene, imbedded in paraffin, sectioned at 6 µ and stained with hematoxylin and eosin.

RESULTS

1. Maze Behavior of Mice after Trauma

Intracerebral (i.c.) and intraperitoneal (i.p.) inoculation techniques were used in the studies discussed in the present article. The use of these techniques involved the injection of relatively large volumes at a period of a few seconds into the brain (0,03 ml) and into the abdominal cavity (0,3 ml). In preliminary experiments we studied behavioral effects produced by traumas associated with the inoculation of different noninfectious fluids by different routes: physiological saline (i.c.), 0,01% suspension of brain obtained from intact mice (i.c. and i.p.), commercial inactivated tissue–culture TBE vaccine with Al(OH)$_3$ (i.p.). The vaccine was introduced in 2 injections at an interval of 1 day, physiolohical saline and brain suspension were introduced in a single injection.

These experiments revealed that the i.c. inoculation of physiological saline or brain suspension did not affect the activity of mice in the linear maze. 48 hours and 7 days after the trauma intact mice found food 4.5–4.8 min. after being placed in the open field (Figure 3).

Table 2 contains the summary of the behavioral testing of intact and traumatized mice in the $\vdash\!\!\perp\!\!\dashv$ – shaped maze. The data demonstrating a considerable behavioral effect produced by the i.c. inoculation of brain suspension are presented. In the traumatized mice latent time was found to decrease after 10–12 sessions; the number of sets of correct movements did not increase during a 10–min. session, the number of false movements increased. From sessions 3–10 perseverations (the multiple repetition of a single false movement) was observed.

After the i.p. inoculation of brain suspension or inactivated TBE virus (vaccine) behavioral alterations were not observed. The latent period, the activity level, the number of false movements were the same as in the intact animals.

Considering the presented data, we used the linear maze only to study the search activity of mice after i.c. inoculation. The $\vdash\!\!\perp\!\!\dashv$ – shaped maze was used for the evaluation of behavioral changes in mice receiving the i.p. inoculation of the virus.

2. Behavioral Abnormalities in Mice after the i.c. Inoculation of Yellow Fever Virus (Strain 17D) or Langat Virus

The behavior of mice with asymptomatic yellow fever infection on day 7 after inoculation was no different from the behavior of intact mice or mice inoculated intracerebrally with brain suspension (Figure 3). However, on day 15 serious CNS

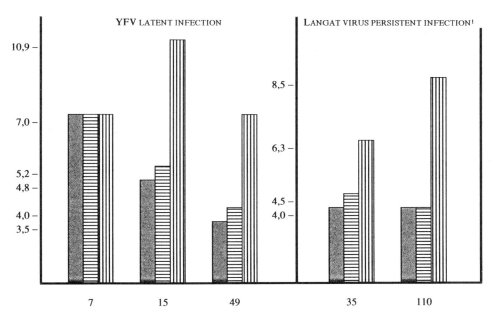

Figure 3. Alteration of the search activity of mice after the i.c. inoculation of yellow fever virus (YFV) and Langat virus.
On the ordinate: time of the search for food (min.) by infected mice (▥), mice receiving brain suspension (▤) or intact mice (▨). On the abscissa: postinoculation days.
[1] Langat virus was isolated from the brain of clinically healthy mice on the next day (36 and 111 respectively) after the completion of behavioral testing.

Table 2. Effect of trauma on the maze behavior of mice

Treatment	Route	L[1]	A[2]	Mf[3]	Me[3]	Perseveration
no		$\dfrac{12\text{-}14^4}{70}$	$\dfrac{17\text{-}26}{4}$	$\dfrac{0.009}{0.10}$	$\dfrac{0}{0.1}$	no
physiological saline	i.c.	$\dfrac{12}{50}$	$\dfrac{12\text{-}15}{2}$	$\dfrac{0}{0.5}$	$\dfrac{0}{0.2}$	no
0.01% brain suspension	i.c.	$\dfrac{170}{85}$	$\dfrac{3\text{-}4}{2}$	$\dfrac{0.8}{0.15}$	$\dfrac{0}{0.1}$	Sessions 3-10
	i.c.	$\dfrac{10\text{-}14}{90}$	$\dfrac{20\text{-}22}{2\text{-}3}$	$\dfrac{0}{0.15}$	$\dfrac{0}{0.2}$	no
TBE vaccine[5]	i.c.	$\dfrac{12}{21}$	$\dfrac{25}{22}$	$\dfrac{0.0093}{0.16}$	$\dfrac{0}{1}$	no

[1] L – latent period (sec.).
[2] A – activity level: the number of sets of behavior tasks during a 10-minute session.
[3] Mf and Me – the freguency of falce and erroneous movements.
[4] In the numerator: the results determined after 10-12 sessions.
   In the denominator: the results determined after 1-2 sessions.
[5] Mice were inoculated with the vaccine after undergoing 10-20 maze sessions.

impairment was found in the infected animals. To find food, they spent 10.9 min., whereas the subjects of both control groups spent 3.5–5.2 and 4.8 min. Behavioral abnormalities were detected till day 49 (the term of observation) with no visual signs of brain infection (paresis, paralysis) being registered.

Experiments with Langat virus also showed that virusinduced behavioral disorders could appear in the absence of clinically overt symptoms. By day 110 of observation the time spent for the search of a food cell increased to 8.5 min. in comparison with 6.5 min. on day 35. For the control animals, 4–4.5 min. were necessary to solve this problem. It should be pointed out that behavioral abnormalities were associated with asymptomatic Langat virus persistence in the CNS. The virus was isolated from the mouse brain, removed 24–48 hours after the completion of behavioral testing (on days 36–37 and days 111–112 respectively).

3. Behavioral Effects of TBE Virus or Langat Virus Latent Infections in Mice

Table 3 summarizes the results of behavioral testing on mice in the ⊢⊥⎯|– shaped maze. It shows that viral infections were associated with numerous patterns of behavioral abnormalities.

The signs of the deficit of maze learning were a considerable latent period of response and a low activity level, characterized by the number of sets of correct movements during a 10-min. session (the shelf - entrance - a refreshed food cell - exit). The deficit of maze learning was retained in the infected animals for 10–12 sessions. Figure 4 shows the dynamics of changes in the latent period and activity levels in infected and control mice. During the whole maze–learning time the infected mice exhibited high values of latent response, equal to 280–300 sec. even after sessions 14–15. In the control animals after sessions 11—12 the latent period of 90 to 15—20 sec. was retained.

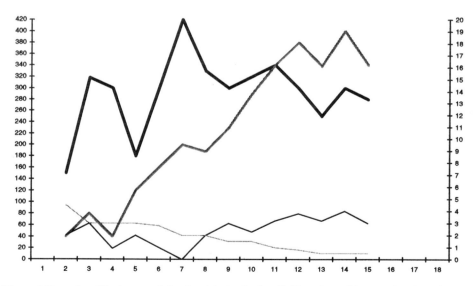

Figure 4. Dynamics of the latent period and activity levels after 10-12 sessions of the maze learning of infected and control mice. On the ordinate: left – the latent activity (sec.) of infected (⬛) and control (⬛) mice; right – the activity level (the number of sets of correct movements within 10 min.) of infected (⎯⎯⎯) and control (⎯ ⎯ ⎯) mice. On the abscissa: No. of the sessions.

Table 3. Behavioral effects of TBE virus and Langat virus
latent infections in mice[1]

1.	Deficient maze learning.
2.	Alterations in maze–acquired experience.
3.	Impairment in the proseccing of proprioceptive information.
4.	Other behavioral disorders:

       – search activity deficit,

       – disorderly hyperactivity,

       – freezing,

       – grooming (sheezing, cleaning, scratching).

[1] The summary of behavior testing in the ⊢—⊥—⊢ shaped maze.

The infected mice made not more than 4 sets of correct movements during a single session; sometimes they could not find the exit from the maze after 10–min. search (see Figure 4, session № 7).

In experiments with the use of the ⊢—⊥—⊣ – shaped maze the mice with virus infection were found to have considerable alterations in the processing of proprioceptive information. The term "alterations in the processing of proprioceptive information" refers to the incessant process of the insertion of new false movements into the closed chain of

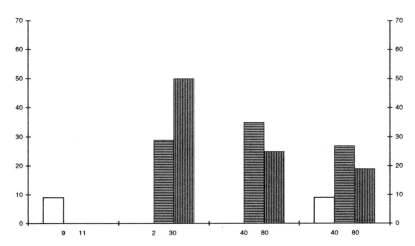

Figure 5. Alteration in processing proprioceptive information in mice with latent Langat virus infection. On the ordinate: left – the average number of 4-v-grams used be the control (▢) and infected (▤) mice; right – the percentage of false movements in infected mice (▥). On the abscissa: postinoculation days.

129

movements (entrance to the maze – a refreshed cell – exit – entrance). The unit of measure for this set of movements is the combination of 4 consecutive movements, denoted as 4 v-grams (1 v-gram is 1 movement).[11]

Figure 5 demonstrates the number of 4 v-grams used by the infected and control mice. The intact animals built up the stereotype of the chain of movements, not containing false movements, after 4–5 sessions (on days 9–11). The stereotype consisted of 9±3 4 v-grams of correct movements. The infected mice included 26±8 or 29±2 4 v-grams into the chain of movements where the proportion of false movements was 19–50%.

In the ├──┴──┤ – shaped maze the infected mice demonstrated numerous different behavioral abnormalities:

– disorderly hyperacivity: the animals moved from one food cell to another, returned from the food cell to the entrance, left the maze without reinforcement;

– freezing: a mouse stands still before a food cell or the entrance/exit for 30 or more sec., sometimes as long as 120 sec.;

– space disorientation: the animal first comes up to a false food cell (erroneous movement), then to a refreshed food cell, or the animal first moves to other sections of the maze (false movement), then finds the exit;

– grooming: the subject stands before a food cell or the exit/entrance and starts washing, scratching; these movements go on for 30–150 sec.;

– perseveration: multiple repetition of the same false movement, e.g. n (AERREA) or n₁ (DKMOOMKD) (see Figure 2).

According to our data, latent virus infection impaired the experience acquired by the animals after 10–20 maze learning sessions. Figure 6 demonstrates the latent period and activity levels of trained mice before and after infection. These data indicate that the latent period increased from 12 sec. (2 days before infection) to 180 or 60 sec. (on days 21 and 27 after infection respectively), while the number of sets of correct movements essentially decreased (from 22 to 3–4 within 10 min.).

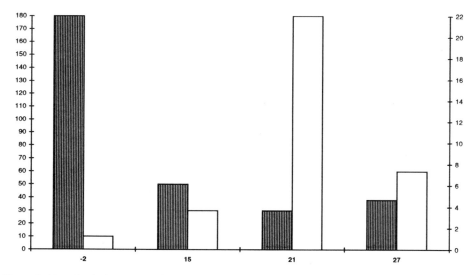

Figure 6. Alteration in the maze-acquired experience after infection with TBE virus (strain Scalica). On the ordinate: left – the latent period (☐) (sec); right – the activity level (▤) (the number sets of correct movements within 10 min.). On the abscissa: days before (–) or after (+) i.p. infection.

Table 4. Behavioral abnormalities (BA) in mice
after the i.p. inoculation to TBE virus (Scalica)

LD$_{50}$	Days p.i.[1]	% of mice with BA	Me[2]	Mf[2]	Freezing	Grooming	Perseveration	Altered processing of proprioceptive information
1,000,000	2–30	61	0.5	0.64	+[3]	+	+	+
	30–40	42	0.28	0.52	+	+	+	+
	40–80	32	0.34	0.6	+	+	+	+
100,000	2–30	57	0.47	0.62	+	+	+	+
	3–40	38	0.28	0.55	+	+	+	+
	40–80	28	0.34	0.6	+	+	+	+

[1] p.i. – post inoculation; [2] Me and Mf – the freguency of false and erroneous movements; [3] – BA were observed.

The results of our experiments, summarized in Table 4, show that disorders of the behavior develop not in all infected mice. During the first 8 days after the inoculation of the virus only 28–61% of the studied animals exhibited functional CNS pathology.

The occurrence of behavioral alterations seemed to depend on the virulence of the virus under study and the chosen experimental protocol. Low-virulent strains Scalica and TP-21 induced the deficit of maze learning in 57–61% and 47% of cases respectively. Infection caused by strain TP-21 was associated with alterations in maze–acquired experience in 57–61% of the animals. After the inoculation of strain Scalica these alterations were found 28–32% of mice. High–virulent strain Sofyin induced the deficit of maze learning and alterations in maze–acquired experience in 28% of the studied subjects.

According to our data, virus–induced behavioral abnormalities in mice were retained till day 80 after inoculation (the term of observation). Still the number of animals exhibiting disorders of the behavior constantly decreased (Table 4). Thus, during the first month after inoculation 57–61% of the tested mice exhibited the impairment of the CNS function. In 30–40 days after inoculation the number of animals with learning deficits decreased to 38–42%. During the period of days 40–80 (the term of observation) behavioral abnormalities were observed in 28–32% of the tested subjects.

Table 5. The onset and frequency of maze behavioral abnormalities occurring in mice
infected with viruses having different virulence

Virus	Strain	Virulence	Mice	Day of onset	Frequency (%)
TBE	Sofyin	high	infected (1)[1]	4	28
			acquiring experience, then infected (2)[1]	8	28
	Scalica	low	1	4	57 – 61
			2	4 – 6	28 – 32
Langat	TP–21	low	1	4	47
			2	2 – 4	57 – 61

[1] For details of the experiments see the text.

The picture of behavioral abnormalities remained practically unchanged during 80 days after inoculation (Table 4).

Behavioral alterations in mice inoculated with different doses of TBE virus were studied (Table 5). Similar results were observed after the injection of 100,000 and 1,000,000 LD_{50} of strain Scalica. The inoculation of 10,000 and 100,000 LD_{50} of Langat virus showed similar effects on maze learning.

Interesting information was provided by the comparison of the data on the development of behavioral abnormalities with the data on pathological changes in the brain. The results of experiments with TBE virus (strain Scalica) and Langat virus indicate that the manifestations of behavioral alterations were not always associated with CNS pathology. And vice versa, brain lesions were found in mice with the normal type of behavior, similar to that of the control animals. The frequency of the appearance of maze–learning deficit, not accompanied by apparent CNS pathology, increased with the increase of the term of observation. In 40–80 days after the inoculation of strain Scaleca functional brain disorders were registered in 90% of the mice. On days 20 and 40 of the experiment only 27% and 50% of the inoculated subjects respectively exhibited the same type of CNS patholgy. Langat virus also induced behabioral changes, not associated with the appearance oa brain lesions (the detailed description of these data is presented for publication[15]).

CONCLUSIONS

Animal models of virus–indiced behavioral disorders are at present regarded as a potential tool for the study of CNS disturbances.[7, 9] These are numerous communications concerning behavioral abnormalities in animals with acute infections caused by, e.g., LCM virus,[5] West Nile virus,[3] 17D strain of yellow fever virus,[4, 8] herpes simplex type virus,[7] Langat virus,[10] etc. A number of reports have been concerned with the effects of perinatal infection on the later behavior of animals.[1, 2] Behavioral abnormalities have been observed relatively early in scrapie–affected mice.[6, 12]

We studied behavioral disorders in mice with latent infections caused by TBE virus, Langat virus and 17D vaccine strain of yellow fever virus. The study revealed that the inoculation of these viruses was accompanied by the development of various patterns of behavioral alterations: the deficit of maze learning, alterations in maze–acquired experience, impairment in the processing of proprioceptive information, the deficit of search activity, alteration in space orientation, desorderly hyperactivity, freezing, grooming.

Behavioral abnormalities appeared on days 2–8 after inoculation and remained for 80 days (the term of observation). The following data should be paid special attention to. Depending on the conditions of the experiment, behavioral changes developed in 28–61% of the tested animals. In some of the mice they disappeared after days 30–80 of the experiment. Maze–learning deficit observed in mice was often not associated with the morphological pathology of the CNS. On day 80 of observation 90% of the mice infected with TBE virus (strain Scalica) exhibited only functional alterations of the CNS.

The main information on virus–induced behavioral impairments was obtained with the use of the ├──┤– shaped maze, developed at the Lomonosov State University, Moscow, Russia.[13, 14] We believe that this maze will prove useful for the examination of early behavioral disorders in prion-inoculated mice.

REFERENCES

1. Anderson, A.A., and Hanson, R.P. Intrauterine infection with St. Louis encephalitis virus: immunological, physiological and behavioral effects on progeny. Infec. Immun. 12: 1173 (1975)

2. Duffy, C.E., and Murphree, O.D. Maze performance of mature rats recovered form early postnatal infection with Murray Valley encephalitis virus. J. Comp. Physiol. Psychol. 52: 175 (1959)

3. Duffy, C.E., and Murphree, O.D., and Morgan, P.N. Learning deficit in early post–natal infection with West Nile virus. Proc. Soc. Exp. Biol. Med. 98: 242 (1958)

4. Hanneberg, L., Haase, J., Museteanu, J., et al. Abnormality in the behavior of mice after experimental encephalitis by yellow fever virus. Zbl. Bakt. Hyg. I Abt. Orig. 234: 1 (1976)

5. Hotchin, J., and Seegal, R. Virus–induced behavioral alteration in mice. Science 196: 671 (1977)

6. McFarland, D., and Hotchin, J. Early behavioral abnormalities in mice due to scrapie virus encephalopathy. Biol. Psychiatry 15: 37 (1980)

7. McFarland, D., and Hotchin, J. Animal models in behavioral neurovirology. In: Viruses, Immunity and Mental Disorders (Kurstak, E., Lipowski, Z.J., Morozov, P.V. eds. Plenum Medical Book Company, New York and London) (1987)

8. Ozherelkov, S.V., Semenov, B.F. Influence of fever virus vaccine, strain 17D, on the funciton of the nervous system of mice. Zh. Mikrobiol. 9: 83 (1980) (in Russian)

9. Ozherelkov, S.V., Semenov, B.F. Use of the methods for the evaluation of animal behavior in the study of the pathogenesis of neurovirus infections. Vopr. Virusol. 5: 516 (1980) (in Russian)

10. Semenov, B.F., Vargin, V.V., Ozherelkov, S.V. The influence of Tahyna virus on Langat virus persistence in the central nervous system of mice. Vopr. Virusol. 6: 724 (1981) (in Russian)

11. Shannon, K. Works on the Theory of Information and Cybernetics, Foreign Literature Publishing House, Moscow, 668 pp. (1963) (in Russian)

12. Suckling, A.J., Bateman, S., Waldron, C.B., et al. Motor activity changes in scrapie–affected mice. Brit. J. Exp. Pathol. 57: 742 (1970)

13. Voronin, L.G., Nikolskaya, K.A., Sagimbaeva, Sh.K. Physiological analysis of experience obtained by white rats in a maze. Dokl. AN SSSR 217: 1225 (1970) (in Russian)

14. Voronin, L.G., Semenov, B.F., Nikolskaya, K.A., et al. Influence of the vaccine strain 17D of yellow fever virus on processing proprioceptive information in the process of the maze learning of BALB/c mice. Vopr. Virusol. 6: 107 (1982) (in Russian)

15. Ozherelkov, S.V., Ravkina, L.I., Semenov, B.F., Morphological and behavioral changes in mice in latent virus infection. Zh. Mikrobiol., in press (in Russian)

PRION BIOLOGY AND DISEASES - FATAL CONFORMATIONS OF PROTEINS DURING A JOURNEY FROM HERESY TO ORTHODOXY

Stanley B. Prusiner

Departments of Neurology and of Biochemistry and Biophysics
University of California
San Francisco, CA 94143-0158

Prions, once dismissed as an impossibility, have now gained wide recognition as extraordinary pathogens that cause a number of infectious, genetic and sporadic disorders. Prions are composed only of a protein which is encoded by a chromosomal gene. This protein, designated PrP, is converted from a normal benign form into a disease causing form by a change in conformation. Prions cause bovine spongiform encephalopathy in cattle, scrapie in sheep, and four fatal CNS diseases in humans. Currently, there is no known treatment for prion diseases, and the fear that prions have passed from cattle to humans may be justified.

A good deal of skepticism was evoked when I proposed that the infectious agents causing certain degenerative disorders of the central nervous system in animals and, more rarely, in humans might consist of protein and nothing else. At that time, the notion was heretical. Dogma held that the conveyers of transmissible diseases required genetic material, composed of nucleic acid (DNA or RNA), in order to establish an infection in a host. Even viruses, among the simplest microbes, rely on such material to direct synthesis of the proteins needed for survival and replication.

Later, many scientists were similarly dubious when my colleagues and I suggested that these "proteinaceous infectious particles"—or "prions," as I called the disease-causing agents—could underlie inherited, as well as communicable, diseases. Such dual behavior was then unknown to medical science. We met resistance again when we concluded that prions (pronounced "preeons") multiply in an incredible way: they convert normal protein molecules (PrP^C) into dangerous ones (PrP^{Sc}) simply by inducing the benign molecules to change their shape.

A wealth of experimental and clinical data has made a convincing case that we are correct on all three counts. Prions are indeed responsible for transmissible and inherited disorders of protein conformation. They can also cause sporadic disease, in which neither transmission between individuals nor inheritance is evident. Moreover, there are hints that the prions causing the diseases explored thus far may not be the only ones. Prions made of different proteins may contribute to other neurodegenerative diseases that are quite prevalent in humans.

Prions and Brain Diseases in Animals and Humans
Edited by Morrison, Plenum Press, New York, 1998

135

The known prion diseases, all fatal, are sometimes referred to as spongiform encephalopathies, so named because they frequently cause the brain to become riddled with holes. These ills, which can brew for years (or even for decades in humans) are widespread in animals.

The most common form is scrapie, found in sheep and goats. Afflicted animals lose coordination and eventually become so incapacitated that they cannot stand. They also become irritable and, in some cases, develop an intense itch that leads them to scrape off their wool or hair (hence the name "scrapie"). The other prion diseases of animals go by such names as transmissible mink encephalopathy, chronic wasting disease of mule deer and elk, feline spongiform encephalopathy, and bovine spongiform encephalopathy. The last, often called mad cow disease, is the most worrisome.

English scientists identified mad cow disease in 1986, after it began striking cows in Great Britain, causing them to become uncoordinated and unusually apprehensive. The source of the emerging epidemic was soon traced to a food supplement that included meat and bone meal from dead sheep. The methods for processing sheep carcasses had been changed in the late 1970s. Where once they would have eliminated the scrapie agent in the supplement, now they apparently did not. The British government banned the use of animal-derived feed supplements in 1988, and the epidemic among cattle peaked in 1992. Nevertheless, many people continued to be concerned that bovine prions might kill humans, a fear that now seems justified.

The human prion diseases are more obscure. Kuru has been seen only among the Fore highlanders of Papua New Guinea where they call it the "laughing death." In 1957, physicians noted that many highlanders became afflicted with a strange, fatal disease marked by loss of coordination (ataxia) and often later by dementia. The affected individuals probably acquired kuru through ritual cannibalism: the Fore tribe reportedly honored the dead by eating their brains. The practice has since stopped, and kuru has virtually disappeared.

Creutzfeldt-Jakob disease, in contrast, occurs worldwide and usually becomes evident as dementia. Most of the time it appears sporadically, striking one person in a million, typically around age 60. About 10 to 15 percent of cases are inherited, and a small number are, sadly, iatrogenic—spread inadvertently by the attempt to treat some other medical problem. Iatrogenic Creutzfeldt-Jakob disease has apparently been transmitted by corneal transplantation, implantation of dura mater or electrodes in the brain, use of contaminated surgical instruments, and injection of growth hormone derived from human pituitaries (before recombinant growth hormone became available).

Despite the decrease in the number of cattle dying from bovine spongiform encephalopathy, many people continue to worry that they will eventually fall ill as a result of having consumed tainted meat. Investigators at the University of Edinburgh reported in the spring of 1996 that three teenagers and seven young adults in Britain had died during the past year of a new variant of Creutzfeldt-Jakob disease. The unusually young age of the patients and the multilobulated plaques in the brain composed of prion protein surrounded by spongiform degeneration made them think that this is a new disease, possibly related to the consumption of bovine prions.

The two remaining human prion disorders are Gerstmann-Sträussler-Scheinker disease which is manifest as ataxia and other signs of damage to the cerebellum and fatal familial insomnia in which dementia follows difficulty sleeping. Both these conditions are usually inherited and typically appear in mid-life. Fatal familial insomnia was discovered at the University of Bologna with help from investigators at Case Western Reserve University.

Understanding how PrP^{Sc} multiplies and causes disease has been all the more difficult due to the enigma of prion strains. The long-standing mystery of how prions consisting of a single kind of protein can vary markedly in their effects is beginning to be explained. Ian

Pattison working at the Institute for Animal Health at Compton, England, initially called attention to the phenomenon of strains in studies where he obtained prions from two separate sets of goats. One isolate made inoculated animals drowsy, whereas the second made them hyperactive. Similarly, it is evident that some prions cause disease quickly, whereas others do so slowly.

Investigators at the Institute for Animal Health in Edinburgh examined the differential effects of varied isolates in mice noting that only pathogens containing nucleic acids are known to occur in multiple strains. Hence, they and others assert, the existence of prion "strains" indicates the prion hypothesis must be incorrect and viruses must be at the root of scrapie and its relatives. Yet because efforts to find viral nucleic acids have been unrewarding, the explanation for the differences must lie elsewhere.

One possibility is that prions can adopt multiple conformations. Folded in one way, a prion might convert PrP^C to PrP^{Sc} efficiently, giving rise to short incubation times. Folded another way, it might be less efficient. Similarly, one "conformer" might be attracted to neuronal populations in one part of the brain, whereas another might be attracted to neurons elsewhere, thus producing different symptoms. Considering that PrP can fold in at least two ways, it would not be surprising to find it can exhibit other structures as well.

In work on prion strains isolated from mink, Richard Marsh working at the University of Wisconsin found evidence that PrP^{Sc} might adopt two different conformations. They called their mink strains "drowsy" and "hyper" as the hamsters they studied exhibited symptoms reminiscent of the goats observed by Pattison. Marsh believed the drowsy and hyper prions isolated from mink originated in a cow that developed a sporadic case of prion disease, as no sheep live in the area of Wisconsin where this outbreak of transmissible mink encephalopathy occurred. Such an argument begs the question of whether or not bovine spongiform encephalopathy exists in the United States. The drowsy strain was more sensitive to digestion by proteases than the hyper, and furthermore, proteolytic digestion cut off about 15 more amino acids from scrapie PrP in "drowsy infected" hamsters than in hyper. I was puzzled by these results, as all strains of prions we had studied showed similar resistance to digestion by proteases.

Recently, we gained substantial insight into prion strains from studies of fatal familial insomnia and familial Creutzfeldt-Jakob disease. We transmitted these inherited human prion diseases to mice carrying chimeric PrP composed of segments of mouse and human PrP, and we found the conformation of PrP^{Sc} depended upon which inoculum was used. PrP^{Sc} in the fatal familial insomnia inoculum and in the recipient mice was similar in molecular size ($\sim$19 kDa) after digestion with proteases; in contrast, PrP^{Sc} from both familial (codon 200 mutation) and sporadic Creutzfeldt-Jakob disease and in the recipient mice inoculated with these prions was larger ($\sim$21 kDa) after digestion with proteases.

For the first time, we had a marker that could be used to trace two different strains from their origin through transmission into animals. These studies demonstrated that the structure of PrP^{Sc} in the brains of patients with fatal familial insomnia is different from that found with familial Creutzfeldt-Jakob disease. More importantly, these structural differences were transferred to chimeric PrP harbored in mice. While in human PrP^{Sc} these results can be explained in terms of the differences in structure resulting from a mutation at codon 178 and codon 200 in fatal familial insomina and familial Creutzfeldt-Jakob disease, respectively, this does not account for the differences observed with chimeric PrP^{Sc} in mice. The production of different chimeric PrP^{Sc} in the brains of inoculated mice can only be explained by human PrP^{Sc} imparting specific conformations to chimeric PrP, a process that many scientists thought does not occur in nature.

As with many other prion strains, specific patterns of PrP^{Sc} accumulation were found in the brains of mice harboring chimeric PrP following inoculation with preparations from the brains of patients who died of fatal familial insomnia or familial Creutzfeldt-Jakob disease.

While the distribution of PrP^{Sc} deposition can be directed by the strain of prions, the distribution pattern can be greatly modified by other factors such as the amino acid sequence and level of PrP^C. While PrP^{Sc} distribution can be useful in tracking strains of prions, it does not help us decipher the basis of prion diversity.

With the demonstration that strains of prions possess different PrP^{Sc} molecules as detected by the size of the protein after exposure to proteases, we no longer need to invoke elusive DNAs or RNAs of the enigmatic scrapie virus as the molecules that carry the biological information of the prion. Instead, it is the conformation of PrP^{Sc} into which such information is encrypted.

Unravelling the mechanism of prion strains is not only of theoretical importance but it also has practical consequences, particularly with respect to the BSE epidemic. If sheep were the source of prions how did all the sheep have the same strain? If cattle were the source of the prions how did they all have the same strain? Perhaps, this strain was the hardiest and outgrew all the others, particularly in cattle. Since the inoculum which initiates the conversion of PrP^C into PrP^{Sc} in cattle essentially disappears as prions replicate, it is not possible for us to trace it. Unlike viruses which carry their own genomes that code for progeny viruses, prions copy the cellular PrP of the host and cannot be traced.

Assessing the likelihood of transmission of bovine prions to humans requires a consideration of prion strains. The passage of human prions into mice carrying chimeric PrP shows both the amino acid sequence of PrP and the conformation of scrapie PrP conspire to produce new PrP^{Sc}. Such a scenario is consistent with but does not establish that variant Creutzfeldt-Jakob disease in young people originated in cattle.

REFERENCES

Alper, T., Cramp, W.A., Haig, D.A. and Clarke, M.C. Does the agent of scrapie replicate without nucleic acid? *Nature* 214: 764-766 (1967).

Anderson, R.M., Donnelly, C.A., Ferguson, N.M., Woolhouse, M.E.J., Watt, C.J., Udy, H.J., MaWhinney, S., Dunstan, S.P., Southwood, T.R.E., Wilesmith, J.W., Ryan, J.B.M., Hoinville, L.J., Hillerton, J.E., Austin, A.R. and Wells, G.A.H. Transmission dynamics and epidemiology of BSE in British cattle. *Nature* 382: 779-788 (1996).

Bessen, R.A. and Marsh, R.F. Identification of two biologically distinct strains of transmissible mink encephalopathy in hamsters. *J Gen Virol* 73: 329-334 (1992).

Bolton, D.C., McKinley, M.P. and Prusiner, S.B. Identification of a protein that purifies with the scrapie prion. *Science* 218: 1309-1311 (1982).

Bruce, M., Chree, A., McConnell, I., Foster, J., Pearson, G. and Fraser, H. Transmission of bovine spongiform encephalopathy and scrapie to mice: strain variation and the species barrier. *Phil Trans R Soc Lond B* 343: 405-411 (1994).

Büeler, H., Aguzzi, A., Sailer, A., Greiner, R.-A., Autenried, P., Aguet, M. and Weissmann, C. Mice devoid of PrP are resistant to scrapie. *Cell* 73: 1339-1347 (1993).

Cousens, S.N., Vynnycky, E., Zeidler, M., Will, R.G. and Smith, P.G. Predicting the CJD epidemic in humans. *Nature* 385: 197-198 (1997).

Dickinson, A.G., Meikle, V.M.H. and Fraser, H. Identification of a gene which controls the incubation period of some strains of scrapie agent in mice. *J Comp Pathol* 78: 293-299 (1968).

Gajdusek, D.C. Unconventional viruses and the origin and disappearance of kuru. *Science* 197: 943-960 (1977).

Hsiao, K., Baker, H.F., Crow, T.J., Poulter, M., Owen, F., Terwilliger, J.D., Westaway, D., Ott, J. and Prusiner, S.B. Linkage of a prion protein missense variant to Gerstmann-Sträussler syndrome. *Nature* 338: 342-345 (1989).

Huang, Z., Prusiner, S.B. and Cohen, F.E. Scrapie prions: a three-dimensional model of an infectious fragment. *Folding & Design* 1: 13-19 (1996).

Lasmézas, C.I., Deslys, J.-P., Demaimay, R., Adjou, K.T., Lamoury, F., Dormont, D., Robain, O., Ironside, J. and Hauw, J.-J. BSE transmission to macaques. *Nature* 381: 743-744 (1996).

Prusiner, S.B. Novel proteinaceous infectious particles cause scrapie. *Science* 216: 136-144 (1982).

Prusiner, S.B. Prions. *Sci Am* 251: 50-59 (1984).

Prusiner, S.B. The prion diseases. *Sci Am* 272: 48-57 (1995).

Prusiner, S.B, Prions, in: *Fields Virology*, B.N.Fields, D.M. Knipe and P.M. Howley, eds., Raven Press, New York (1996).

Riek, R., Hornemann, S., Wider, G., Billeter, M., Glockshuber, R. and Wüthrich, K. NMR structure of the mouse prion protein domain PrP(121-231). *Nature* 382: 180-182 (1996).

Telling, G.C., Scott, M., Mastrianni, J., Gabizon, R., Torchia, M., Cohen, F.E., DeArmond, S.J. and Prusiner, S.B. Prion propagation in mice expressing human and chimeric PrP transgenes implicates the interaction of cellular PrP with another protein. *Cell* 83: 79-90 (1995).

Telling, G.C., Haga, T., Torchia, M., Tremblay, P., DeArmond, S.J. and Prusiner, S.B. Interactions between wild-type and mutant prion proteins modulate neurodegeneration in transgenic mice. *Genes & Dev* 10: 1736-1750 (1996).

Telling, G.C., Parchi, P., DeArmond, S.J., Cortelli, P., Montagna, P., Gabizon, R., Mastrianni, J., Lugaresi, E., Gambetti, P. and Prusiner, S.B. Evidence for the conformation of the pathologic isoform of the prion protein enciphering and propagating prion diversity. *Science* 274: 2079-2082 (1996).

Westaway, D., Goodman, P.A., Mirenda, C.A., McKinley, M.P., Carlson, G.A. and Prusiner, S.B. Distinct prion proteins in short and long scrapie incubation period mice. *Cell* 51: 651-662 (1987).

Wickner, R.B. [URE3] as an altered URE2 protein: evidence for a prion analog in Saccharomyces cerevisiae. *Science* 264: 566-569 (1994).

Will, R.G., Ironside, J.W., Zeidler, M., Cousens, S.N., Estibeiro, K., Alperovitch, A., Poser, S., Pocchiari, M., Hofman, A. and Smith, P.G. A new variant of Creutzfeldt-Jakob disease in the UK. *Lancet* 347: 921-925 (1996).

NEW VARIANT CREUTZFELDT-JAKOB DISEASE

Robert Will

National Creutzfeldt-Jakob Disease Surveillance Unit
Western General Hospital
Edinburgh EH4 2XU, UK

INTRODUCTION

The identification in the United Kingdom of a novel form of Creutzfeldt-Jakob disease (CJD), designated new variant Creutzfeldt-Jakob disease (nvCJD), led to the proposal that these cases might be causally linked to the occurrence of bovine spongiform encephalopathy (BSE). Since the publication of an article describing the cases and the predominantly epidemiological evidence implying a causal link (1), there has been extensive comment and debate on the strength of this data and it is perhaps timely to review the hypothesis in the light of subsequent scientific developments.

CLINICO-PATHOLOGICAL PHENOTYPE

Ten cases of nvCJD were described in the original publication. These cases were first linked because of a remarkably early age of onset (mean 29 years) in comparison to previous experience of CJD from systematic surveys which consistently indicated that CJD was predominantly a disease of late middle-age. Clinically the cases were also unusual because of a consistent phenotype, including early psychiatric symptoms and a prolonged duration of illness averaging over 12 months in comparison to 4 months in classical cases. Review of case records held at the CJD surveillance Unit (CJDSU) and of the previous literature indicated that this clinical phenotype was highly unusual in CJD, although rarely there were young patients with classical CJD with similar features. However as a group these cases exhibited a strikingly similar clinical phenotype and the hypothesis of the novelty of this form of CJD would be supported if the clinical phenotype were maintained in subsequent cases.

By July 1997 10 further cases of nvCJD have been identified in the UK and a further case in France (2) and all have a clinical phenotype consistent with the original description. The incidence of CJD in cases aged less than 35 years at death, based on the original 10 cases, was statistically clearly distinct from previous experience. The current situation in the UK is that 15 of the 19 cases confirmed cases of nvCJD were aged less than 35 years at death and 18 of 20 cases (including one probable case) were aged 35 years or less at disease onset. The duration of illness in the 20 UK cases averages 16 months with a range of 9-38 months. Evidence from the cases identified since the original publication is consistent with the continuing occurrence of a novel form of CJD.

The most powerful evidence linking the first 10 cases and indicating the potential novelty of nvCJD was the neuropathological phenotype. All cases exhibited widespread florid plaque deposition in addition to the spongiform change, astrocytic gliosis and neuronal loss typical of classical CJD. Quantitative image analysis confirmed prion protein (PrP) deposition in the brain which exceeded by a factor of two or three the amount of PrP seen in previous cases of classical CJD. There were two possible caveats to this evidence: would the neuropathological features have been identified in the past without modern immunocytochemical techniques and were there cases of CJD with this neuropathological appearance either currently or in the past in countries without BSE? Review of material from past cases of CJD held at the CJDSU had been carried out prior to the announcement of the identification of nvCJD and has continued but no similar past case has been identified. Crucially, extensive review of neuropathological archives in other countries has failed to lead to the identification of even a single similar case, and this includes a review of European CJD neuropathology, utilising standardised immunocytochemical techniques coordinated by Prof H Budka. Single florid plaques have been identified in two cases of CJD, one historical case and one recent iatrogenic case, but in neither was the overall pathology consistent with nvCJD. On current evidence the neuropathology of nvCJD is indeed novel as originally proposed.

EPIDEMIOLOGY

The identification of a new clinico-pathological form of CJD does not in itself lead to any inference regarding aetiology. Variants of CJD have been identified in the past, for example the Heidenhein, Brownell-Oppenheimer and pan-encephalopathic types of CJD. Evidence of a causal link with any putative novel risk factor such as BSE depends on the timing of the occurrence of any related human disease, the geographical co-localisation of such cases with the risk factor and the identification of a common exposure in affected patients.

BSE was first identified in 1986, although in retrospect there were almost certainly cases in 1985, and with an incubation period averaging 5-6 years the exposure of the UK cattle population is likely to have started early in the 1980s (4). The proposed mechanism of cross-contamination was through animal feed in the form of meat and bone meal contaminated initially with sheep scrapie and later with BSE itself. The feeding of ruminant protein to ruminants was banned in July 1988 and the BSE epidemic peaked in 1992/3 and has subsequently significantly declined, consistent with the proposed mechanism of infection. Exposure of the human population to the BSE agent may have started in the mid 1980s and continued to increase until affected animals were excluded from the human food chain in 1988 and a ban on the use of a range of bovine tissues, including brain and spinal cord from all cattle, in human (and animal) food in 1989. If BSE were a human pathogen, an important question is when the first cases of an associated human disease might develop in relation to the BSE epidemic.

There is of course no direct experimental evidence on the incubation period of BSE in humans and speculation on the likely timing of the occurrence of BSE-related human disease is based on inference from data on the incubation period in kuru and CJD. In kuru the incubation period ranges from 4.5 to over 30 years, but there is no hard data on the mean incubation period because of multiple potential exposures and because there is only limited information on incubation periods in individuals who attended single cannabilistic rituals. Evidence on incubation periods in CJD comes from iatrogenic cases (5). With central inoculation through neurosurgical procedures or dura mater grafts, the incubation periods are relatively short ranging from 18 months to a few years, although there are occasional cases of dura mater related CJD with incubation periods exceeding 10 years. Transmission of BSE to humans would not involve direct implantation of infectivity in the CNS but oral exposure through dietary contamination or perhaps some other peripheral route. The closest analogy in iatrogenic CJD is the occurrence of cases of CJD following human pituitary hormone therapy administered by peripheral injection. In these cases there is a wide range of incubation periods from about 5 years to over 25 years, with a mean of 13 years. In prion diseases the incubation period is shortest with intra-species transmission and is usually extended when crossing from one species to another. The incubation period of BSE in humans is therefore likely to be longer than in pituitary hormone related CJD and with human exposure to BSE potentially starting in the mid 1980s it is reasonable to propose that a BSE related human disease might first occur in the mid 1990s. The disease onsets of the first cases of nvCJD were in 1994.

BSE has largely been a disease of British cattle. By October 1996 there had been over 165,000 cases of BSE in the UK in comparison to small numbers of cases in other countries: 129 in the Republic of Ireland, 224 in Switzerland, 47 in Portugal, 23 in France and cases in single figures in Denmark, Germany, Italy and the Netherlands. If BSE were pathogenic to humans any associated disease should occur predominantly in the UK. In 1993 a collaborative system for the study of the epidemiology of CJD was established in Europe, funded through the Biomed system. Surveillance methodologies in France, Germany, Italy, The Netherlands, Slovakia and the UK were harmonised in order to obtain comparable data and to identify any change in the characteristics of CJD that might be linked to BSE. In early 1996 it became important to determine whether cases of nvCJD were occurring in countries other than the UK and detailed case review indicated that by March 1996 no similar case had been identified in any other collaborating country. Since then 10 further cases of nvCJD have been identified in the UK and only one case in continental Europe. This single case does not refute the hypothetical causal link with BSE as significant quantities of meat products were exported from the UK to France in the 1980s as well as cattle and cattle feed. With effective surveillance of CJD in a large proportion of the continental European population and currently no other cases of nvCJD, it is apparent that nvCJD is occurring almost exclusively in the UK, the country with by far the highest incidence of BSE. This geographical co-localisation of BSE and nvCJD is only circumstantial evidence that the two conditions are linked, but evolving epidemiological evidence reinforces this hypothesis. It would be impossible to interpret the occurrence of nvCJD in the UK without comparative data from other countries.

Detailed information on past medical history, occupation, and dietary history has been obtained in all cases of nvCJD in order to identify any common linking factor and in particular any common exposure to BSE. No case has been treated with human pituitary hormones and none has a history of previous neurosurgery or dura mater graft. Some cases have undergone surgical procedures, often minor, but with a similar frequency to age matched control subjects and there is no common occupational history, although one patient with nvCJD had worked for a 2 year period in a butcher's shop. After the identification of the first cases of nvCJD, one hypothesis was that these cases might represent genetic forms of CJD associated with mutation of the prion protein gene (PRNP). However full sequencing of the open reading frame of PRNP has excluded any of the mutations known to be associated with hereditary forms of

CJD. All cases tested so far are homozygous for methionine at codon 129 of PRNP, which compares with a similar frequency of this genotype in sporadic CJD (6). In the French cohort of human growth hormone related CJD, the codon 129 genotype appears to influence the incubation period with early disease in methionine homozygotes, followed by valine homozygotes and finally heterozygotes (7). In nvCJD it is possible that cases with genotypes other than methionine/methionine at codon 129 may occur in the future with more prolonged incubation periods.

The likeliest mechanism of transmission of the BSE agent to the human population is through dietary exposure and an important question is whether such exposure took place. A range of tissues from BSE affected animals have been inoculated into mice in order to determine the tissue distribution of infectivity. These experiments are not fully sensitive because they involve crossing a species barrier between the source species and the indicator species. In BSE infectivity in natural disease has been found only in brain, spinal cord and retina, with high titres of infectivity in CNS tissues. Although these experiments cannot exclude low levels of infectivity in apparently negative tissues, the level of infectivity in the CNS tissues is orders of magnitude higher than in any other tissue. If BSE has caused nvCJD, the likeliest mechanism is through dietary exposure to bovine brain or spinal cord. A case-control study of past dietary history has not demonstrated any dietary factor consumed more frequently by nvCJD cases in comparison to age-matched controls. The failure, so far, to identify a common and distinct dietary factor in nvCJD cases should not be interpreted as refuting a causal link between BSE and nvCJD. There are a number of methodological problems with this type of dietary case-control study. Information on dietary history is necessarily obtained from a surrogate witness, details of past dietary exposures (particularly years in the past) are known to be unreliable, and there is evidence of respondent bias in previous dietary studies in CJD. Exposure to bovine brain and/or spinal cord in human food may have been intermittent, a range of food products may have contained infectivity, and there may have been variation with time in the level of contamination and the products contaminated. It is possible that single meat products only rarely contained sufficient infectivity to initiate agent replication and subsequent disease. The identification of specific causal exposures may be impossible through the retrospective study of dietary histories years later.

In the absence of evidence from individual cases implicating dietary exposure as the cause of nvCJD, it is important to consider whether there was a population dietary exposure to the BSE agent, and in particular exposure to high titres of infectivity through consumption of bovine brain and spinal cord. Currently there is no direct information on past inclusion rates of these tissues in human food products in the UK. However an investigation on the contemporary use of these products has been carried out in Australia and New Zealand, and this may provide some indication of past practices in the UK as dietary habits in these countries are similar to the UK and no legislative action has been necessary in these countries which are free of all animal spomgiform encephalopathies. With rare exceptions bovine brain does not enter the human food chain in either Australia or New Zealand. A detailed investigation in New Zealand has however indicated that cattle remains, including vertebral column, are used in the production of mechanically recovered meat (MRM). Spinal cord is removed from the majority of carcases, but spinal cord remnants, including occasional whole cords, are present in a significant proportion of vertebral columns used in the production of MRM. In the UK prior to 1988, spinal cord from affected cattle may have been included in MRM and, up until the introduction of the specified bovine offals ban in November1989, spinal cord from animals incubating the disease may also have been included. Mathematical modelling has indicated that large numbers of cattle incubating BSE, but clinically unaffected, may have been slaughtered in the late 1980s (8). It is likely that there was exposure, and potentially extensive exposure, of the population in the UK to CNS titres of BSE infectivity through the use of MRM in the 1980s. This product is said to have been used in a variety of meat products, including burgers, pies, sausages and pate, but there is no hard data on which products contained MRM, nor on inclusion rates. This type of information would be of particular use in targetting questions on dietary history in nvCJD. However current evidence indicates thet there is likely to have been significant dietary exposure to high titre BSE infectivity in the UK in the 1980s.

CONCLUSION

In summary, nvCJD is almost certainly a novel disease. The appearance of this disease in the mid 1990s would be consistent with exposure to BSE in the 1980s and cases of nvCJD are occuring almost exclusively in the UK, co-localising with the epidemic of BSE. There is no direct evidence of common specific exposures to BSE in nvCJD, but these may never be identified because of the limitations of retrospective investigation. This evidence is not conclusive, but in the absence of any other reasonable explanation, it is probable that BSE and nvCJD are causally linked. The consistency of the clinico-pathological phenotype and of the epidemiological evidence since April 1996 reinforces the likelihood of such a link.

REFERENCES

1. R.G. Will, J.W. Ironside, M. Zeidler, S.N. Cousens, K. Estibeiro, A. Alperovitch, S. Poser, M. Pocchiari, A. Hofman, and P.G. Smith, A new variant of Creutzfeldt-Jakob disease in the UK, Lancet. 347:921-925 (1996).
2. G. Chazot, E. Broussole, C.I. Lapras, T. Blattler, A. Aguzzi, and N. Kopp, New variant of Creutzfeldt-Jakob disease in a 26-year old French man, Lancet. 347:1181 (1996).
3. J.W. Wilesmith, G.A.H. Wells, M.P. Cranwell and J.B. Ryan, Bovine spongiform encephalopathy: epidemiological studies, Veterinary Record. 123:638-644 (1988).
4. P. Brown, M.A. Preece, R.G. Will, 'Friendly fire' in medicine: hormones, homografts and Creutzfeldt-Jakob disease, Lancet. 340:24-27 (1992).
5. O. Windl, M. Dempster, J.P. Estibeiro, R. Lathe, R. De Silva, T. Esmonde, R. Will, A. Sprigbett, T.A. Campbell, K.C.L. Sidle, M.S. Palmer and J. Collinge, Genetic basis of Creutzfeldt-Jakob disease in the United Kingdom: a systematic analysis of predisposing mutations and allelic variation in the PRNP gene, Human Genetics. 98:259-264 (1996).
6. T. Billette de Villemeur, J-P. Deslys, A. Pradel, C. Soubrie, A. Alperovitch, M. Tardieu, J-L. Chaussain, J-J. Hauw, D. Dormont, M. Ruberg, and Y. Agid, Creutzfeldt-Jakob disease from contaminated growth hormone extracts in France, Neurology, 47:690-695 (1996).
7. R. M. Anderson, C.A. Donnelly, N.M. Ferguson, M.E.J. Woolhouse, C.J. Watt, H.J. Udy, S. MaWhinney, S.P. Dunstan, T.R.E. Southwood, J.W. Wilesmith, J.B.M. Ryan, L.J. Hoinville, J.E. Hillerton, A.R. Austin, and G.A.H. Wells, Transmission dynamics and epidemiology of BSE in British cattle, Nature. 382:779-788 (1996).

THE MOLECULAR BASIS OF CELLULAR DYSFUNCTION IN PRION DISEASES

Randal R. Nixon[1], Yin Qiu[1], William Hyun[2], Stanley B. Prusiner[3,4]
William C. Mobley[4], and Stephen J. DeArmond[1,4]

[1]Department of Pathology, Neuropathology Unit
[2]Laboratory for Cell Analysis
[3]Department of Biochemistry and Biophysics
[4]Department of Neurology
University of California, San Francisco, San Francisco, CA, 94143

INTRODUCTION

Our Neuropathology Research Laboratory at the University of California San Francisco has collaborated with Stanley Prusiner's laboratory for almost 14 years. The vast majority of the laboratory work to be reported here was done and/or initiated by Krister Kristensson, M.D., a neuropathologist from Stockholm who spent a year in our laboratory; by four of our neuropathology research fellows, Hans Kretzschmar, M.D., Klaus Jendroska, M.D., Kondi Wong, M.D. and Randal Nixon, M.D., Ph.D.; and by a research associate in our laboratory, Yin Qiu, M.D., Ph.D. Our focus has been on the molecular and cellular mechanisms which cause nerve cell dysfunction, neuropathological changes, and nerve cell death in prion diseases. We began our studies following our first immunohistochemical study done in conjunction with Stanley Prusiner's laboratory which showed that the amyloid plaques in scrapie-infected hamster brains and in human prion diseases contain protease resistant prion protein (PrPSc)[1-4]. From those early studies, it became clear that amyloidogenesis, which is very limited in amount in prion diseases and fails to occur in most cases of sporadic and familial Creutzfeldt-Jakob disease (CJD), does not account for neuronal dysfunction and degeneration. However, finding that protease resistant PrP can deposit as amyloid in prion diseases raised the possibility that PrPSc may be the cause of neurodegeneration in prion diseases. Consequently, we set out to test whether or not the accumulation of PrPSc in the brain is associated with the neuropathological changes characteristic of prion diseases and, if so, by what mechanisms could it cause neuronal dysfunction and death.

PRION DISEASES IN ANIMALS: PrPSc ACCUMULATION CAUSES NEURODEGENERATION

The conversion of PrPC to PrPSc and/or the accumulation of PrPSc causes the clinically relevant neuropathology. The most characteristic neuropathological changes in prion diseases which correlate best with clinical signs are spongiform (vacuolar) degeneration of neurons and nerve cell loss[5,6]. Astrocytic gliosis has been described as "hyperreactive" in prion diseases[7]; however, it appears to be secondary to nerve cell degeneration[6]. Although spongiform degeneration of neurons and nerve cell loss correlate well with dementia and motor signs in prion diseases, their presence is not a constant. There

Prions and Brain Diseases in Animals and Humans
Edited by Morrison, Plenum Press, New York, 1998

are many examples of patients with little spongiform degeneration, no detectable nerve cell loss, and no reactive astrocytic gliosis who present with severe dementia with myoclonus. This suggests there is a more fundamental functional abnormality which causes nerve cell dysfunction and results in vacuolation of neurons and their processes.

Several lines of indirect evidence have argued that local PrPSc accumulation in the central nervous system causes the neuropathological changes. First, our earliest immuno-histochemical studies during the terminal stages of scrapie in Syrian hamsters revealed that most of the PrPSc in the brain accumulates in the grey matter neuropil away from amyloid plaques and that spongiform degeneration and reactive astrocytic gliosis colocalize with sites of PrPSc deposition[3]. Interpretation of those early immunohistochemical studies, however, were hampered by marked variation in the intensity of immunostaining for PrPSc in aldehyde-fixed tissue sections and by the uncertainty that the normal, constitutively expressed prion protein isoform, PrPC, was completely eliminated from the sections by pretreatment with proteinase K. To overcome the problems presented by standard immunohistochemical techniques, brain regions from Syrian hamsters were microdissected at different times post-intracerebral inoculation with scrapie prions, homogenized, and the relative concentration of PrPSc determined by quantitative Western analysis[8]. Limited exposure to proteinase K eliminated PrPC with certainty from the homogenates without loss of PrPSc, which was partially digested to PrP 27-30. PrP 27-30 was first detected unilaterally in the thalamus at the site of intracerebral inoculation with scrapie prions at about 14 days post-inoculation. The amount of PrPSc increased in concentration and spread from the thalamus in a stereotypical pattern. The timing and pattern of spread suggested that it occurred mostly by orthograde axonal transport from one brain region to another; however, it also appeared to spread to some regions by transport in the extracellular space[9]. Once PrPSc began to accumulate in a grey matter region, its concentration increased exponentially. Neurohistological examination of tissue sections during the course of scrapie in the Syrian hamster showed that spongiform degeneration followed the accumulation of PrPSc by one-two weeks as did reactive astrocytic gliosis.

Following these studies, the histoblot technique was developed by Albert Taraboulos, who was then a research fellow in Stanley Prusiner's laboratory. This has proven to be the most sensitive and consistent method for localization and quantification of PrPSc in tissue sections[10]. Histoblots are made by pressing cryostat sections of unfixed brain tissue onto nitrocellulose paper where the tissue is treated with proteinase K to eliminate PrPC and with guanidinium to denature PrPSc (PrP 27-30) to make the latter's epitopes accessible to PrP-specific antibodies. By combining histoblot localization of PrPSc with standard neurohistological sections stained by the hematoxylin and eosin method for evaluation of the degree of spongiform degeneration and by GFAP immunohistochemistry for reactive astrocytic gliosis, a precise correlation between PrPSc deposition and the neuropathologic lesions was clearly demonstrated[11]. For example, in hamsters inoculated with the Sc237 scrapie prion strain, the outer cortical layers (Figure 1, region A) were devoid of PrPSc while the adjacent deeper cortical layers (Figure 1, region B) stained intensely for PrPSc. Correspondingly, spongiform degeneration and reactive astrocytic gliosis were confined to the deeper layers of the cerebral cortex.

Independent evidence that the formation and accumulation of PrPSc causes nerve cell dysfunction and degeneration has been the failure of intracerebral prion inoculation to cause disease or neuropathological changes in PrP knockout mice which do not express PrPC and, therefore, do not form PrPSc[13-15]. Furthermore, continuous exposure of the brain of PrP knockout mice to PrPSc by intracerebral implantation of scrapie-infected wild-type mouse brain has no effect on the receipient brain, whereas, the neurons in the transplant, which express PrPC, degenerate[16]. This observation argues that the mere presence of PrPSc in the central nervous system is not sufficient to cause neuronal dysfunction; rather, it appears that the conversion of PrPC to nascent PrPSc, following the exposure of the former to exogenous PrPSc, is necessary. Whether the conversion of PrPC to nascent PrPSc is pathogenic or whether accumulation of nascent PrPSc is pathogenic remains to be determined.

Direct evidence that abnormal PrP causes neuronal dysfunction. Our current line of studies have focused on prion diseases acquired by infection with prions; however, investigations of dominantly inherited prion diseases have yielded the strongest evidence that abnormal PrP is pathogenic. More than 15 mutations of the human PrP gene

Sc237

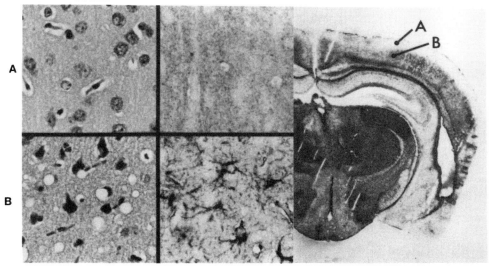

Figure 1. PrP^{Sc} colocalizes with spongiform degeneration and reactive astrogliosis in scrapie infected Syrian Hamster brain. Syrian hamsters inoculated with prion isolate Sc237 were sacrificed following development of clinical signs (approximately 70 days post-inoculation). Brains were harvested and processed for histoblot and neurohistological evaluation[12]. The left panels are photomicrographs of H&E stained sections and the middle panels photomicrographs of immunohistochemical stains for GFAP from regions A and B as shown on the histoblot at right. The histoblot demonstrates the distribution of PrP^{Sc} in a coronal section of the Syrian hamster brain.

have been genetically linked to the spontaneous development of a variety of clinically and neuropathologically defined prion diseases including several forms of familial CJD, Gerstmann-Sträussler syndrome (GSS), and fatal familial insomnia (FFI)[17]. The codon 102 mutation, which has been linked to the original GSS family[18], results in a leucine for proline substitution at residue 102[19]. When this mutation is made at the homologous site in a mouse PrP transgene, the resulting transgenic (Tg)-GSS mice spontaneously develop a neurodegenerative disease with widespread spongiform degeneration of grey matter, astrocytic gliosis, and GSS-type amyloid plaque formation[20]. Infectious prion particles also form spontaneously in the Tg(GSS) mice. Therefore, regardless of whether abnormal PrP is formed as a result of infection with prions or as the result of a dominantly inherited PrP gene mutation, the result is propagation of prions, neuronal dysfunction and death.

Spongiform degeneration of neurons indicates abnormal ion and water homeostasis. Spongiform (vacuolar) degeneration of neurons occurs principally in synaptic regions and is characterized by focal pre- and post-synaptic neuritic swelling, loss of internal organelles, and accumulation of abnormal membranes[21,22]. These ultrastructural characteristics suggest focal abnormalities of ion and water transport across the plasma membrane of neurons. Consistent with this possibility, glutamate toxicity, which results in receptor mediated increased ion and water transport into post-synaptic regions, has virtually identical ultrastructural features[23]. For these reasons, we began our current line of studies by testing whether or not receptor mediated Ca^{2+} regulation is abnormal in scrapie infected cell lines[24]. Four clonal cell lines were used: two uninfected lines, the mouse derived neuroblastoma cell line (N2a) and a cell line derived from scrapie-infected hamster brain (HaB), and their corresponding scrapie-infected counterparts, ScN2a and ScHaB respectively.

Cells were exposed to bradykinin (Bk) and the intracellular Ca^{2+} ([Ca^{2+}]$_i$) concentration monitored by the fluorescence ratio method using the dye Indo-1. Following exposure of N2a cells to bradykinin (Bk), a transient increase in [Ca^{2+}]$_i$ occurred in the N2a cells with a peak at approximately 20 seconds (Figure 2A). In ScN2a cells, the Bk stimulated [Ca^{2+}]$_i$

increase was either absent or markedly decreased (Figure 2B). Similar results were obtained with the HaB and ScHaB lines (data not shown). During the course of these experiments, the ScHaB hamster cell line unexpectedly and spontaneously ceased converting PrP^C to PrP^{Sc} and lost the ability to infect animals. Simultaneously, a normal Bk-stimulated intracellular Ca^{2+} response returned to these cells which argued that scrapie infection and not the clonality of the cell line was the cause of the attenuated Ca^{2+} responses (data not shown). The ScHaB cell line has since resisted reinfection with scrapie prions.

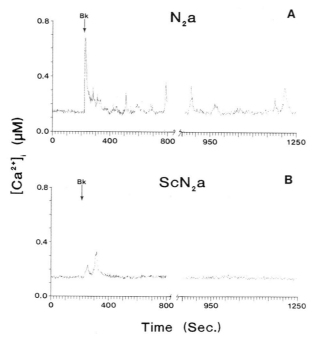

Figure 2. ScN2a cells have a diminished $[Ca^{2+}]_i$ flux in response to Bradykinin. N2a cells (A) show an immediate, rapid, transient increase in $[Ca^{2+}]_i$ followed by multiple smaller transient increases. ScN2a cells (B) have a markedly attenuated flux in $[Ca^{2+}]_i$; monitoring of the cells for greater than 15 min failed to identify any increase. Similar responses were seen in HaB and ScHaB cells (data not shown)[24].

PRION INFECTED CELL LINES: PHYSIOLOGICAL ABNORMALITIES

Decreased Bk-stimulated Ca^{2+} responses correlate with decreased Bk receptor binding affinity. The next series of studies were designed to determine the mechanism of decreased Bk-stimulated Ca^{2+} responses in ScN2a cells. A cascade of events links the Bk-receptor in the plasma membrane with the Ca^{2+} stores in the endoplasmic reticulum. Binding of Bk to its receptor is followed by coupling to a G-protein which then activates phospholipase-C and leads to the production of inositol triphosphate (IP_3)[25, 26]. The generated IP_3 then binds to receptors on the endoplasmic reticulum which open Ca^{2+} channels and release stored Ca^{2+} into the cytoplasm[27]. Because IP_3 is the link between the plasma membrane and endoplasmic reticulum, we compared the intracellular IP_3 response to Bk-stimulation in N2a and ScN2a cells[28]. A normal IP_3 response would implicate an abnormality of the Ca^{2+} release mechanisms in the endoplasmic reticulum, whereas, a decreased IP_3 response would argue for an abnormality in the plasma membrane. The mean and standard deviation of the IP_3 response in N2a cells 20 sec after exposure to Bk was 15 ± 7.6 pM. In contrast, the response was markedly reduced in ScN2a cells, 1 ± 0.6 pM ($p < 0.01$), a value not significantly different from resting concentrations in either N2a or ScN2a cells. These results argued that scrapie alters the IP_3 formation mechanisms in the plasma membrane; however, they do not exclude the possibility that there are also abnormalities of the endoplasmic reticulum. The latter has not yet been tested.

To test whether or not the decreased Bk-stimulated IP_3 response was due to a loss of Bk receptors or a change in their Bk-binding affinity in scrapie, a series of competitive and saturation binding studies were performed with Bk and Bk analogues[28]. The competitive binding assays indicated that both N2a and ScN2a cells express Bk-2 receptors exclusively (data not shown) and that both the N2a and ScN2a cells demonstrated similar competitive binding characteristics, although subtle differences were present (Figure 3). Compared to N2a cells, ScN2a cells consistently showed greater binding of ^{3}H-Bk in the presence of the lower concentrations of unlabeled Bk and less binding at the highest concentrations of unlabled Bk. The half maximal inhibitory concentrations were comparable for the two cell lines, 3.2 nM and 2.1 nM for N2a and ScN2a respectively, similar to values reported for Bk receptors on astrocytes[29]. It was concluded that Bk receptors are not decreased on ScN2a cells, in fact, these results suggested an increase compared to N2a. Scatchard plots of saturation binding assays yielded markedly different X-intercepts and slopes for N2a and ScN2a cells (Figure 4). The greater B_{max} (e.g. X-intercept) for ScN2a cells supported the conclusion from the competitive binding assays of more Bk-receptors on ScN2a cells. However, the slopes indicated that the binding affinity of ScN2a Bk-receptors for their ligand was 5-13 fold less than that seen with N2a cells.

Plasma membrane abnormalities are generalized with scrapie infection. Because of the abnormal properties of Bk-receptors and their inability to initiate a normal IP_3 response, the possibility was raised that scrapie causes a generalized abnormality of the plasma membrane. The plasma membrane is a heterogeneous semi-solid composed of a complex mixture of proteins, lipids, and carbohydrates. Qualitative and/or quantitative alterations in any one of these components can have significant effects on membrane function. Plasma membrane fluidity (or stiffness) is a quantifiable parameter which reflects changes in its composition[30]. Furthermore, increases or decreases in fluidity are known to affect the functions of many of the plasma membrane components, including enzyme systems and receptors. Fluidity has at least three components; lateral, rotational, and transitional or "flip-flop". In our studies, we measured lateral fluidity in the plane of the plasma membrane by the technique of Fluorescence Recovery After Photobleaching (FRAP). In brief, the fluorescent phospholipid analogue, nitrobenz-oxadiazole labeled phosphocholine (NBD-C6-PC) was incorporated into the outer leaflet of the plasma membrane. A 2.6 μm diameter spot was bleached on the cell surface with a brief pulse of laser light using a

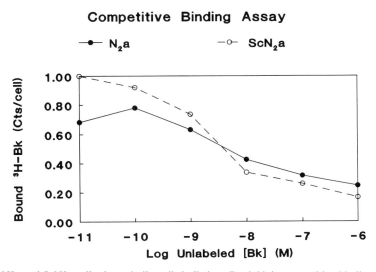

Figure 3. N2a and ScN2a cells show similar, albeit distinct, Bradykinin competitive binding assay curves. ScN2a cells bound more ^{3}H-Bk at the lowest concentration of unlabeled ligand and less ^{3}H-Bk at the highest amounts of added competitor than N2a cells. The half maximal inhibitory concentration was similar for N2a and ScN2a cells[28]. These results imply binding of Bk to both cell types.

Bradykinin (Bk) Receptor

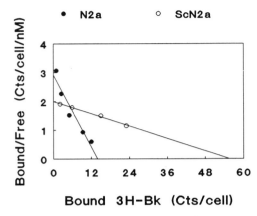

Figure 4. Scatchard Analysis of Bradykinin binding to N2a and ScN2a cells. N2a or ScN2a cells were incubated with increasing ^{3}H-Bk (ranging from 0.1 to 20 nM) and the amounts of free and bound ^{3}H-Bk determined after correcting for non-specific binding[28]. Data were converted to molar concentrations and plotted according to the method of Scatchard. The x-intercept corresponds to the receptor number (B_{Max}) and the slope to the negative reciprocal of the dissociation constant ($-1/K_D$). Each data point is the result of assays performed in triplicate. The ScN2a cells demonstrate increased numbers of receptors but with decreased affinity for the ^{3}H-Bk ligand.

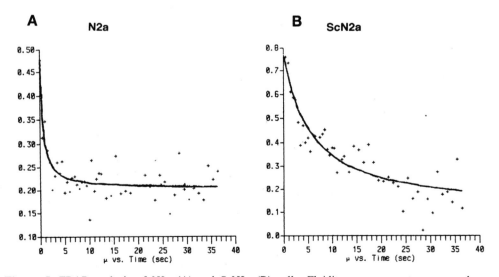

Figure 5. FRAP analysis of N2a (A) and ScN2a (B) cells. Fluidity measurements were made on subconfluent cultures labeled with the fluorescent phospholipid analogue NBD-C6-PC and a ACAS 570 laser photocytometer[28].

152

Meridian ACAS 570 laser photocytometer (Meridian Instruments, Inc., Okemos, MI). The rate of diffusion of unbleached NBD-C6-PC into the photobleached area from the surrounding membrane was measured by the ACAS 570. Examples of the type of data obtained from single N2a (Figure 5A) and ScN2a (Figure 5B) cells are shown. The inverse of the fluorescence intensity is shown on the ordinate and time on the abscissa. Recovery of fluorescence took longer in the ScN2a cell in this example than in an N2a cell. The rate of lateral diffusion (cm²/sec) was calculated for each cell analyzed by the ACAS 570. The mean and standard deviation for 18 N2a cells was $8.04 \pm 6 \times 10^{-9}$ cm²/sec and for 22 ScN2a cells, $1.13 \pm 1 \times 10^{-9}$ cm²/sec ($p < 0.01$) or a 7-fold decrease in plasma membrane fluidity. About 100 N2a and 100 ScN2a cells were analyzed; however, most measurements had to be discarded because of cell membrane movement, floating debris which blocked measurements, or incomplete recovery of fluorescence. The values for lateral membrane mobility obtained for the N2a and ScN2a cells are compatible with those reported for other cells[31]. In those studies, a 3-fold change in fluidity was considered significant because of profound effects on membrane function[32, 33]. Our results argue that the conversion of PrP^C to PrP^Sc and/or the accumulation of PrP^Sc in the cell markedly alters the composition of the plasma membrane which in turn alters the function of its components.

PATHOGENIC MECHANISMS: HYPOTHESES

These studies have identified one possible mechanism by which prion disease might cause cell dysfunction, namely by altering the composition/properties of the plasma membrane. How could the conversion of PrP^C to nascent PrP^Sc and/or the accumulation of PrP^Sc do this? There are at least three possibilities: 1) PrP^Sc itself accumulates in the plasma membrane and directly alters its properties; 2) The conversion of PrP^C to PrP^Sc and the shunting of nascent PrP^Sc to lysosomes alters normal steady state endocytic recycling, which maintains the composition of the plasma membrane constant, and creates a non-steady state condition which removes proteins and/or lipids faster than they can be replaced; and 3) Intracellular accumulation of PrP^Sc sequesters cellular chaperones and alters chaperone responses which, in turn, allows proteins destined for the plasma membrane to become mal-folded during their synthesis thus altering protein-protein and protein-lipid interactions in the membrane. The first possibility is supported by recent evidence which indicates that PrP^Sc becomes concentrated in caveolae-like detergent-insoluble complexes purified from plasma membranes[34].

The second possibility suggested is consistent with our knowledge of PrP^C and PrP^Sc metabolism. PrP^C and PrP^Sc are phosphatidylionositol glycolipid (GPI) anchored, sialoglyco-proteins[35-37]. The GPI anchor is attached to a serine at residue 231. They also contain two asparagine (N)-linked oligosaccharide side-chains at residues 181 and 197. Most of our knowledge about the metabolism and trafficking of PrP^C and PrP^Sc in cells comes from investigations of scrapie infected cell lines. Only a small number of cell lines, such as the mouse neuroblastoma N2a line, appear to be infectable; however, once infected, the resulting ScN2a cell line retains infectivity and on all subsequent passages in culture without the necessity for reinfection[38]. Evidence for scrapie infection in these cultures is the conversion of PrP^C to PrP^Sc and the formation of infectious prion particles which cause disease in animals. PrP^C is constitutively expressed in both N2a and ScN2a cells. Ninety percent of the cell's PrP^C becomes attached to the outer layer of the plasma membrane by its GPI anchor following its synthesis and passage through the Golgi complex[39]. Like other GPI anchored proteins, PrP^C re-enters the cell through caveolae[40]; however, it also appears to be capable of internalization via clathrin-coated pits[41]. About 10% of the endocytosed PrP^C is degraded while the remainder is recycled to the cell surface along with other plasma membrane-derived proteins and lipids of the endosome. Constitutive synthesis of PrP^C plus endocytic recycling maintains both the plasma membrane concentration of PrP^C and the lipid-protein composition of the plasma membrane constant. In pulse-chase experiments, PrP^C in uninfected cells is rapidly labelled by a radioactive amino acid tracer and appears to be completely degraded with loss of the tracer in about six hours[39, 42]. In scrapie infected ScN2a cells, labelling of PrP^Sc is delayed by about one hour after the pulse and increases to a maximum during the time period when the PrP^C pool loses its tracer[39]. Over the next 24 hours, there is little or no loss of tracer from the PrP^{Sc} pool. These results argue that PrP^{Sc} is derived from PrP^C but, unlike PrP^C, PrP^{Sc} is not degraded and accumulates in the cell. The PrP^C precursor pool from which PrP^{Sc} is formed must reach the cell surface before it can be converted to PrP^{Sc} since blocking PrP^C export from the ER-Golgi complex to the plasma membrane with brefeldin[43] or release of PrP^C from the plasma membrane of scrapie-infected cells with phosphatidylinositol-specific phospholipase C inhibits formation of PrP^{Sc}[44]. The sites of PrP^{Sc} distribution in cells is still being determined. Immunoelectron microscopy suggests that most of the PrP^{Sc} accumulates in structures containing "myelin" figures (e.g. layers of phospholipid rich membranes) believed to be secondary lysosomes[45]. Another location definitely known to contain PrP^{Sc} is the conditioned culture medium which appears to accumulate PrP^{Sc} as a result of excretion from infected cells or, more likely, as a result of cell death. More recently, we have found that as of yet, an unknown amount of PrP^{Sc} becomes attached to the plasma membrane[34]. The site where PrP^C is converted to PrP^{Sc} is not known although it is suspected to occur in a cholesterol-rich compartment. It is likely that the continuous conversion of PrP^C to PrP^{Sc} followed by sorting of nascent PrP^{Sc} in endosomes disrupts normal steady state endosomal recycling which is necessary for homeostasis of plasma membrane composition.

As for the third possibility, Tatzelt et. al. have recently reported abnormal heat shock protein responses and aberrant cellular localization in scrapie-infected cell lines[46]. This raises the possibility that proteins targeted to the plasma membrane, such as the Bk receptor we investigated in our studies, are misfolded during synthesis. The arrival of abnormally folded Bk-receptor protein in the plasma membrane might explain why its binding constants were markedly decreased. It could also explain decreased membrane fluidity particularly if many membrane proteins are abnormally folded during their synthesis. An intriguing variation on this possibility has been suggested by Liautard to explain the origin of the three forms of prion diseases: infectious, genetic, and sporadic[47]. The foundation of the argument is that a misfolded chaperone can induce misfolding of proteins and, in the case of autofolding (auto-chaperone), may lead to new misfolded chaperones. From some structural similarities between the prion protein and chaperones and from the effects of pathogenic mutations of the prion protein on its structure, Liautard proposes that PrP may itself be a chaperone and perform an auto-chaperone function. The recently described abnormal Hsp responses in scrapie-infected cells and the possibility that the prion protein is an auto-chaperone are particularly interesting because the conversion of PrP^C to PrP^{Sc} involves a change in molecular conformation from an α-helical to ß-sheet configuration[48-51]. Nevertheless, it is likely that the abnormal heat shock protein responses and locations in scrapie-infected cells are secondary to decreased membrane fluidity. Thus, modulation of membrane fluidity in dimorphic fungi by exposure to unsaturated or saturated fatty acids has a profound effect on the transcription of the Hsp70 and Hsp82 genes[52]. Futhermore, an association between membrane fluidity and Hsp regulation could also be via the former's affect on Ca^{2+} signalling given that the calcium ionophore, A23187, induces the 80 and 100 kDa heat shock proteins in tracheal epithelium [53].

The possibility that all of the known physiological abnormalities in scrapie-infected cells are secondary to a primary abnormality of plasma membrane composition and function is very attractive. First, it is consistent with an earlier spin-label study of synaptosomal membranes from scrapie-infected mice which showed alterations in the rotational fluidity of membrane proteins and the superficial layers of the membrane bilayer[54]. Secondly, it provides a rational explanation for neuronal vacuolation in prion diseases. Thus, alterations of plasma membrane fluidity would have a profound effect on those regions of the nerve cell membrane rich in receptors and ion-channels, particularly, pre- and post-synaptic regions. In fact, the vacuolation in prion diseases appears to affect synaptic regions preferentially, although not exclusively. Vacuolation itself suggests focal disturbances in both water and ion transport across the neuronal membrane. Additionally, changes in the binding affinity of trophic factor receptors for their ligands could deprive neurons of stimuli required to maintain their viability and may contribute to nerve cell degeneration and death in prion diseases. A change in membrane fluidity could explain why some patients with CJD have severe dementia but only mild neuropathological changes since altered fluidity has the potential to disturb synaptic transmission and, in doing so, functionally disconnect neural networks.

Finally, the results we presented here are consistent with the evidence that PrP^Sc is the sole functional component of prions, with our understanding of the steps required for propagation of prions, and with the roles played by PrP^Sc and PrP^C in the pathogenesis of nerve cell dysfunction. But perhaps more importantly, they provide insight into how the propagation of prions and cell dysfunction are related.

ACKNOWLEDGMENTS

The authors wish to thank Ms. Marlyna R. Stewart for her help preparing the manuscript.

REFERENCES

1. P.E. Bendheim, R.A. Barry, S.J. DeArmond, D.P. Stites, and S.B. Prusiner. Antibodies to a scrapie prion protein. Nature; 310: 418-421 (1984).

2. S.J. DeArmond, M.P. McKinley, R.A. Barry, M.B. Braunfeld, J.R. McColloch, and S.B. Prusiner. Identification of prion amyloid filaments in scrapie-infected brain. Cell; 41: 221-235 (1985).

3. S.J. DeArmond, W.C. Mobley, D.L. DeMott, R.A. Barry, J.H. Beckstead, and S.B. Prusiner. Changes in the localization of brain prion proteins during scrapie infection. Neurology; 37: 1271-1280 (1987).

4. A.D. Snow, R. Kisilevsky, J. Willmer, S.B. Prusiner, and S.J. DeArmond. Sulfated glycosaminoglycans in amyloid plaques of prion diseases. Acta Neuropathol. (Berl.); 77: 337-342 (1989).

5. E. Beck, P.M. Daniel, and H.B. Parry. Degeneration of the cerebellar and hypothalamo-neurohypophysial systems in sheep with scrapie; and its relationship to human system degenerations. Brain; 87: 153-176 (1964).

6. C.L. Masters and E.P. Richardson Jr. Subacute spongiform encephalopathy Creutzfeldt-Jakob disease - the nature and progression of spongiform change. Brain; 101: 333-344 (1978).

7. D. Dormont, A. Delpech, M.N. Courcel, J. Viret, P. Markovitz, and L. Court. Hyperproduction de proteine glio-fibrillarie acide (GFA) au cours de l'evolution de la tremblante de la soruis. C. R. Acad. Sci. (Paris); 293: 53-56 (1981).

8. K. Jendroska, F.P. Heinzel, M. Torchia, L. Stowring, H.A. Kretzschmar, A. Kon, A. Stern, S.B. Prusiner, and S.J. DeArmond. Proteinase-resistant prion protein accumulation in Syrian hamster brain correlates with regional pathology and scrapie infectivity. Neurology; 41: 1482-1490 (1991).

9. S.J. DeArmond and S.B. Prusiner. The neurochemistry of prion diseases. J. Neurochem.; 61: 1589-1601 (1993).

10. A. Taraboulos, K. Jendroska, D. Serban, S.L. Yang, S.J. DeArmond, and S.B. Prusiner. Regional mapping of prion proteins in brains. Proc. Natl. Acad. Sci. USA; 89: 7620-7624 (1992).

11. R. Hecker, A. Taraboulos, M. Scott, K.M. Pan, M. Torchia, K. Jendroska, S.J. DeArmond, and S.B. Prusiner. Replication of distinct prion isolates is region specific in brains of transgenic mice and hamsters. Genes Dev.; 6: 1213-1228 (1992).

12. R. Hecker, N. Stahl, M. Baldwin, S. Hall, M.P. McKinley, and S.B. Prusiner. Properties of two different scrapie prion isolates in the Syrian hamster. VIIIth Intl. Congr. Virol., Berlin; 284, Aug. 26-31 (1990).

13. H. Büeler, M. Fischer, Y. Lang, H. Bluethmann, H.P. Lipp, S.J. DeArmond, S.B. Prusiner, M. Aguet, and C. Weissmann. Normal development and behaviour of mice lacking the neuronal cell-surface PrP protein. Nature; 356: 577-582 (1992).

14. H. Büeler, A. Aguzzi, A. Sailer, R.A. Greiner, P. Autenried, M. Aguet, and C. Weissmann. Mice devoid of PrP are resistant to scrapie. Cell; 73: 1339-1347 (1993).

15. S.B. Prusiner, D. Groth, A. Serban, R. Koehler, D. Foster, M. Torchia, D. Burton, S.L. Yang, and S.J. DeArmond. Ablation of the prion protein (PrP) gene in mice prevents scrapie and facilitates production of anti-PrP antibodies. Proc. Natl. Acad. Sci. USA; 90: 10608-10612 (1993).

16. S. Brandner, S. Isenmann, A. Raeber, M. Fischer, A. Sailer, Y. Kobayashi, S. Marino, C. Weissmann, and A. Aguzzi. Normal host prion protein necessary for scrapie-induced neurotoxicity. Nature; 379: 339-343 (1996).

17. S.J. DeArmond and S.B. Prusiner. Etiology and pathogenesis of prion diseases. Am. J. Path.; 146: 785-811 (1995).

18. H.A. Kretzschmar, G. Honold, F. Seitelberger, M. Feucht, P. Wessely, P. Mehraein, and H. Budka. Prion protein mutation in family first reported by Gerstmann, Straussler, and Scheinker. Lancet; 337: 1160 (1991).

19. K. Hsiao, H.F. Baker, T.J. Crow, M. Poulter, F. Owen, J.D. Terwilliger, D. Westaway, J. Ott, and S.B. Prusiner. Linkage of a prion protein missense variant to Gerstmann-Sträussler syndrome. Nature; 338: 342-345 (1989).

20. K. Hsiao, D. Groth, M. Scott, S.L. Yang, H. Serban, D. Rapp, D. Foster, M. Torchia, S.J. DeArmond, and S.B. Prusiner. Serial transmission in rodents of neurodegeneration from transgenic mice expressing mutant prion protein. Proc. Natl. Acad. Sci. USA; 91: 9126-9130 (1994).

21. P.W. Lampert, D.C. Gajdusek, and C.J. Gibbs Jr. Subacute spongiform virus encephalopathies. Scrapie, kuru and Creutzfeldt-Jakob disease: a review. Am. J. Pathol.; 68: 626-652 (1972).

22. S.M. Chou, W.N. Payne, C.J. Gibbs Jr., and D.C. Gajdusek. Transmission and scanning electron microscopy of spongiform change in Creutzfeldt-Jakob disease. Brain; 103: 885-904 (1980).

23. J.W. Olney. Inciting excitotoxic cytocide among central neurons. Adv Exp Med Biol; 203: 631-645 (1986).

24. K. Kristensson, B. Feuerstein, A. Taraboulos, W.C. Hyun, S.B. Prusiner, and S.J. DeArmond. Scrapie prions alter receptor-mediated calcium responses in cultured cells. Neurology; 43: 2335-2341 (1993).

25. T.M. Keravis, H. Nehlig, M.F. Delacroix, D. Regoli, C.R. Hiley, and J.C. Stoclet. High-affinity bradykinin B2 binding sites sensitive to guanine nucleotides in bovine aortic endothelial cells. Eur J Pharmacol; 207: 149-55 (1991).

26. J.L. Bascands, C. Emond, C. Pecher, D. Regoli, and J.P. Girolami. Bradykinin stimulates production of inositol (1,4,5) triphosphate in cultured mesangial cells of the rat via a BK2-kinin receptor. Brit. J. Pharmacol.; 102: 962-966 (1991).

27. M.J. Berridge. Inositol trisphosphate and diacylglycerol: two interacting second messengers. Annu Rev Biochem; 56: 159-193 (1987).

28. K. Wong, Y. Qiu, W. Hyun, R. Nixon, J. VanCleff, J. Sanchez-Salazar, S. Prusiner, and S. DeArmond. Decreased receptor-mediated calcium response in prion-infected

cells correlates with decreased membrane fluidity and IP3 release. Neurology; 47: 741-750 (1996).

29. A.J. Cholewinski, G. Stevens, A.M. McDermott, and G.P. Wilkin. Identification of B2 bradykinin binding sites on cultured cortical astrocytes. J Neurochem; 57: 1456-1458 (1991).

30. J. Szollosi. Fluidity/viscosity of biological membranes. in Mobility and Proximity in Biological Membranes. (eds. Damjanovich S) CRC Press, Bocaraton, FL; pp 137-208 (1994).

31. S. Ladha, A.R. Mackie, and D.C. Clark. Cheek cell membrane fluidity measured by fluorescence recovery after photobleaching and steady-state fluorescence anisotropy. J Membr Biol; 142: 223-8 (1994).

32. C.J. Field, E.A. Ryan, A.B. Thomson, and M.T. Clandinin. Diet fat composition alters membrane phospholipid composition, insulin binding, and glucose metabolism in adipocytes from control and diabetic animals. J Biol Chem; 265: 11143-11150 (1990).

33. H.A. Lehr, J.P. Zimmer, C. Hubner, M. Ballmann, W. Hachmann, W. Vogel, H. Baisch, P. Hartter, M. Albani, A. Kohlschutter, et al. Decreased binding of HIV-1 and vasoactive intestinal peptide following plasma membrane fluidization of CD4+ cells by phenytoin. Virology; 179: 609-617 (1990).

34. M. Vey, S. Pilkuhn, H. Wille, R. Nixon, S.J. DeArmond, E.J. Smart, R.G.W. Anderson, A. Taraboulos, and S.B. Prusiner. Subcellular colocalization of cellular and scrapie prion proteins in caveolae-like membranous domains. Proc. Natl. Acad. Sci. USA; 93:14945-14949 (1996).

35. D.C. Bolton, R.K. Meyer, and S.B. Prusiner. Scrapie PrP 27-30 is a sialoglycoprotein. J. Virol.; 53: 596-606 (1985).

36. T. Endo, D. Groth, S.B. Prusiner, and A. Kobata. Diversity of oligosaccharide structures linked to asparagines of the scrapie prion protein. Biochemistry; 28: 8380-8388 (1989).

37. N. Stahl, D.R. Borchelt, K. Hsiao, and S.B. Prusiner. Scrapie prion protein contains a phosphatidylinositol glycolipid. Cell; 51: 229-240 (1987).

38. D.A. Butler, M.R.D. Scott, J.M. Bockman, D.R. Borchelt, A. Taraboulos, K.K. Hsiao, D.T. Kingsbury, and S.B. Prusiner. Scrapie-infected murine neuroblastoma cells produce protease-resistant prion proteins. J. Virol.; 62: 1558-1564 (1988).

39. D.R. Borchelt, M. Scott, A. Taraboulos, N. Stahl, and S.B. Prusiner. Scrapie and cellular prion proteins differ in their kinetics of synthesis and topology in cultured cells. J. Cell Biol.; 110: 743-752 (1990).

40. Y.S. Ying, R.G.W. Anderson, and K.G. Rothberg. Each caveola contains multiple glycosyl-phosphatidylinositol-anchored membrane proteins. Cold Spring Harb. Symp. Quant. Biol.; 57: 593-604 (1992).

41. S.L. Shyng, J.E. Heuser, and D.A. Harris. A glycolipid-anchored prion protein is endocytosed via clathrin-coated pits. J. Cell Biol.; 125: 1239-1250 (1994).

42. B. Caughey, R.E. Race, D. Ernst, M.J. Buchmeier, and B. Chesebro. Prion protein biosynthesis in scrapie-infected and uninfected neuroblastoma cells. J. Virol.; 63: 175-181 (1989).

43. A. Taraboulos, A.J. Raeber, D.R. Borchelt, D. Serban, and S.B. Prusiner. Synthesis and trafficking of prion proteins in cultured cells. Mol. Biol. Cell; 3: 851-863 (1992).

44. B. Caughey and G.J. Raymond. The scrapie-associated form of PrP is made from a cell surface precursor that is both protease- and phospholipase-sensitive. J. Biol. Chem.; 266: 18217-18223 (1991).

45. M.P. McKinley, A. Taraboulos, L. Kenaga, D. Serban, A. Stieber, S.J. DeArmond, S.B. Prusiner, and N. Gonatas. Ultrastructural localization of scrapie prion proteins in cytoplasmic vesicles of infected cultured cells. Lab. Invest.; 65: 622-630 (1991).

46. J. Tatzelt, J. Zuo, R. Voellmy, M. Scott, U. Hartl, S.B. Prusiner, and W.J. Welch. Scrapie prions selectively modify the stress response in neuroblastoma cells. Proc. Natl. Acad. Sci. USA; 92: 2944-2948 (1995).

47. Liautard J-P. Prions and molecular chaperones. Arch. Virol.; 7 [Suppl.]: 227-243 (1993).

48. B.W. Caughey, A. Dong, K.S. Bhat, D. Ernst, S.F. Hayes, and W.S. Caughey. Secondary structure analysis of the scrapie-associated protein PrP 27-30 in water by infrared spectroscopy. Biochemistry; 30: 7672-7680 (1991).

49. M. Gasset, M.A. Baldwin, R.J. Fletterick, and S.B. Prusiner. Perturbation of the secondary structure of the scrapie prion protein under conditions associated with changes in infectivity. Proc. Natl. Acad. Sci. USA; 90: 1-5 (1993).

50. K.M. Pan, M. Baldwin, J. Nguyen, M. Gasset, A. Serban, D. Groth, I. Mehlhorn, Z. Huang, R.J. Fletterick, F.E. Cohen, and S.B. Prusiner. Conversion of α-helices into β-sheets features in the formation of the scrapie prion proteins. Proc. Natl. Acad. Sci. USA; 90: 10962-10966 (1993).

51. J. Safar, P.P. Roller, D.C. Gajdusek, and C.J. Gibbs Jr. Conformational transitions, dissociation, and unfolding of scrapie amyloid (prion) protein. J. Biol. Chem.; 268: 20276-20284 (1993).

52. B. Maresca and G. Kobayashi. Changes in membrane fluidity modulate heat shock gene expression and produced attenuated strain in the dimorphic fungus *Histoplasma capsulatum*. Arch Med Res; 24: 247-249 (1993).

53. P.L. Smith, M.J. Welsh, J.S. Stoff, and R.A. Frizzell. Chloride secretion by canine tracheal epithelium: I. Role of intracellular c AMP levels. J Membr Biol; 70: 217-226 (1982).

54. J. Viret, D. Dormont, D. Molle, L. Court, F. Leterrier, F. Cathala, C.J. Gibbs, and D.C. Gajdusek. Structural modifications of nerve membranes during experimental scrapie evolution in mouse. Biochem Biophys Res Commun; 101: 830-836 (1981).

PRESENILIN PROTEINS AND THE PATHOGENESIS OF EARLY-ONSET FAMILIAL ALZHEIMER'S DISEASE: β-AMYLOID PRODUCTION AND PARALLELS TO PRION DISEASES

D. Westaway[1], G. A. Carlson[2], C. Bergeron[1], G. Levesque[1],
R. Sherrington[1], H. Yao[1], R. Strome[1], B. Perry[2], A. Davies[2],
S. Gandy[3], C. Weaver[4], P. Davies[4], D. Shenk[5], J. Rommens[6],
J. Roder[7], P.E. Fraser[1], and P. St. George-Hyslop[1]

1 Centre for Research in Neurodegenerative Diseases
University of Toronto, Ontario, Canada
2 McLaughlin Research Institute, Great Falls, Montana, USA
3 Cornell University Medical College, NY, USA
4 Albert Einstein College of Medicine, Bronx, NY, USA
5 Athena Neurosciences, South San Francisco, CA
6 Hospital for Sick Children, Toronto, Ontario, Canada
7 Samuel Lunenfeld Research Institute, Mount Sinai Hospital
Toronto, Ontario, Canada

SUMMARY

In contrast to rare mutations in the amyloid presursor protein (APP) gene, missense mutations in the presenilin 1 (PS1) and presenilin 2 (PS2) genes, on chromosomes 14 and 1 respectively, are the most common causes of early-onset familial Alzheimer's disease (AD)(Sherrington et al. 1995)(Rogaev et al. 1995)(Levy-Lahad et al. 1995). Presenilin genes encode proteins with at least seven putative transmembrane domains and an extruded cytoplasmic "loop", the latter with a proponderance of acidic amino acid residues: these proteins are expressed in a variety of cell types. While the physiologic function of these genes is unknown, their protein products have been demonstrated to accumulate in intracellular sites including the endoplasmic reticulum, and the Golgi apparatus. Similarly, the mechanism by which the 30 different point mutations have been identified in PS1 and PS2 to date cause the clinical and neuropathological hallmarks of Alzheimer disease is unknown. However, fibroblasts from heterozygous carriers of PS1 and PS2 mutations secrete increased levels of the amyloidogenic long-tailed amyloid β-peptides ending at residues 42 or 43 (Aβ42)(Martin 1995)(Scheuner et al. 1996). Increased levels of Aβ42 and other Aβ-peptides can also be measured in postmortem brain tissue from human patients dying with early-onset FAD associated with PS1 mutations(Lemere et al. 1996). To determine whether overproduction of Aβ peptides occurs in brain as an early biochemical event prior to the onset of neurodegeneration, we constructed transgenic mice with either mutant or wild-type human PS1 and mated them with another line of transgenic mice overexpressing wild-type human βAPP695 under the control of the same transcriptional regulatory element. These studies reveal that mutant PS1 transgenes but not wild-type PS1 transgenes act in a dominant fashion to programme over-production of long-tailed Aβ42 peptides in brain, and that this biochemical difference is present by at least 2-4 months of age and in the absence of any detectable neuropathologic lesions. These advances in our understanding of presenilin function are discussed in relation to the two schools of thought on AD pathogenesis, "tau-ist" and "ßaptist", and also with regards the hypothesis that similarities between AD and prion diseases reflect the existence of shared pathogenic pathways.

INTRODUCTION

Alzheimer's disease has a well-defined neuropathologic profile which includes a surfeit of extracellular amyloid deposits (comprised mainly of the Aß peptide), accretion of intracellular neurofibrillary tangles, comprised mainly of hyperphosphorylated forms of the microtubule-associated protein tau (τ), reduced synaptic density(Masliah et al. 1991)(Masliah et al. 1991), and loss of cholinergic neurons of the basal forebrain. The disease clearly has a complex aetiology, existing in early and late-onset familial forms in addition to manifestation in a common sporadic (idiopathic) form. Together, these observations have been taken to indicate both environmental and genetic risk-factors triggering a chain of events which converge at a final pathogenic pathway leading to the stereotypic neuropathology(Hardy and Higgins 1992). The relative importance of amyloid deposition and tangle formation in AD pathogenesis have been intensely debated, with proponents of two camps punningly dubbed "ßaptists" and "tauists", respectively. ßaptists emphasize mutations in the amyloid precursor protein (APP) gene in familial Alzheimer's disease kindreds and the early accumulation of amyloid deposits in Down's syndrome patients prior to the onset of frank dementia(Bugiani et al. 1991), indicative of an early, crucial pathogenic step while tauists cite poor correlations between amyloid burden and cognitive decline in the elderly and a good correlation with tangle formation(Arriagada et al. 1992). Though space precludes a detailed elaboration of these two schools of thought, a convergence of genetic and biochemical studies (vide infra) make it unlikely that the Aβ peptide will be unseated as a key player in AD pathogenesis.

Because of the power of recombinant technologies, genetic approaches have played a key role in dissecting this complex disease. Such studies have revealed at least four loci which either cause, or strongly predispose to the development of AD. These include the APP gene located on chromosome 21(Goate et al. 1991), and the e4 allele of the apoE gene(Strittmatter et al. 1993), a modifier associated with late-onset AD, and located on chromosome 19. More recently, kindreds carrying the AD3 sub-type of the early-onset FAD have been exploited to positionally clone(Sherrington et al. 1995) a locus previously located on chromosome 14(St George-Hyslop et al. 1992)(Mullan et al. 1992)(Van Broeckhoven et al. 1992). The presenilin 1 (PS1) membrane protein encoded by this locus is related to the presenilin 2 (PS2) gene on chromosome 1. PS2 missense mutations are found in the "Volga-German" kindred and also an Italian FAD pedigree(Rogaev et al. 1995)(Levy-Lahad et al. 1995).

CREATION OF MICE HARBOURING APP AND PS1 TRANSGENES

Since mutations in the PS1 gene cause AD with onset as early as 30 years, we reasoned that these dominant mutations would comprise a good starting-point for transgenic modeling of AD, especially since mice have a life-expectancy of only 2-3 years. Accordingly, transgenic mice were created expressing either wild-type or mutant PS1 cDNAs (see below). These were then interbred with transgenic mice bearing a human wild type βAPP695 transgene to yield litters of mice expressing wild type human βAPP695 (single transgenic mice) or expressing both human wild type βAPP695 and either mutant or wild type human PS1 (double transgenic mice). This strategy was chosen for several reasons. First, we have developed a reliable and accurate assay for human Aβ peptides, allowing us to determine whether transgenic mice can model the elevation in human Aβ42 seen in postmortem brain and in cultured fibroblasts of PS1 carriers and in transfected cells. Second, endogenous mouse βAPP, differing from human Aβ-peptides at three N-terminal residues (Arg5Gly,Tyr10Phe, and His13Arg) is not associated with Aβ deposition under physiological or naturally occurring pathological conditions. Although murine Aβ is capable of folding into potentially amyloidogenic β-sheets, these substitutions alter the aggregation properties of the murine β-peptide such that it is less able to assemble into well define fibrils(Fraser et al. 1992). However, Aβ derived from human βAPP is capable of forming amyloid deposits in the brain of mice with human βAPP transgenes(Games 1995)(Hsiao et al. 1996). Third, although no direct interaction between βAPP and presenilin proteins had been formally shown at the time of initiating these experiments (see however(Weidemann et al. 1997)), we were concerned that if such putative direct interactions did occur, then the subtle amino acid sequence differences between human and murine βAPP, and between human and murine PS1 might be functionally important. Furthermore, processing of murine βAPP might differ quantitatively or qualitatively from human βAPP, regardless of whether or not there is a direct interaction between PS1 and βAPP.

Transgenic mice over-expressing wild-type human βAPP695 were constructed using the prion protein-derived cosmid expression vector cos.Tet, in which the human βAPP695 cDNA was cloned downstream of the promoter, a 5' untranslated region (UTR) exon, and an intron derived from the Syrian hamster PrP gene(Scott et al. 1992)(Fig.1). This vector directs transgene expression in most CNS neurons and some astrocytes(Prusiner et al. 1990)(Moser 1995), and has been used to create transgenic mice over-expressing human βAPP695 constructs(Hsiao et al. 1995). Transgenic mice over-expressing either mutant or wild-type human PS1 were created using the same cos.Tet vector system. Additionally, to facilitate cloning of cDNAs which include naturally-occuring Not 1 restriction endonuclease sites (e.g. PS2), the cos.Tet vector was modified to include Fse1 sites at the borders of the prokaryotic vector sequences (Fig.1, bottom panel). In general, use of the same promoter element for two transgene constructs offers the advantage that both transgenes are likely to be expressed in adequate quantities in the same cell, a likelihood that is enhanced in the particular case of PrP cosmid vector by a documented ability to direct transgene expression largely independent of chromosomal integration site(Prusiner et al. 1990)(Scott et al. 1993)(Telling et al. 1995): this special property of cos.Tet presumably reflects the presence of sequences functionally analogous to a "locus control region"(Grosveld et al. 1987).

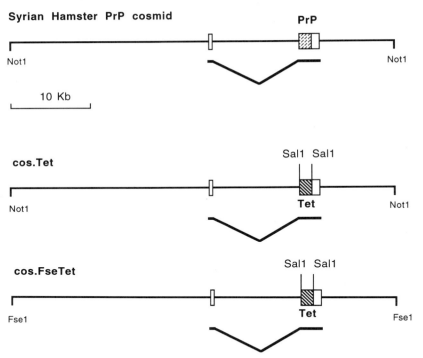

Figure 1. Derivation and features of the PrP cosmid expression vectors used in the presenilin protein research project. The parental Syrian hamsrer PrP cosmid is shown (after Scott et. al.). cos.FseTet was derived by filling-in the Not1 sites of cos.Tet with *pfu* DNA polymerase, followed by religation (H. Yao and D. Westaway, unpublished).

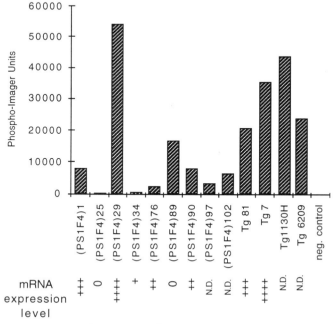

Figure 2. Transgene copy-numbers. Taq 1-digested genomic DNA was transferred to a nylon membrane and probed with a fragment derived from the 3'UTR of the SHaPrP cosmid. Two phophorimage analyses of different gel-loadings are superimposed. Estimates of transgene-encoded mRNA levels in adult brain are presented below the bar-graph (see also Fig. 3).

PS1 cDNAs were derived from the clone CC33 which contains a full length PS1 ORF including the alternatively spliced residues 324-335 in Exon 4 and residues 1018-1116 in Exon 9(Sherrington et al. 1995)(Rogaev et al. 1997). Two different mutant transgenes were created, namely the Met146Leu and the Leu286Val missense mutations, which are both associated with FAD but have slightly different phenotypes (Met146Leu, "FAD4", age of onset = 40 years; Leu286Val, "FAD2", onset = 50 years) (Foncin et al. 1985)(Frommelt et al. 1991). Several different founders were obtained for each construct, and these exhibited a range of copy-numbers. A Southern blot of Taq1-digested genomic DNA from representative Tg(M146L) animals was probed with a DNA fragment from the PrP 3' untranslated region, and a phosphorImage analysis of this analysis is presented in Figure 2. For purposes of comparison, DNA samples from previously characterized transgenic lines (TgSHaPrP: Tg81, Tg7(Prusiner et al. 1990) and Tg APP695: TG1130H,Tg6209(Hsiao et al. 1995)) are also presented.

mRNA and protein encoded by PS1 Transgenes

Expression of the PS1 transgenes in brain was determined by Northern blot analysis (Fig. 3A): with the single exception of the Tg(M146L)89 line (mRNA undetectable: not shown), expression of transgene-encoded PS1 mRNA in the adult brain was generally found to parallel transgene copy-number (Fig.2 and data not presented).

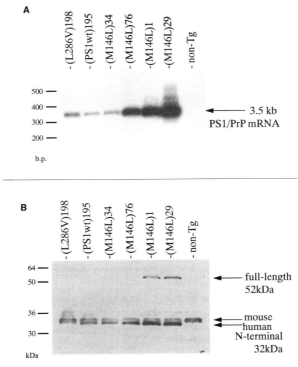

Figure 3. **Panel A:** Northern blot of mRNA from brain of non-transgenic (non-Tg), wild type PS1 [(PS1wt)195], Met146Leu PS1 [(M146L)1 and (M146L)29], and Leu286Val PS1 [(L286V)198] transgenic mice using a human PS1 cDNA hybridization probe. **Panel B:** Western blot of brain lysates from the same mice using the PS1 specific antibody Ab14. Positions of the putative human PS1 holoprotein, and of the endogenous mouse and human N-terminal PS1 endoproteolytic cleavage products are denoted by arrows.

Western blot analyses were also performed using an N-terminal directed polyclonal antibody-Ab14(Thinakaran et al. 1996) (Fig. 3B). In accord with the results of others the predominant transgene-encoded protein was found to correspond to an N-terminal proteolytic fragment of PS1(Thinakaran et al. 1996), which on this tris-glycine gel system has a different electrophoretic mobility from the putative endogenous mouse PS 1 fragment. In the highest copy-number lines, Tg(M146L)1 and Tg(M146L)29, putative uncleaved PS1 holoprotein was detected. Analyses with C-terminal directed anti-PS1 antibodies confirm the presence of transgene encoded C-terminal ~19 kDa fragments in the Tg(M146L)1, Tg(M146L)29, Tg(PS1wt)195 and Tg(L286V)198 lines (data not shown).

Based upon the above analyses, four independent PS1 transgenic lines were selected for cross-breeding with the wild-type human βAPP695 transgenic line Tg6209, which contains approximately 28 ± 6.1 βAPP copies and produces 1.6 ± 0.43-fold more human βAPP than endogenous murine βAPP. Offspring were selected either for the presence of both the βAPP and the PS1 transgenes (double transgenics), or were selected for the presence of the βAPP transgene alone (single transgenic controls). Because all of the parental lines were produced using the same inbred FVB/N mice, the only difference between double and single transgenic animals is the presence or absence of the human PS1 transgene array and any associated insertional mutation.

Aβ-peptide in transgenic mice

Animals were killed at 38-133 days of age (median 75.5 days), and the hippocampi were quick frozen. Two sandwich ELISAs were used to measure the concentration of both total Aβ human peptides (Aβ1-x, where x > residue 28) and the concentration of Aβ1-42 peptides in hippocampus, an area of brain typically heavily involved by AD neuropathology. Aβ levels were measured simultaneously for the Tg(M146L) 29 double transgenic, Tg(L286V)198 double transgenic, Tg(PS1wt)195 double transgenic animals and their single transgenic control littermates harbouring

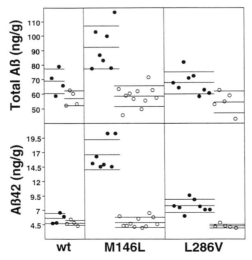

Figure 4. Aβ42 and total Aβ levels in the brain of double transgenic (human PS1 and human βAPP695) mice (filled symbols) compared to littermates with the human βAPP695 transgene only (open symbols). The PS1 genotype of the double transgenic mice is denoted below each panel. The Tg1 Met146Leu line is indicated by diamond symbols because the Aβ assays were performed at a different time.

only APP transgenes. Aβ levels in the Tg1 (Met146Leu double transgenic) line and their single transgenic control littermates were measured together in an independent assay. These results showed small but significant increases in total human Aβ peptides in the hippocampus of the Tg(M146L)29 (P<0.001) and Tg(L286V)198 (P<0.01) double transgenic mice compared to their single transgenic littermates expressing only the human wild-type APP695 (Fig. 4). In contrast, the hippocampi from the Tg(Ps1wt)195 double transgenic mice displayed only a very small increase in total Aβ peptide (P=0.0574).

More impressively, there was a very significant increase in the amount of long-tailed Aβ42 peptide in the brain of double transgenic mice expressing mutant PS1 (P<0.0001). In contrast, the amount of Aβ42 peptide in the brain of wild type PS1 double transgenic mice did not differ from that in the brain of the single transgenic βAPP695 control mice (Fig. 5). Furthermore, the ratio of Aβ42 to total Aβ (Aβ42/Aβtotal ratio), a measure of the relative utilization of the 42-cleavage pathway which eliminates differences in total Aβ production and APP expression, was significantly increased in the mutant PS1 double transgenic mice (p < 0.001), but not in the wild type PS1 double transgenic mice (Fig. 5). These results were confirmed by an independent assay of Aβ in the second Met146Leu double transgenic line, Tg(M146L)1 (diamond symbols in Fig. 5), which was run using slightly different methods.

There are three further observations worthy of comment. First, both the magnitude of the increase in Aβ42 and the increase in the Aβ42/Aβtotal ratio was less in the Leu286Val mice compared to both of the Met146Leu mice (Tg1 and Tg29). At least two potential explanations can be envisaged for this difference. First, the level of expression of the PS1 transgenes was higher in both of the PS1 Met146Leu animals than in the PS1 Leu286Val and the wild type PS1 animals (Fig. 1). Alternatively, the PS1 Leu286Val mutation in humans is associated with a milder phenotype (onset at approximately 50 years compared to approximately 40 years)[Foncin, 1985 #3563](Frommelt et al. 1991). Thus the difference may reflect different biological activities of the mutations rather than disparities in the level of transgene expression. The difference in Aβ levels between the two Met146Leu double transgenic lines Tg1 and Tg29, which express slightly different levels of PS1 (Fig.4), does not help resolve this because the differences in the Aβ levels for Tg1 and Tg29 could reflect that fact that the Aβ assays were conducted using slightly different conditions. However, the argument that the lower levels of Aβ42 secretion in the TG(L286V) line could reflect

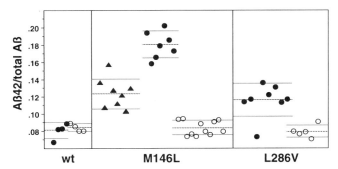

Figure 5. Aβ42/total Aβ ratio in the brain of double transgenic (human PS1 and human βAPP695) mice (filled symbols) compared to littermates with the human βAPP695 transgene only (open symbols). The PS1 genotype of the double transgenic mice is denoted below each panel. The Tg1 Met146Leu line is indicated by diamond symbols because the Aβ assays were performed at a different time.

differences in the biological effect of the mutations is supported by the observation of comparatively lower levels of Aβ42 secretion from cultured fibroblasts from human carriers of the PS1 Leu286Val mutation compared to fibroblasts of subjects heterozygous for the PS1 Met146Leu mutation(Scheuner et al. 1996).

Second, in contrast to native fibroblasts and the majority of βAPP/PS1 double transfected 293 cell lines, where total Aβ peptide concentrations were not elevated ((Martin 1995)(Scheuner et al. 1996)and Citron et. al. submitted), there was an increase in total Aβ peptide concentration in the brains of double transgenic mice bearing mutant PS1 sequences. A similar small increase in total Aβ peptide relative to a larger increase in Aβ42 has been observed in postmortem brain tissue from human subjects clinically affected with PS1 mutation linked FAD (Tamaoka et al, unpublished). The significance of this observation remains to be determined, but might indicate that there are subtle differences in the way in which mutant human PS1 alters the processing of human βAPP proteins in brain as compared to non-neural cell lines.

Finally, although not statistically significant, both the total Aβ-peptide and the Aβ42 peptide concentrations were slightly and proportionately elevated in the wild-type PS1 double transgenic mice. It is not clear whether this slight increase reflects the cumulative effect of increased murine and human PS1 level, or a preference for species identity in the putative βAPP/PS1 processing pathways.

Transgenic modeling of Alzheimer's disease

Our results and also those obtained by others, using Tg mice or transfected cells (Duff et al. 1996)(Borchelt et al. 1996)(Citron et al. 1997) support the contention that one effect of PS1 mutations is to alter the catabolism of βAPP by directly or indirectly alter the activity of γ-secretase (but not α- or β- secretases), resulting in the production of long-tailed Aβ-peptides. Insofar as young TgPS1/TgHuAPP mice described here have no overt signs of AD-like neuropathology, these changes in βAPP metabolism appear to antedate neuropathological evidence of neurodegeneration. The mechanism by which PS1 mutations lead to an aberration in βAPP metabolism remains to be worked out, although the localization of PS1 immunoreactivity in the endoplasmic reticulum, Golgi apparatus, and some uncharacterized intracytoplasmic vesicles *in vivo* in primate neurons (Levesque *et al.*, unpublished observations) supports the hypothesis that PS1 may have a role in the intracellular events regulating trafficking and degradation of transmembrane proteins like βAPP. A similar function has been proposed for SPE4, a weak homologue of human presenilins present in spermatocytes of *C.elegans*(L'Hernault and Arduengo 1992).

In sum, our current observations strongly suggest that transgenic models expressing mutant human presenilin transgenes can recapitulate at least some of the biochemical phenotypes of Alzheimer disease. It will be of considerable interest to seek behavioural, electrophysiological, and pathological changes consistent with clinical AD as these mice approach middle-age.

ALZHEIMER'S DISEASE AND CJD

In contrast to AD, prion diseases are quite rare. The incidence of the commonest prion disease, Creutzfeldt-Jakob disease (CJD) has been estimated at approximately one per million. Other prion diseases such as Gerstmann-Sträussler-Scheinker syndrome and Fatal Familial Insomnia (FFI) are rarer still. A cardinal feature of prion diseases, distinguishing them from AD(Brown et al. 1994) is their transmissibility, either in an experimental setting or iatrogenically. However, unlike viral diseases, the major component of the <u>pro</u>teinaceous <u>in</u>fectious particles (prions) that cause diseases such as CJD, and scrapie in sheep and goats, is thought to be an abnormal conformer of the prion protein, a host glycoprotein known as PrPCJD or PrPSc, respectively. The cerebral amyloid found in some cases of CJD, GSS, and scrapie disease is almost invariably composed of the prion protein (PrP), rather than Aß. Despite these crucial differences, many have commented on the intriguing parallels that nonetheless remain between AD and prion diseases (DeArmond 1993) (Table 1). Additionally, features of AD and CJD may even coexist within some patients.

Table 1. Neuropathology and epidemiology of AD, CJD, and GSS

Feature	AD	CJD	GSS
Progressive dementia?	yes	yes	yes
Transmissible?	no†	yes	yes, but less efficient than CJD
cerebral amyloidosis	yes (major component = Aß)	yes, in 5-15% cases (major component = PrP)	yes, major component = PrP
neurofibrillary tangles	yes	no	sometimes present (see Table 2)
spongiform change	sometimes	invariably	yes
loss of cholinergic neurons	yes	no	no
familial and sporadic manifestation	yes	yes	yes
site of causative mutation in familial forms of disease (see also Table 2)	presenilin 1 (chr. 14) presenilin 2 (chr. 1) APP (chr. 21)	PRNP (chr. 20)	PRNP(chr.20)

†reports of transmissions from the blood of AD patients, or their relatives (Manuelidis et al. 1988), must be treated with a degree of scepticism as the pathological phenotype in inoculated animals was reminiscent of CJD or scrapie, not AD

The perceived overlap between AD and prion diseases begs the question of origin: why should such parallels exist? A trivial explanation for coincident AD and CJD is along the lines of a sampling-error: superimposition of AD, a comparatively common disease in the elderly, upon the much rarer disease, CJD. However, the alternative viewpoint that the similarities and occasional coexistence of these diseases reflects fundamental shared elements in pathogenesis cannot be readily dismissed. Indeed, as our understanding of CJD and AD has increased, further similarites have come to light. These overlaps become even more striking when GSS is taken into consideration as well as CJD, and can be organized into four categories

Familial prion diseases resembling AD. Purification and molecular cloning of the prion protein lead to the realization that familial "clustering" of prion disease in Western societies had a fundamentally different mechanism from infectious spread of kuru among tribesmen and women in New Guinea(Gajdusek 1977). Such instances were instead found to represent germline transmission of pathogenic missense mutations in the prion protein gene (PRNP) on chromosome 20(Hsiao et al. 1989)(Hsiao et al. 1991)(Goldfarb et al. 1990). With the advent of the polymerase chain reaction, genomic sequences were obtained from many GSS and familial CJD patients, confirming and extending the original description of a proline to leucine mutation at codon 102 in two GSS kindreds(Hsiao et al. 1989). As neurologists cast their nets further afield to include clinical diagnoses of atypical CJD and atypical dementias, new disease entities defined upon molecular genetic criteria became apparent(Parchi and Gambetti 1995), some of which show blending or overlap with features of AD.

Perusal of Table 2 reveals that cases of AD-like familial prion diseases identified to date (denoted with asterisks) were diagnosed as GSS or AD, and exhibit deposition of neurofibrillary tangles.

Interestingly, three of these familial diseases (F198S, "Indiana GSS kindred"(Hsiao et al. 1992); Q217R , "Swedish GSS kindred"(Hsiao et al. 1992); Y145stop, "PrP-cerebral amyloid angiopathy"(Ghetti et al. 1996)) are characterized by accumulation of amyloidogenic fragments of PrP with discrete termini which differ from that of wild-type PrP, a feature not generally seen in other forms of GSS or CJD. Whether formation of neurofibrillary tangles in the Indiana and Swedish GSS kindreds is caused by a biologic effect of intact, mutant PrPC molecules or the derived PrP peptide fragments is unknown: however, observations of NFT's in the cerebral grey matter of a patient bearing a *PRNP* termination codon at position 145(Ghetti et al. 1996) is in accord with the latter hypothesis, and tallies with neurotoxic effects of PrP peptides observed by others(Forloni et al. 1993). Remarkably, double-immunostaining of material from Indiana kindred GSS families and from a patient with ungenotyped GSS has revealed Aß-specific plaque-like lesions which sometimes occur in "hybrid" plaques, surrounding PrP-amyloid cores (Bugiani et al. 1993)(Ikeda et al. 1994).

Familial AD with prion disease-like features. At this time there is a single example, a kindred described by Foncin and co-workers, where an FAD mutation appears to programme prion-like pathology (in addition to causing features typical of AD). In this family two siblings carrying an H163R presenilin 1 mutation were found at autopsy to display independent amyloid deposits composed of either PrP, or of βA4, but rarely of both (Table 2). In other regards disease presentation was indistinguishable from that caused by other mutations in presenilin 1 such as M146L. One brother was heterozygous for methionine and valine at *PRNP* codon 129 while the other was a methionine homozygote and also deleted for a single copy of the octarepeat array(El Hachimi et al. 1996).

This finding is remarkable because it indicates that presenilins (predominantly located in the ER) can facilitate amyloid formation from not one but two cell-surface proteins, namely APP and PrP. It also begs the questions of whether PrP deposits are a general feature of the AD3 subtype of FAD or only associated with certain mutations in PS1: given the intense pathological scrutiny applied to most of these kindreds, the latter situation seems more likely. In this regard it will be of interest to determine if prion-like features are present in a Finnish presenilin 1 H163Y kindred(Haltia et al. 1994)(Alzheimer 1995).

Sporadic AD with prion-like features.   On rare occasions patients have been described with sporadic neurologic disease bearing pathological and/or clinical hallmarks of both CJD and AD(Brown et al. 1990)(Powers et al. 1991)(Muramoto et al. 1992).   CJD diagnosis was verified by western blot analysis for PrP or by transmission studies(Brown et al. 1990). As mentioned earlier, given that AD is a common disease it is not surprising that that some sporadic CJD patients exhibit AD-like features, and the advanced ages of the above patients (73, 69, and 75 years) is consistent with the hypothesis of "coincidence".

Table 2.   Pathologic features in familial prion disease and some overlaps with FAD

Disease	Gene	mutation	Aβ plaques	PrP plaques	NFTs	spongiform change	Comments
F-CJD	*PRNP - 129M*	E200K	absent	absent	absent	present	most common F-CJD mutation: transmissible
GSS	*PRNP- 129M*	P102L	absent	present	absent	variable	
GSS	*PRNP- 129V*	A117V	absent	present	absent	variable	
GSS* (Indiana kindred)	*PRNP- 129V*	F198S	absent	present	present	mild	accumulation of an internal PrP amyloid fragment, residues ~81 to ~150. Non-transmissible?
GSS* (Swedish kindred)	*PRNP- 129V*	Q217R	absent	present	present	mild	accumulation of an internal PrP amyloid fragment, residues ~81 to ~146. Non-transmissible?
GSS*	*PRNP- 129M*	145 stop	absent	present	present	absent	"PrP cerebral amyloid angiopathy" Non-transmissible?
"FAD"	*PS1*	H163R	abundant neuritic plaques	kuru-like and diffuse	present	"micro-spongiosis"	PrP and Aβ plaques usually distinct but occasionally coincide. βA4 amyloid angiopathy
Early-onset FAD	*PS1 or PS2*	numerous	present	absent	present	not a consistent feature	Different mutations may be associated with intrinsic differences in rates of Aβ deposition
FAD	*APP*	V717F, "Swedish" mutation	present	absent	present	not a consistent feature	

Data from (Hsiao et al. 1991)(Hsiao et al. 1989)(Hsiao et al. 1992)(Ghetti et al. 1989)(Ghetti et al. 1996)(El Hachimi et al. 1996)(Parchi and Gambetti 1995)(Citron et al. 1997)

Patients with earlier disease onset (e.g. 39 years, 38 years(Gaches and Supino-Viterbo 1977)(Liberski et al. 1987)) are potentially difficult to reconcile with the hypothesis of coincidence, since these ages lie decades before the typical onset of sporadic AD. However, the prion disease component of the diagnoses in these patients can be called into question, e.g. there are no immunoblots for protease-resistant PrP and in the instance of the 38-year old patient transmissions into non-human primates were negative (P. Liberski, personal communication).

Aβ , PrP , and Inclusion Body Myositis. Approaching the potential pathogenic convergence of AD and prion diseases from the "molecular" perspective of the implicated proteins brings to mind one last, surprising area of overlap - the muscle disease inclusion body myositis (IBM). IBM is a common myopathy with onset in middle-age and has recently been reviewed by Carpenter(Carpenter 1996). Rimmed vacuoles are a prominent feature in light microscopic analyses and electron micrographs reveal cytoplasmic (and less frequently, nuclear) aggregates of unusual filaments. In 1991 Mendell and co-workers used the Congo-red stain to reveal green-birefringent amyloid deposits in vacuolated fibres(Mendell et al. 1991). Others went on to demonstrate staining for Aβ, indicating that IBM constitutes the only known occurence of intracellular amyloid deposits of Aβ peptide. However, since the filamentous structures were not co-localized with immunogold deposits it should be clearly understood that they are not themselves composed of Aβ subunits(Askanas et al. 1992): instead it has been suggested that they correspond to paired helical filaments(Askanas et al. 1996), closely resembling those of AD brains (however see also(Carpenter 1996)).

Askanas and co-workers have also demonstrated accumulations of PrP (detected with the 3F4 monoclonal antibody) in IBM-affected muscle fibres(Askanas et al. 1993). Increases in the abundance of PrP mRNA were inferred from in situ hybridizations(Sarkozi et al. 1994). Intriguingly, in transfected cells and transgenic mice, overexpression of either APP and PrPC appears to cause muscle fibre necrosis(Askanas et al. 1996)(Westaway et al. 1994). Of course, whether accumulations of either or both proteins actually play a causal role in IBM is far from proven and other workers have stressed nuclear changes in the pathogenesis of this disease, as well as the paucity of the birefringent material(Nalbantoglu et al. 1994)(Sherriff et al. 1995)(Carpenter 1996). Nonetheless, the numerous parallels to AD, which also include immunopositivity for α-antichymotrypsin, ubiquitin, and apolipoprotein E (Askanas and Engel 1995) are intriguing. If they can be substantiated(Sherriff et al. 1995) investigation of the biochemistry of PS1 - which is expressed in skeletal muscle and known to affect the metabolism of APP and perhaps also of PrP - may be worthwhile in future studies of this disease.

CONCLUDING REMARKS

With regards AD, analyses of mice expressing presenilin and APP transgenes lend further support to the notion that accumulation of fibrillogenic derivatives of Aβ may be an important pathogenic event in AD, as previously suggested by the accumulation of amyloid deposits in Down's syndrome patients prior to the onset of dementia. While it could be argued that the "amyloid versus neurofibrillary tangle" debate is largely over, the balance having tipped in favour of the ßaptists, in a sense this issue is passé. What is of more interest is defining the biochemical events upstream of Aß deposition and discerning whether pathogenesis in the most common forms of AD truly proceeds via a single unidirectional pathway. Insofar as age-at-onset in presenilin FAD kindreds may precede those of APP kindreds by decades, and presenilins control Aß deposition, these membrane proteins are good candidates for a pathogenetic nodal point that might be altered in sporadic AD, and might be reset therapeutically. Positioning of presenilins within the ER and with an extruded cytoplasmic loop might facilitate meaningful interactions with a plethora of proteins: consequently, the pathogenic potential of presenilins might extend beyond APP. This possibility can now be tested transgenetically. These other hypothetical substrates might account for the "non-amyloid" features of AD pathology, namely accumulation of tangles, loss of cholinergic neurons, and reduced synaptic density. For example, molecular genetic studies of a PS1 H163R kindred advance PrPC as a potential substrate(El Hachimi et al. 1996).

Table 3. Pathogenic proteins in AD and CJD

Property	PS 1	PS2	APP	PrP
location	ER	ER	cell-surface	cell-surface
membrane associated?	yes	yes	yes	yes
accumulation of stable, endoproteolytically cleaved products	yes	yes	yes	yes
protease-inhibitor domain?	unknown	unknown	yes (kunitz)	unknown
GAG-binding?	unlikely	unlikely	yes	yes
amyloidogenic?	unknown	unknown	yes	yes
metal binding site?	unknown	unknown	discrete Cu and Zn sites.	Cu?
Comments				may interact with the APP-like protein PLP (Yehiely et al. 1997)

Similarly, despite ambiguities concerning IBM and some "coincident" cases of AD and CJD, molecular genetic studies of variant GSS and PS1 H163R kindreds argue that the complexities of AD and prion diseases will be found to converge in some shared pathogenetic pathways. Such overlaps indicate a continuum of pathologic changes extending from classical AD to classical CJD, with the familial prion disease Gerstmann-Sträussler-Scheinker syndrome (GSS) occupying an intermediate position. Documented biochemical similarities between presenilins, APP, and PrPC (Table 3) are absolutely in accord with this notion and hint at cellular processes that might be perturbed in the two types of disease (e.g. protein trafficking, protein misfolding, refolding salvage pathways, endoproteolysis and degradation). It is likely that further transgenetic studies of presenilin and prion protein function will permit these diverse and overlapping pathogenic processes to be arranged into a meaningful hierarchy.

ACKNOWLEDGEMENTS

This work was supported by grants from the Medical Research Council of Canada, Canadian Genetic Diseases Network - NCE Program, Alzheimer Association of Ontario, EJLB Foundation, American Health Assistance Foundation, the Bayer/Canadian Red Cross Society Research Fund, the National Institute on Aging (AG106812), US Public Health Service, and the Fraternal Order of Eagles.

REFERENCES

Alzheimer, D. C. G. (1995). The structure of the presenilin 1 (S182) gene and identification of six novel mutations in early onset AD families. *Nature Genetics* **11**: 219-222.

Arriagada, P. V., J. H. Growdon, E. T. Hedley-Whyte and B. T. Hyman (1992). Neurofibrillary tangles but not senile plaques parallel duration and severity of Alzheimer's Disease. *Neurology* **42**: 631-639.

Askanas, V., R. B. Alvarez, M. Mirabella and W. K. Engel (1996). Use of Anti-Neurofilament antibody to identify paired-helical filaments in Inclusion-Body Myositis. *Ann. Neurol* **39**: 389-391.

Askanas, V., M. Bilak, W. K. Engel, R. B. Alvarez, F. Tomé and A. Leclerc (1993). Prion protein is abnormally accumulated in inclusion-body myositis. *NeuroReport* **5**: 25-28.

Askanas, V. and W. K. Engel (1995). New advances in the understanding of sporadic inclusion-body myositis and hereditary inclusion-body myopathies. *Curr. Opinion Rheumatol.* **7**: 486-496.

Askanas, V., W. King Engel and R. B. Alvarez (1992). Light and electron microscopic localization of β-amyloid protein in muscle biopsies of patients with inclusion-body myositis. *Am. J. Pathol.* **141**: 31-36.

Askanas, V., J. McFerrin, S. Basque, R. B. Alvarez, E. Sarkozi and W. K. Engel (1996). Transfer of beta-amyloid precursor protein gene using adenovirus vector causes mitochondrial abnormalities in cultured normal human cells. *PNAS* **93**: 1314-1319.

Borchelt, D. R., G. Thinakaran, C. B. Eckman, M. K. Lee, F. Davenport, T. Ratovitsky, C.-M. Prada, G. Kim, S. Seekins, D. Yager, H. H. Slunt, R. Wang, M. Seeger, A. I. Levey, S. E. Gandy, N. G. Copeland, N. A. Jenkins, D. L. Price, S. G. Younkin and S. S. Sisodia (1996). Familial Alzheimer's Disease-Linked presnilin 1 Variants elevate Aβ 1-42/1-40 ratio In Vitro and In Vivo. *Neuron* **17**: 1005-1013.

Brown, P., C. J. Gibbs, Jr., P. Rodgers-Johnson, D. M. Asher, M. P. Sulima, A. Bacote, L. G. Goldfarb and D. C. Gajdusek (1994). Human spongiform encephalopathy: the National Institutes of Health series of 300 cases of experimentally transmitted disease. *Ann. Neurol.* **35**: 513-529.

Brown, P., F. Jannotta, C. J. Gibbs, Jr., H. Baron, D. C. Guiroy and D. C. Gajdusek (1990). Coexistence of Creutzfeldt-Jakob disease and Alzheimer's disease in the same patient. *Neurology* **40**: 226-228.

Bugiani, O., G. Giaccone, L. Verga, B. Pollo, B. Frangione, M. R. Farlow, F. Tagliavini and B. Ghetti (1993). βPP participates in PrP-amyloid plaques of Gerstmann-Straussler-Scheinker disease, Indiana kindred. *J. Neuropathol. Exp. Neurol.* **52**: 64-70.

Bugiani, o., F. Tagliavini and G. Giaccone (1991). Amyloid deposits and senile plaques in Alzheimer's disease, Down Syndrome, and Aging. *Ann NY Acad Sci* **640**: 122-127.

Carpenter, S. (1996). Inclusion Body Myositis, a review. *J. Neuropathology and Experimental Neurology* **55**: 1105-1114.

Citron, M., D. Westaway, W. Xia, G. A. Carlson, T. Diehl, G. Levesque, K. Johnson-Wood, M. Lee, P. Seubert, A. Davis, D. Kholodenko, R. Motter, R. Sherrington, B. Perry, H. Yao, R. Strome, I. Lieberburg, J. Rommens, S. Kim, D. Schenk, P. Fraser, StGeorge-Hyslop and D. Selkoe (1997). Mutant presenilins of Alzheimer's Disease increase production of 42-residue amyloid β-protein in both transfected cells and transgenic mice. *Nature Medicine* **3**: 67-72.

DeArmond, S. J. (1993). Alzheimer's disease and Creutzfeldt-Jakob disease: overlap of pathogenic mechanisms. *Current Opinion in Neurobiology* **6**: 872-881.

Duff, K., C. Eckman, C. Zehr, X. Yu, C.-M. Prada, J. Perez-tur, M. Hutton, L. Buee, Y. Harigaya, D. Yager, D. Morgan, M. Gordon, L. Holcomb, L. Refolo, B. Zenk, J. Hardy and S. Younkin (1996). Increased amyloid-β42(43) in brains of mice expressing presenilin 1. *Nature* **383**: 710-713.

El Hachimi, K. H., L. Cervenakova, P. Brown, L. Goldfarb, R. Rubenstein, D. C. Gajdusek and J.-F. Foncin (1996). Mixed features of Alzheimer disease and creutzfeldt-Jakob disease in a family with a presenilin 1 mutation in chromosome 14. *Amyloid: Int. J. Exp. Clin. Invest.* **3**: 223-233.

Foncin, J.-F., D. Salmon, V. Supino-Viterbo, R. G. Feldman, G. Macchi, P. Mariotti, C. Scopetta, G. Caruso and A. C. Bruni (1985). Alzheimer's Presenile dementia transmitted in an extended kindred. *Rev. Neurol.(Paris)* **141**: 194-202.

Forloni, G., N. Angeretti, R. Chiesa, E. Monzani, M. Salmona, O. Bugiani and F. Tagliavini (1993). Neurotoxicity of a prion protein fragment. *Nature* **362**: 543-546.

Fraser, P., J. Nguyen, H. Inouye, W. K. Surewicz and D. J. Selkoe (1992). Fibril formation by primate rodent and Dutch-hemorrhagic analogues of Alzheimer amyloid B-protein. *Biochemistry* **31**: 10716-10723.

Frommelt, P., R. Schnabel, W. Kuhne, L. E. Nee and R. J. Polinsky (1991). Familial Alzheimer Disease: a large multigenerational German kindred. *Alzheimer Dis. Assoc. Disorders* **5**: 36-43.

Gaches, J. and V. Supino-Viterbo (1977). Association de maladies d'Alzheimer et de Creutzfeldt-Jakob. *Acta Neurol. Belg.* **77**: 202-212.

Gajdusek, D. C. (1977). Unconventional viruses and the origin and disappearance of kuru. *Science* **197**: 943-960.

Games, D., et al. (1995). Alzheimer-type neuropathology in transgenic mice overexpressing V717F beta-amyloid precursor protein. *Nature* **373**: 523-527.

Ghetti, B., P. Piccardo, M. G. Spillantini, Y. Ichimiya, M. Porro, F. Perini, T. Kitamoto, J. Tateishi, C. Seiler, B. Frangione, O. Bugiani, G. Giaccone, F. Prelli, M. Goedert, S. R. Dlouhy and F. Tagliavini (1996). Vascular variant of prion protein cerebral amyloidosis with τ-positive neurofibrillary tangles: The phenotype of a stop codon145 mutation in *PRNP. Proc. Natl.Acad. Sci. U.S.A.* **93**: 744-748.

Ghetti, B., F. Tagliavini, C. L. Masters, K. Beyreuther, G. Giaccone, L. Verga, M. R. Farlo, P. M. Conneally, S. R. Dlouhy, B. Azzarelli and O. Bugiani (1989). Gerstmann-Sträussler-Scheinker disease. II. Neurofibrillary tangles and plaques with PrP-amyloid coexist in an affected family. *Neurology* **39**: 1453-1461.

Goate, A. M., M.-C. Chartier-Harlin, M. Mullan, J. Brown, F. Crawford, L. Fidani, L. Guiffra, A. Haynes and J. A. Hardy (1991). Segregation of a missense mutation in the amyloid precursor protein gene with Familial Alzheimer Disease. *Nature* **349**: 704-706.

Goldfarb, L. G., E. Mitrova, P. Brown, B. H. Toh and D. C. Gajdusek (1990). Mutation in codon 200 of scrapie amyloid protein gene in two clusters of Creutzfeldt-Jakob disease in Slovakia. *Lancet* **336**: 514-515.

Grosveld, F., G. B. van Assendelft, D. R. Greaves and G. Kollias (1987). Position-independent, high-level expression of the human ß-globin gene in transgenic mice. *Cell* **51**: 975-985.

Haltia, M., M. Viitanen, R. Sulkava, V. Ala-Hurula, M. Poyhonen, L. G. Goldfarb, P. Brown, E. Levy, H. Houlden, R. Crook, A. Goate, R. Clark, K. Korenblat, s. Pandir, D. Keller, L. Lilius, L. Lui, K. Axelman, L. Forsell, B. Winblad, L. Lannfelt and J. Hardy (1994). Chromosome14-encoded Alzheimer's disease; Genetic and clincopathological description. *Ann. neurol* **36**: 362-367.

Hardy, J. A. and G. A. Higgins (1992). The amyloid cascade hypothesis. *Science* **256**: 184-85.

Hsiao, K., H. F. Baker and T. J. Crow (1989). Linkage of a prion protein missense variant to Gerstmann-Straussler Syndrome. *Nature* **338**: 342-345.

Hsiao, K., H. F. Baker, T. J. Crow, M. Poulter, F. Owen, J. D. Terwilliger, D. Westaway, J. Ott and S. B. Prusiner (1989). Linkage of a prion protein missense variant to Gerstmann-Sträussler syndrome. *Nature* **338**: 342-345.

Hsiao, K., S. Dlouhy, M. R. Farlow, C. Cass, M. Da Costa, M. Conneally, M. E. Hodes, B. Ghetti and S. B. Prusiner (1992). Mutant prion proteins in Gerstmann-Sträussler-Scheinker disease with neurofibrillary tangles. *Nat. Genet.* **1**: 68-71.

Hsiao, K., Z. Meiner, E. Kahana, C. Cass, I. Kahana, D. Avrahami, G. Scarlato, O. Abramsky, S. B. Prusiner and R. Gabizon (1991). Mutation of the prion protein in Libyan Jews with Creutzfeldt-Jakob disease. *N. Engl. J. Med.* **324**: 1091-1097.

Hsiao, K. H., P. Chapman, S. Nilsen, C. Eckman, Y. Harigawa, S. Younkin, F. S. Yang and G. Cole (1996). Correlative memory deficits, a-beta elevation, and amyloid plaques in transgenic mice. *Science* **274**: 99-102.

Hsiao, K. K., D. R. Borchelt, K. Olson, R. Johannsdottir, C. Kitt, W. Yunis, S. Xu, C. Eckman, S. Younkin, D. Price, C. Iadecola, H. B. Clark and G. A. Carlson (1995). Age-related CNS disorder and early death in transgenic FVB/N mice overexpressing Alzheimer amyloid precursor proteins. *Neuron* **15**: 1203-1218.

Ikeda, S.-i., N. Yanagisawa, D. Allsop and G. G. Glenner (1994). Gerstmann-Straussler-Scheinker diease showing β-protein type cerebellar and cerebral amyloid angiopathy. *Acta Neuropathol.* **88**: 262-266.

L'Hernault, S. W. L. and P. M. Arduengo (1992). Mutation of a putative sperm membrane protein in Caenohabitis elegans prevents sperm differentiation but not its associated meiotic divisions. *J.Cell Biol.* **119**: 55-69.

Lemere, C. A., F. Lopera, K. Kosik, C. L. Lendon, J. Ossa, T. C. Saido, H. Yamaguchi, A. Ruiz, A. Martinez, L. Madrigal, L. Hincapie, J. C. Arango I, D. C. Anthony, E. H. Koo, A. M. Goate, D. J. Selkoe and J. C. Arango V (1996).

The E280A presenilin 1 Alzheimer mutation produces increased Abeta42 deposition and severe cerebellar pathology. *Nature medicine* **2**: 1146-1150.

Levy-Lahad, E., W. Wasco, P. Poorjaj, D. M. Romano, J. Oshima, W. H. Pettingell, C.-e. Yu, P. D. Jondro, S. D. Shmidt, K. Wang, A. C. Crowley, Y.-H. Fu, S. Y. Guenette, D. Galas, e. Nemens, E. M. Wijsman, T. D. Bird, G. D. Schellenberg and R. E. Tanzi (1995). Candidate gene for the chromosome 1 familial Alzheimer's Disease locus. *Science* **269**: 973-977.

Liberski, P. P., W. Papeerz and J. Alwasiak (1987). Creutzfeldt-Jakob disease with plaques and paried helical filaments. *Acta Neurol. Scand.* **76**: 428-432.

Manuelidis, E. E., J. M. de Figueiredo, J. H. Kim, W. W. Fritch and L. Manuelidis (1988). Transmission studies from blood of Alzheimer disease patients and healthy relatives. *Proc. Natl. Acad. Sci. USA* **85**: 4898-4901.

Martin, R. N. e. a. (1995). High levels of amyloid beta-protein from S182 (Glu246) familial Alzheimer's cells. *Neuroreport* **7**: 217-220.

Masliah, E., L. Hansen, T. Albright, M. Mallory and R. D. terry (1991). Immunoelectron microscopic study of synaptic pathology in Alzheimer's disease. *Acta Neuropathol (Berl)* **81**(428-433).

Masliah, E., R. D. Terry, M. Alford, R. M. DeTeresa and L. A. Hansen (1991). Cortical and subcortical patterns of synaptohysin-like immunoreactivity in Alzherimer's disease. *Am. J. Pathol.* **138**: 235-246.

Mendell, J. R., Z. Sahenk, T. Gales and L. Paul (1991). Amyloid filaments in inclusion body myositis. *Arch. Neurol.* **48**: 1229-1234.

Moser, M., Colello, R.J., Pott, U., and Oesch, B. (1995). Developmental Expression of the Prion Protein in Glial Cells. *Neuron* **14**: 509-517.

Mullan, M., H. Houlden, M. Windelspecht, L. Fidani, C. Lombardi, P. Diaz, M. Rossor, R. Crook, J. Hardy, K. Duff and F. Crawford (1992). A locus for familial early-onset Alzheimer's disease on the long arm of chromosome 14, proximal to the α-1-antichymotrypsin gene. *Nat. Genet.* **2**: 340-342.

Muramoto, T., T. Kitamoto, H. Koga and J. Tateishi (1992). The coexistence of Alzheimer's disease and Creutzfeldt-Jakob disease in a patient with dementia of long duration. *Acta Neuropathol.* **84**: 686-689.

Nalbantoglu, J., G. Karpati and S. Carpenter (1994). Conspicuous accumulation of a singel-stranded DNA binding protein in skeletal muscel fibers in Inclusio Body Myositis. *Am. J. Pathol.* **144**: 874-882.

Parchi, P. and P. Gambetti (1995). Human prion diseases. *Current Opinion in Neurobiology* **8**: 286-293.

Powers, J. M., Y. Liu, S. Hair, R. J. Kascsak, L. D. Lewis and L. A. Levy (1991). Concomitant Creutzfeldt-Jakob and Alzheimer diseases. *Acta Neuropathol.* **83**: 95-98.

Prusiner, S. B., M. Scott, D. Foster, K.-M. Pan, D. Groth, C. Mirenda, M. Torchia, S.-L. Yang, D. Serban, G. A. Carlson, P. C. Hoppe, D. Westaway and S. J. DeArmond (1990). Transgenetic studies implicate interactions between homologous PrP isoforms in scrapie prion replication. *Cell* **63**: 673-686.

Rogaev, E. I., R. Sherrington, E. A. Rogaeva, G. Levesque, M. Ikeda, Y. Liang, H. Chi, C. Lin, K. Holman, T. Tsuda, L. Mar, S. Sorbi, B. Nacmias, S. Piacentini, L. Amaducci, I. Chumakov, D. Cohen, L. Lannfelt, P. E. Fraser, J. M. Rommens and P. St George-Hyslop (1995). Familial Alzheimer's disease in kindreds with missense mutations in a novel gene on chromosome 1 related to the Alzheimer's Disease type 3 gene. *Nature* **376**: 775-778.

Rogaev, E. I., R. Sherrington, C. Wu, G. Levesque, Y. Liang, E. A. Rogaeva, M. Ikeda, K. Holman, C. Lin, W. J. Lukiw, P. J. de Jong, P. E. Fraser, J. M. Rommens and P. St George-Hyslop (1997). Analysis of the 5' sequence, genomic structure, and alternative splicing of the *presenilin-1* gene (*PSEN1*) associated with Early Onset Alzheimer Disease. *Genomics* **40**: 415-424.

Sarkozi, E., V. Askanas and W. K. Engel (1994). Abnormal accumulation of prion protein mRNA in muscle fibers of patients with sporadic Inclusion-Body Myositis and hereditary Inclusion-Body Myopathy. *Am. J. Pathol.* **145**: 1280-1284.

Scheuner, D., J. Eckman, M., X. Song, M. Citron, N. Suzuki, T. D. Bird, J. Hardy, M. Hutton, W. Kukull, E. Larson, E. Levy-Lahad, M. Viitanen, E. Peskind, P. Poorkaj, G. Schellenberg, R. Tanzi, W. wasco, L. Lannfelt, D. Selkoe and S. Younkin (1996). Secreted amyloid beta-protein similar to that in the senile plaques of Alzheimer's disease is increased in vivo by the presenilin 1 and 2 and APP mutations linked to familial Alzheimer's disease. *Nature medicine* **2**: 864-870.

Scott, M., D. Groth, D. Foster, M. Torchia, S.-L. Yang, S. J. DeArmond and S. B. Prusiner (1993). Propagation of prions with artificial properties in transgenic mice expressing chimeric PrP genes. *Cell* **73**: 979-988.

Scott, M. R., R. Köhler, D. Foster and S. B. Prusiner (1992). Chimeric prion protein expression in cultured cells and transgenic mice. *Protein Sci.* **1**: 986-997.

Sherriff, F., C. L. Joachim, M. V. Squier and M. M. Esiri (1995). Ubiquitinated inclusions in inclusion-body myositis patients are immunoreactive for cathepsin D but not β-amyloid. *Neuroscience Letters* **194**: 37-40.

Sherrington, R., E. Rogaev, Y. Liang, E. Rogaeva, G. Levesque, M. Ikeda, H. Chi, C. Lin, K. Holman, T. Tsuda, L. Mar, P. Fraser, J. M. Rommens and P. St George-Hyslop (1995). Cloning of a gene bearing missense mutations in early onset familial Alzheimer's disease. *Nature* **375**: 754-760.

St George-Hyslop, P., et al. (1992). Genetic Evidence for a novel Familial Alzheimer Disease gene on chromosome 14. *Nature Genetics* **2**: 330-334.

Strittmatter, W. J., A. M. Saunders, D. Schmechel, D. Goldgaber, A. D. Roses and e. al. (1993). Apolipoprotein E: high affinity binding to B/A4 amyloid and increased frequency of type 4 allele in Familial Alzheimers Disease. *Proc.Natl.Acad.Sci.USA* **90**: 1977-1981.

Telling, G. C., M. Scott, J. Mastrianni, R. Gabizon, M. Torchia, F. E. Cohen, S. J. DeArmond and S. B. Prusiner (1995). Prion Propagation in Mice Expressing Human and Chimeric PrP transgenes implicates the Interaction of Cellular PrP with Another Protein. *Cell* **83**: 79-90.

Thinakaran, G., D. R. Borchelt, M. K. Lee, H. H. Slunt, L. Spotzer, G. Kim, T. Ratovitsky, F. Davenport, C. Nordstedt, M. Seeger, J. Hardy, A. I. Levey, S. E. Gandy, N. A. Jenkins, N. G. Copeland, D. L. Price and S. S. Sisodia (1996). Endoproteolysis of Presenilin 1 and accumulation of processed derivatives in vivo. *Neuron* **17**: 181-190.

Van Broeckhoven, C., H. Backhovens, M. Cruts, G. De Winter, M. Bruyland, P. Cras and J.-J. Martin (1992). Mapping of a gene predisposing to early-onset Alzheimer's disease to chromosome 14q24.3. *Nat. Genet.* **2**: 335-339.

Weidemann, A., K. Paliga, U. Durrwang, C. Czech, G. Evin, C. L. Masters and K. Beyreuther (1997). Formation of stable complexes between two Alzheimer's disease gene products: Presenilin-2 and β-amyloid precursor protein. *Nature medicine* **3**.

Westaway, D., S. J. DeArmond, J. Cayetano-Canlas, D. Groth, D. Foster, S.-L. Yang, M. Torchia, G. A. Carlson and S. B. Prusiner (1994). Degeneration of skeletal muscle, peripheral nerves, and the central nervous system in transgenic mice overexpressing wild-type prion proteins. *Cell* **76**: 117-129.

Yehiely, F., P. Bamborough, M. Da Costa, B. J. Perry, G. Thinakaran, F. E. Cohen, G. A. Carlson and S. B. Prusiner (1997). Identification of candidate proteins binding to prion protein. *Neurobiology of Disease* **3**: 339-355.

POLYENE ANTIBIOTICS IN EXPERIMENTAL TRANSMISSIBLE SUBACUTE SPONGIFORM ENCEPHALOPATHIES

Vincent Beringue,[1] Rémi Demaimay,[1] Karim T. Adjou,[1]
Séverine Demart,[1] François Lamoury,[1] Michel Seman,[2]
Corinne I. Lasmézas,[1] Jean-Philippe Deslys,[1]
and Dominique Dormont[1]

[1]Service de Neurovirologie, DRM/DSV/SSA
Commissariat à l'Energie Atomique
B.P. 6, 92 265 Fontenay aux Roses Cedex, France

[2]Laboratoire d'Immunodifférenciation
Institut Jacques Monod, Université de Paris VII
Paris, France

INTRODUCTION

Transmissible subacute spongiform encephalopathies (TSSE) form a group of fatal neurodegenerative diseases which include Creutzfeldt-Jakob disease in humans, scrapie in sheep and goats and bovine spongiform encephalopathy (BSE) in cattle. They are characterized by a long incubation period which precedes clinical symptoms related to the central nervous system (CNS) degeneration. In experimental models, a modified proteinase-resistant form (PrPres) of a host-encoded protein (PrP) accumulates in the brain proportionally to infectivity[1,2]. A transcriptional overexpression of glial fibrillary acidic protein (GFAP), a specific marker of astrocytes, has also been observed[3,4,5].

Several treatments have been performed in TSSE[6]. Only three classes of drugs increase survival time of rodents experimentally infected with TSSE agents : polyanions,[7,8] the amyloid-binding dye Congo red,[9] and a polyene antibiotic amphotericin B, which is the most efficient one[10,11,12,13]. Amphotericin B (AmB) prolongs incubation period of scrapie infected hamsters, although the duration of the clinical phase and the clinical signs remain identical, and delays PrPres accumulation in the brain. The anti-scrapie effect of AmB is dose-dependant and is correlated to the period of treatment. AmB administration during the early steps of infection seems to be the best regimen : a treatment performed 4 weeks after the inoculation is unefficient in prolonging survival time of intracerebrally (ic) infected hamsters[11,13]. These previous reports indicated that AmB effects were restricted to the hamster model infected with the 263K scrapie strain. It may be due to the toxicity of AmB, which limits its administration : the maximal tolerated dose of AmB is around 3 mg/kg in rodents[14,15] and most of the experiments in experimental scrapie were done at a dose of 1 mg/kg.

The prion hypothesis supposes an identity between PrPres and infectivity[1]. In scrapie infected hamsters, AmB treatment prolongs survival time, delays significantly PrPres accumulation in the brain but it remains a controversial discrepancy whether AmB delays brain infectivity or not[16,17]. Once again, this doubt could be explained by the low dose of AmB (1 mg/kg) used in the treatment of hamsters.

MS-8209 is a polyene macrolide antibiotic derived from AmB. It exhibits similar antifungal and antiviral properties[18,19] but has a greater water solubility and is at least five times less toxic[20]. Thus, MS-8209 could be a useful drug for chemotherapeutic studies in experimental scrapie.

Experiments reported here were designed 1) to compare the clinical and molecular effects of MS-8209 with that of AmB in different TSSE experimental models (scrapie in hamsters and mice and BSE in mice), 2) to investigate the effects of polyene antibiotics on scrapie agent replication in the brain, 3) to extend AmB derivatives effects to the late stages of experimental TSSE.

Thus, various treatments were performed at the time of inoculation or at the preclinical stages of hamster and mouse experimental TSSE and the major characteristics of TSSE were analyzed : incubation period, PrPres accumulation, GFAP gene expression and infectious titers in the brain.

MATERIALS AND METHODS

Animals

Mice Inoculation. The mouse scrapie strain C506M3 (7.9×10^8 50% lethal dose/g of brain, D.C. Gajdusek, USA) and the mouse-adapted BSE strain[21,22] were obtained from brain homogenates of terminally ill infected animals. C57BL/6 mice (Centre d'Elevage René Janvier, Le Genest-Saint-Isle, France) were infected intracerebrally or intraperitoneally (ip) with 20 µl and 80 µl respectively of 1% (w/v) brain homogenate.

Hamsters Inoculation. The hamster scrapie strain 263K (4.4×10^8 50% lethal doses/g of brain, H. Fraser, UK)) was obtained from brain homogenates of terminally ill infected animals. Female golden Syrian hamsters (Centre d'Elevage René Janvier, Le Genest-Saint-Isle, France) were infected by the ic route with 50 µl of 1% (w/v) brain homogenate.

Mice and hamsters were regularly monitored for the determination of the onset of clinical symptoms. Sacrifices were performed by cervical column disruption. CNS (including the cerebellum) was dissected, rapidly frozen and kept at -80°C until use.

Infectious titers were estimated by the end point dilution method in hamsters[23]. In mice, incubation period was mesured in recipient mice and was compared to a standard curve of incubation time versus infectious titer[24].

Drugs

AmB (Fungizone®, Squibb), MS-8209 (N-methyl glucamine salt of 1-deoxy-1-amino-4,6,O-benzylidine-D-fructosyl-AmB, Mayoly-Spindler laboratories, Chatou, France) were suspended in a 5% glucose (w/v) sterile solution. All animals were treated by the intraperitoneal route 6 days a week.

Protein Analysis

Briefly, PrPres was isolated by proteinase K digestion, as previously described by Adjou *et al.*[25]. Mouse and hamster PrPres were visualized by immunoblot using respectively a polyclonal antiPrP Antibody 007JB[26] and the monoclonal 3F4 (kindly provided by R.J. Kascsak, USA). Immunodetections were carried out with an Enhanced Chemiluminescence kit (ECL®, Amersham) and signals were quantified with a Radio Imager (Molecular Imager®, Biorad).

Northern-blot

Total RNAs were extracted and size fractionated on a 0.6 M formaldehyde / 1% agarose gel. RNAs were blotted onto a nylon membrane (Schleicher & Schuell) using the Vacugene system (Pharmacia, LKB) and cross-linked under 260 nm U.V. irradiation. Blots were hybridized with (α-^{32}P) dCTP (110 MBq/mM, Amersham) labelled glyceraldehyde-3-

phosphate-dehydrogenase (GAPDH), GFAP or PrP probes (10^8 cpm/µg). Stringent washing procedures included three successive 20 min incubations at 42°C in 0.1 % SDS and respectively 2X, 1X and 0.1X SSPE (SSPE 1X : NaCl 174 g/l, Na$_2$HPO$_4$ 27.6 g/l, pH 7.4). Signals were quantified with a Radio Imager (Biorad) and normalized by comparison with the GAPDH signals.

RESULTS

Polyene Antibiotics Treatments at the Early Stages of the Disease

Early Treatments of Hamster Scrapie. Initial treatments with MS-8209 and AmB were performed around the time of infection, in hamster scrapie. AmB effects (at 2.5 mg/kg during one week) were similar to published results in ic 263K infected hamsters[11]. MS-8209, at the same dose is as efficient as AmB in prolonging the incubation time of the disease[27], the duration of the clinical phase remaining unchanged compared with untreated animals. As an improved action of MS-8209 should be possible with higher doses, a long term treatment (i.e. from the day of inoculation until death) with 10 and 25 mg/kg MS-8209 was performed. These MS-8209 concentrations could increase hamster survival time by 100% (table 1), a delay which has never been observed with AmB[25].

Table 1. Polyene antibiotics efficiency in prolonging the incubation time of ic infected rodents. Results are expressed as an increase in percentage of incubation time in infected untreated animals

		treatment schedule	AmB (2.5 mg/kg)	MS-8209 (2.5 mg/kg)	MS-8209 (10 mg/kg)	MS-8209 (25 mg/kg)
Long term treatment						
	hamster	0 - †	+ 66%	+ 75%	+ 100%	+ 100%
	mouse	0 - †	+ 31%	+ 26%	ND	ND
8 weeks treatment						
	mouse (scrapie)	0 - 56 dpi	+ 10%	+ 15%	ND	ND
	mouse (BSE)		+ 3.6%	NS	+ 6%	+ 3%
Preclinical treatment						
	mouse	80 - †	+ 24%	+ 25%	ND	+ 40%
	mouse	110 - †	+ 15%	+ 14%	ND	+ 20%
	mouse	140 - †	+ 5%	+ 4%	ND	+ 9%
	hamster	30 - †	+ 32%	ND	+ 45%	ND

ND : not done.
NS : not significant.
† : death.

dpi : days post-inoculation.
BSE : bovine spongiform encephalopathy.

In these treatments, molecular analyses were done at the time of clinical signs for untreated animals. Both AmB and MS-8209 induced a delay in PrPres accumulation in the brain, which correlated with the delay observed in animals survival time. Furthermore, a delay in GFAP accumulation, proportionately to that of PrPres was noticed[25,27]. Thus, the anti-scrapie effects of AmB have been extended to a new derivative, MS-8209, which is less toxic and permits to double AmB efficiency at high doses (table 1). Then, the use of higher doses of MS-8209 could contribute, better than AmB, to study PrPres accumulation in parallel with TSSE agent infectivity in the brain of infected treated animals.

In hamsters, after a long term treatment with AmB (2.5 mg/kg) and MS-8209 (2.5 to 25 mg/kg), the greatest difference in PrPres accumulation between untreated and treated animals was observed at 40 days post-inoculation (dpi), as previously reported[16]. Brain infectivity was titrated at this time by end point dilution in recipient hamsters[23]. It was significantly reduced in all infected treated groups compared with that of infected untreated or solvent treated groups (table 2).

Table 2. Reduction of infectious titers in scrapie infected rodents, treated with polyene antibiotics, compared with infected untreated animals. Infectivity was analyzed at 40 dpi in the long term treatment protocol for hamsters, where the greatest difference in PrPres accumulation between infected untreated and infected treated animals is observed. Infectious titers were estimated in mice at 110 dpi, the treatment beginning at the time of neuroinvasion (80 dpi)

	Difference in infectious titers	
Treatment schedule	hamster : 0 - †	mouse : 80 - †
Infectivity analyses	40 dpi	110 dpi
AmB (2.5 mg/kg)	-2.4	-2.0
MS-8209 (2.5 mg/kg)	-2.1	-2.1
MS-8209 (10 mg/kg)	-3.4	ND
MS-8209 (25 mg/kg)	-3.0	ND

ND : not done.
NS : not significant.
† : death.
dpi : days post-inoculation.

For each concentration and each drug, this infectivity diminution correlates with the delay observed in PrPres accumulation (K.T. Adjou, unpublished data). The same results were obtained whatever the inoculum used for recipient hamsters inoculation (brain homogenate or purified PrPres). These results are in accordance for an association between PrPres accumulation and infectivity in the brain[17].

It was also interesting to test MS-8209 effects in scrapie mice model, which seemed to be resistant to AmB treatment[16].

Early Treatments of Mouse Scrapie. Treatments with polyene antibiotics at the time of inoculation were performed in mice infected with the C506M3 scrapie strain. Intracerebrally infected mice had a prolonged incubation time when they were treated until death with AmB or MS-8209 at 2.5 mg/kg (table 1), the duration of the clinical phase and the clinical signs remaining unchanged. Higher doses of MS-8209 permitted to prolong the benefits of AmB. Polyene antibiotics effects were also extended to ip infected mice[27]. As in hamster scrapie, PrPres accumulation in the brain was delayed and correlated with the increase in survival time (figure 1).

Moreover, a delay in GFAP accumulation, proportionately to that of PrPres was noticed[27]. For the first time, polyene antibiotics efficiency could be observed in another animal model. MS-8209 seems to be as efficient as AmB in delaying clinical signs, its lower toxicity allowing the use of higher doses, and therefore prolonging experimental scrapie to a larger extend than AmB.

MS-8209 was then tested in a murine BSE model, to get further insight into the strain specificity of polyene antibiotics treatments.

Polyene Antibiotics Effects on a Mouse-Adapted BSE Strain. An eight weeks treatment was performed in ic infected mice with both the BSE and the scrapie agent[28]. The increase in survival time with MS-8209 (at 10 mg/kg) was only 5% in BSE infected mice, whereas only 2.5 mg/kg MS-8209 increased the survival time of 15% in scrapie infected mice (table 1). At the molecular level, PrPres and GFAP accumulation were reduced in the CNS of scrapie infected and treated animals (when clinical signs appear in infected

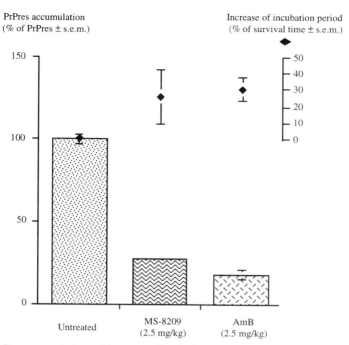

Figure 1. Brain PrPres accumulation and increase of incubation period in C506M3 ic infected mice, treated from the day of inoculation until death by polyene antibiotics.
PrPres accumulation in the brain was quantified at 170 days post-inoculation, at onset of clinical signs in untreated infected animals.

untreated animals) but not in BSE infected mice, which is in agreement with the slight delays observed in survival time[28]. These results underscore the strain specific efficiency of polyene antibiotics treatment at high doses and also show that disease retardation by AmB derivatives is a more general phenomenom.

Polyene Antibiotics Treatments at the Late Stages of the Disease

Pocchiari *et al.* reported that the AmB anti-scrapie effect in infected treated hamsters was limited to the early stages of the infection[11]. It was interesting to look if the unefficiency of AmB at the late stages of the disease was due to its toxicity limiting administration or to a host restriction. We therefore performed treatments with MS-8209 and AmB at preclinical stages of experimental scrapie in mice and hamsters.

Preclinical Treatments of Mouse Scrapie. Treatments were performed until death at 3 different preclinical phases of the ic infected C506M3 murine model : at the time of neuroinvasion when PrPres accumulation is detectable in the brain (80 dpi), at the appearance of histopathological lesions in the brain (110 dpi) and at the prodromic phase of the disease (140 dpi).

All the infected treated groups had a significantly prolonged incubation time compared with the solvent and the infected untreated group whenever the beginning of the treatment (table 1). MS-8209 efficiency was correlated to the length of the treatment and to the dose used (figure 2). It is noteworthy that a preclinical treatment at 80 dpi with MS-8209 (at 2.5 mg/kg) is as efficient as a long term treatment at the same dose beginning at the day of inoculation (table 1).

Molecular analyses were performed at 110 dpi for the treatment with AmB and MS-8209 beginning at 80 dpi. PrPres accumulation was reduced in all infected treated groups as compared with infected untreated and solvent treated animals. This reduction was proportional to the dose used for drug administration and to the delay observed in survival time (figure 3). GFAP gene expression was also reduced (data not shown). At the same time, brain infectivity was also estimated in recipient mice[24] for untreated, AmB and MS-8209 (2.5

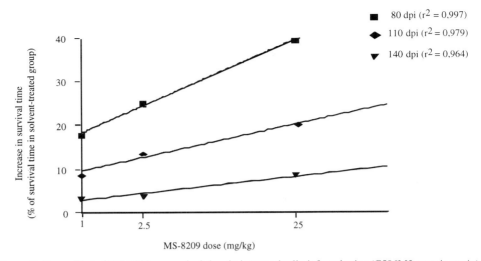

Figure 2. Dose effect of MS-8209 on survival time in intracerebrally infected mice (C506M3 scrapie strain), treated at the preclinical stages of the disease.
Mice were treated at the beginning of detectable PrPres accumulation (80 dpi), at the appearance of histopathological lesions (110 dpi) and at the prodromic phase (140 dpi) of the disease.

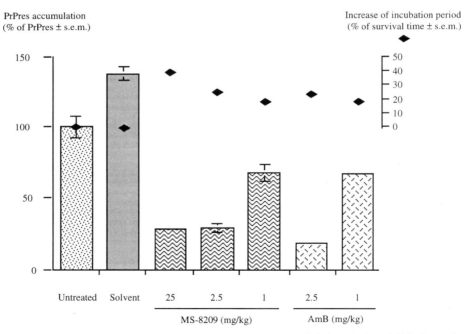

Figure 3. Incubation period and brain PrPres accumulation in intracerebrally infected mice (C506M3 scrapie strain), treated at the preclinical stages of the disease by polyene antibiotics.
Mice were treated until death at the beginning of detectable PrPres accumulation in the brain (80 days post-inoculation). PrPres accumulation in the brain was quantified 30 days after the beginning of the treatment.

mg/kg) treated mice. The apparent infectious titer was significantly lower in treated groups as compared with untreated groups (table 2). Thus, polyene antibiotics may reduce TSSE agent replication in the brain after the onset of neuroinvasion. These results show again that infectivity and PrPres accumulation are tightly associated in experimental scrapie.

Preclinical Treatments of Hamster Scrapie. Experiments are in progress to evaluate the preclinical treatment of hamster scrapie. Intracerebrally infected hamsters have been treated with MS-8209 and AmB at the beginning of detectable PrPres accumulation in the brain (30 dpi). Preliminary results showed that the use of higher doses of both drugs has permitted for the first time to obtain a beneficial anti-scrapie effect (table 1), which has never been observed before[11].

DISCUSSION

Polyene antibiotics prolonged the incubation time both in 263K infected hamsters and in C506M3 infected mice,[25,27] regardless of the inoculation route (ic or ip). More generally, this delay was related to the dose of the drug used and to the length of the treatment. This effect has been extended to another mouse-adapted strain, derived from BSE-infected cattle,[21,22] although the treatment displayed only a low efficiency[28]. These results, obtained particularly with MS-8209, indicate that the interactions between polyene antibiotics and TSSE agents are not limited to one strain in one animal model. Evaluation of the efficiency of AmB derivatives for other scrapie strains, which have been formerly described as being resistant to AmB treatment[29,30,31] would permit completion of this study.

These drugs are efficient, not only when administered close to the experimental infection, but also later during the course of the infection, both in 263K infected hamsters and in C506M3 infected mice models, when neuroinvasion has already occurred and replication of the agent in lymphoid tissues may be no more of pathogenetical significance[32,33,34]. Moreover, polyene antibiotics treatments in hamster and mouse scrapie, when administered around the time of infection or at the neuroinvasion, delay PrPres accumulation, GFAP gene expression and infectious titers in the brain[17,25,26,27]. Polyene antibiotics interfere, by an as yet unknown mechanism, both with the TSSE agent replication and PrPres accumulation in experimentally scrapie infected rodents.

Polyanions are another class of anti-scrapie drugs which have been also extensively studied. They also decrease infectious titer and delay PrPres accumulation in mouse scrapie,[35] by an unknown mechanism. But they are only efficient when their administration occurs prior or very shortly after the time of infection[7,8,35,36,37]. Thus, the efficacy of polyene antibiotics in TSSE is stongly supported by their efficiency during all the preclinical stages of the disease.

These results show that polyene antibiotics could act on the pathogenetical mechanisms leading to TSSE, at the time of inoculation when performed intracerebrally or at the preclinical stages (from the neuroinvasion to the prodromic phase in mice). Single and multiple doses of AmB administered in rodents have shown that the drug concentrates mainly in peripheral organs (liver, spleen and kidney) and is scarcely detectable in the brain and in cerebro-spinal fluid[14,38]. Moreover, the metabolism of AmB has been poorly described and no AmB metabolites have been identified in animals organs[38]. In experimental TSSE, the blood-brain barrier (BBB) is disrupted at the clinical phase of the disease[39]. The pathogenetical processes occuring consequently to neuroinvasion may also damage the BBB and permit AmB or their eventual metabolites to cross the barrier and be efficient. Thus, a further step would be reached by looking if both TSSE agents and polyene antibotics have similar targets in the CNS areas. Then, several hypotheses could be verified for the mechanisms of action of these drugs, based on the properties of AmB : 1) AmB interacts with cholesterol and modifies the lipid composition of mammalian membranes[40,41]. Thus, polyene antibiotics could modify the conformation of a hypothetical receptor molecule for the TSSE agent on the target cell or could interfere with the interaction TSSE agent / target cell. This receptor is unknown but the best candidate is cellular PrP[42], 2) because polyene antibiotics delay PrPres accumulation in the brain, they may have a direct effect on the conversion of PrP into PrPres. This could happen by a decrease in PrP expression at the cell surface, due to the destabilizing properties of AmB for the cell membrane or by a direct interaction during the conversion process. The location of polyene antibiotics and their potential metabolites in the brain of experimentally scrapie infected rodents may permit a better understanding of how they act on TSSE pathophysiology.

Polyene antibiotics represent today a powerful tool to dissect TSSE pathogenesis. They permit the study of relationships between PrPres accumulation, scrapie agent replication and the onset of the disease, in different host-agent strain combinations. Further experiments are needed to look if these major characteristics of experimental TSSE are always strongly associated or not. It will be also interesting to look at their effects *in vivo* on the classical neuropathological changes in the CNS of TSSE infected animals (spongiosis, neuronal loss and astrogliosis). Indeed, *in vitro*, studies with MS-8209, which is less toxic than AmB, may permit the study of interactions between polyene antibiotics, normal PrP and *de novo* synthetized PrPres in order to investigate their relationships with the conversion of PrP into PrPres.

Acknowledgments

We would like to thank Dr K. Cherifi and Mayoly-Spindler laboratories for kindly provided MS-8209, and especially R. Rioux and J-C. Mascaro for animal care. This work was made possible by a grant from DRET (Paris, France) and the support of ACC-SV 10 of the French Ministry of Research.

REFERENCES

1. S. B. Prusiner, Novel proteinaceous infectious particles cause scrapie, *Science*. 216:136 (1982).
2. D. C. Bolton, M. P. McKinley, and S. B. Prusiner, Identification of a protein that purifies with the scrapie prion, *Science*. 218:1309 (1982).
3. D. Dormont, B. Delpech, A. Delpech, M. N. Courcel, J. Viret, P. Markovits, and L. Court, Hyperproduction de protéine gliofibrillaire acide (GFAP) au cours de l'évolution de la tremblante expérimentale de la souris, *C R Acad Sc Paris*. 293:53 (1981).
4. F. Lazarini, J. P. Deslys, and D. Dormont, Variations in prion protein and glial fibrillary acidic protein messenger RNAs in the brain of scrapie-infected newborn mouse, *J Gen Virol*. 73:1645 (1992).
5. A. Mackenzie, Immunohistochemical demonstration of glial fibrillary acidic protein in scrapie, *J Comp Pathol*. 93:251 (1983).
6. P. Brown, A therapeutic panorama of the spongiform encaphalopathies, *Antiviral Chemistry & Chemoth.* 1:75 (1990).
7. R. H. Kimberlin, and C. A. Walker, The antiviral compound HPA-23 can prevent scrapie when administred at the time of infection, *Arch Virol*. 78:9 (1983).
8. R. H. Kimberlin, and C. A. Walker, Suppression of scrapie infection in mice by heteropolyanion 23, dextran sulfate, and some other polyanions, *Antimicrob Agents Chemother*. 30:409 (1986).
9. L. Ingrosso, A. Ladogana, and M. Pocchiari, Congo red prolongs the incubation period in scrapie-infected hamsters, *J Virol*. 69:506 (1995).
10. H. Amyx, A. M. Salazar, D. C. Gajdusek, and C. J. Gibbs, Chemotherapeutic trials in experimental slow virus disease, *Neurology*. 34:149 (1984).
11. M. Pocchiari, S. Schmittinger, A. Ladogana, and C. Masullo, Effects of Amphotericin B in intracerebrally scrapie inoculated hamster, in *Unconventional Virus Diseases of the Central Nervous System*, L.A. Court, D. Dormont, P. Brown and D.T. Kingsbury, eds., C.E.A. diffusion, Paris (1986).
12. M. Pocchiari, S. Schmittinger, and C. Masullo, Amphotericin B delays the incubation period of scrapie in intracerebrally inoculated hamsters, *J Gen Virol*. 68:219 (1987).
13. M. Pocchiari, P. Casaccia, and A. Ladogana, Amphotericin B : a novel class of antiscrapie drugs, *J Infect Dis*. 160:795 (1989).
14. R. T. Proffitt, A. Satorius, S. M. Chiang, L. Sullivan, and J. P. Adler-Moore, Pharmacology and toxicology of a liposomal formulation of amphotericin B (Ambisome) in rodents, *J Antimicrob Chemother*. 28, Suppl. B:49 (1991).
15. E. S. Tabosa Do Egito, M. Appel, H. Fessi, G. Barratt, F. Puisieux, and J. P. Devissaguet, *In-vitro* and *in-vivo* evaluation of a new amphotericin B emulsion-based delivery system, *J Antimicrob Chemother*. 38:485 (1996).
16. Y. G. Xi, L. Ingrosso, A. Ladogana, C. Masullo, and M. Pocchiari, Amphotericin B treatment dissociates *in vivo* replication of the scrapie agent from PrP accumulation, *Nature*. 356:598 (1992).
17. D. McKenzie, J. Kaczkowski, R. Marsh, and J. Aiken, Amphotericin B delays scrapie agent replication and PrP-res accumulation early in infection, *J Virol*. 68:7534 (1994).
18. M. Magierowskajung, D. Cefai, H. Marrakchi, F. Chieze, H. Agut, J. M. Huraux, and M. Seman, *In vitro* determination of antiviral activity of MS-8209, a new amphotericin B derivative, against primary isolates of HIV 1, *Res Virol*. 147:313 (1996).
19. O. Pleskoff, N. Sol, H. Marrakchi, M. Serlin, M. Seman, and M. Alizon, Possible role of the V3 domain of gp120 in resistance to an Amphotericin B derivative (MS-8209) blocking human immunodefiency virus entry, *J Virol*. 70:8247 (1996).

20. L. Saint-Julien, V. Joly, M. Seman, C. Carbon, and P. Yeni, Activity of MS-8209, a nonester Amphotericin B derivative, in treatment of experimental systemic mycoses, *Antimicrob Agents Chemother.* 36:2722 (1992).
21. C. I. Lasmézas, J. P. Deslys, R. Demaimay, K. T. Adjou, J. J. Hauw, and D. Dormont, Strain specific and common pathogenic events in murine models of scrapie and bovine spongiform encephalopathy, *J Gen Virol.* 77:1601 (1996).
22. C. I. Lasmézas, J. P. Deslys, O. Robain, A. Jaegly, V. Beringue, J. M. Peyrin, J. G. Fournier, J. J. Hauw, J. Rossier, and D. Dormont, Transmission of BSE agent to mice in the absence of detectable abnormal prion protein, *Science.* 275:402 (1997).
23. L. J. Reed, and H. Muench, A simple method of estimating fifty percent endpoints, *Amer J Hygiene.* 27:493 (1938).
24. S. B. Prusiner, S. P. Cochran, D. F. Groth, D. E. Downey, K. A. Bowman, and H. M. Martinez, Measurement of the scrapie agent using an incubation time interval assay, *Ann Neurol.* 11:353 (1982).
25. K. T. Adjou, R. Demaimay, C. I. Lasmézas, J. P. Deslys, M. Seman, and D. Dormont, MS-8209, a new amphotericin B derivative, provides enhanced efficacy in delaying hamster scrapie, *Antimicrob Agents Chemother.* 39:2810 (1995).
26. R. Demaimay, K. T. Adjou, V. Beringue, S. Demart, F. Lamoury, C. I. Lasmézas, J. P. Deslys, M. Seman, and D. Dormont, Polyene antibiotics in experimental transmissible spongiform encephalopathies, in *Transmissible Subacute Spongiform Encephalopathies : Prion Diseases*, L. Court and B. Dodet, eds., Elsevier, Paris (1996).
27. R. Demaimay, K. T. Adjou, C. I. Lasmézas, F. Lazarini, K. Cherifi, M. Seman, J. P. Deslys, and D. Dormont, Pharmalogical studies of a new derivative of Amphotericin B, MS-8209, in mouse and hamster scrapie, *J Gen Virol* 75, 2499 (1994).
28. K. T. Adjou, R. Demaimay, C. I. Lasmézas, M. Seman, J. P. Deslys, and D. Dormont, Differential effects of a new Amphotericin B derivative, MS-8209, on mouse BSE and scrapie: Implications for the mechanism of action of polyene antibiotics, *Res Virol.* 147:213 (1996).
29. R. I. Carp, Scrapie, unconventional infectious agent, in *Neuropathogenic Viruses and Immunity*, S. Specter, ed., Plenum Press, New-York (1992).
30. M. Pocchiari, L. Ingrosso, and A. Ladogana, Effect of Amphotericin B on different experimental strains of spongiform encephalopathy agents, in *Bovine Spongiform Encephalopathy : the BSE Dilemma*, C. J. Gibbs Jr, ed., Springer Verlag, New-york (1996).
31. K. T. Adjou, J. P. Deslys, R. Demaimay, and D. Dormont, Probing the dynamics of prion diseases with Amphotericin B, *Trends Microb.* 5:27 (1997).
32. R. H. Kimberlin, and C. A. Walker, Pathogenesis of mouse scrapie: effect of route of inoculation on infectivity titres and dose-response curves, *J Comp Pathol.* 88:39 (1978).
33. R. H. Kimberlin, and C. A. Walker, Pathogenesis of mouse scrapie: dynamics of agent replication in spleen, spinal cord and brain after infection by different routes, *J Comp Pathol.* 89:551 (1979).
34. R. H. Kimberlin, and C. A. Walker, Pathogenesis of scrapie (strain 263K) in hamsters infected intracerebrally, intraperitoneally or intraocularly, *J Gen Virol.* 67:255 (1986b).
35. H. Diringer, and B. Ehlers, Chemoprophylaxis of scrapie in mice, *J Gen Virol.* 72:457 (1991).
36. B. Ehlers, and H. Diringer, Dextran sulphate 500 delays and prevents mouse scrapie by impairment of agent replication in spleen, *J Gen Virol.* 65:1325 (1984).
37. A. Ladogana, P. Casaccia, L. Ingrosso, M. Cibati, M. Salvatore, Y. G. Xi, C. Masullo, and M. Pocchiari, Sulphate polyanions prolong the incubation period of scrapie-infected hamsters, *J Gen Virol.* 73:661 (1992).
38. H. A. Gallis, R. H. Drew, and W. W. Pickard, Amphotericin B : 30 years of clinical experience, *Reviews of Infectious Diseases.* 12:308 (1990).
39. Y. L. Chung, A. Williams, J. S. Beech, S. C. R. Williams, J. D. Bell, J. Cox, and J. Hope, MRI assessment of the blood -brain barrier in a hamster model of scrapie, *Neurodegeneration.* 4:203 (1995).
40. F. C. Szoka, and M. Tang, Amphotericin B formulated in liposomes and lipid based systems : a review, *J Liposome Res.* 3:363 (1993).
41. A. Vertut-Doï, S. I. Ohnishi, and J. Bolard, The endocytic process in CHO cells, a toxic pathway of the polyene antibiotic Amphotericin B, *Antimicrob Agents Chemother.* 38:2373 (1994).
42. B. Chesebro, and B. Caughey, Scrapie agent replication without the prion protein, *Curr Biol.* 3:696 (1993).

SCRAPIE PATHOGENESIS IN BRAIN GRAFTS

Adriano Aguzzi and Sebastian Brandner

Institute of Neuropathology (Dept. of Pathology)
University of Zürich
CH-8091 Zürich, Switzerland

INTRODUCTION

Many compelling questions are still unresolved in the field of prion diseases, two of which have captured much of the attention of researchers. The first and most burning question, obviously, relates to the actual nature of the infectious agent which has been named prion and which replicates in the central nervous system and in certain further tissues of infected animals and humans.

The second question, however, is no less intriguing: by which mechanisms can prions bring about the devastating damage to the central nervous system which is characteristic of almost all transmissible spongiform encephalopathies? This unresolved problem raises many additional related questions. Is the damage related to the actual replication of the prion? Or is the spongiform encephalopathy the result of accumulation of toxic metabolites within or around neurons? One prime candidate for the latter hypothesis of toxicity would certainly be PrP^{Sc}, the pathologically changed isoform of the normal prion protein, PrP^C. If so, would PrP^{Sc} be toxic only if it is generated within cells, or would it damage nervous cells if it acts from without or is internalized? And anyway, since prions appear to replicate, or at least to accumulate, in the organs of the lymphoreticular system, such as spleen, lymph nodes, and Peyer's plaques of the intestine, why is that we do not observe immune deficiencies or structural pathologies of the latter organs after infection with prions? In other words, is susceptibility to prion toxicity a unique property of neural tissue, or is it rather the result of the 100fold higher levels of PrP^{Sc} accumulation seen in brains of terminally scrapie-sick mice as compared to lymphoreticular organs?

The complexity of these questions, along with our limited understanding of the nature of the infectious agent, suggests that it may be very difficult to devise suitable systems to address them experimentally. However, the recent generation of genetic *in vivo* model systems, such as transgenic mice expressing, at various levels, normal and mutated forms of the prion protein, as well as knockout mice which bear hetero- or homozygous ablations of the *Prnp* gene which encodes the prion protein, has now opened new, promising avenues of investigation.

Along these lines, we have taken advantage of various strains of transgenic mice and have asked whether the neurografting technology could be used to address the question of prion neurotoxicity. In this paper, we present a characterization of biological properties such as tissue growth, proliferation and differentiation in neuroepithelial grafts. Special emphasis is laid on the development of the blood-brain barrier (BBB) after grafting. We then discuss how embryonic telencephalic grafting was applied to the study of scrapie pathogenesis.

More than 100 years have passed since the first transplantation studies in the CNS of rodents were undertaken (Das, 1990; Fisher and Gage, 1993; Woerly and Marchand, 1990). In a number of studies, neural grafting has been used to address questions related to developmental neurobiology (Fisher and Gage, 1993; Kromer et al. 1983; O'Leary and Stanfield, 1989; Renfranz et al. 1991). Several studies investigated the establishment of neuronal organization within grafts and interactions with the host CNS (Jaeger and Lund, 1980; Kromer et al. 1983; Lund and Hauschka, 1976). More recently, grafting studies were aimed at questions related to neural plasticity. For example, it was asked whether and to what extent undifferentiated progenitor cells can integrate and take part to the formation of the host CNS (Brüstle, 1995; Campbell et al. 1995; Gage et al. 1995). Other studies were undertaken to address questions related to tumorigenic potential of various oncogenes by grafting retrovirally transduced cells into the rodent CNS (Aguzzi et al. 1991; Brüstle et al. 1992; Wiestler et al. 1992).

In the field of neurodegenerative disorders, grafting studies have been mainly aimed at reconstituting certain pathways or particular functions after surgical or toxic lesions to selected functional systems (Dunnett, 1990; Dunnett et al. 1981; Fisher and Gage, 1993; Lindvall, 1991). In these models, an artificial lesion leads to degeneration of specific neuronal systems. Grafting of neural tissue or genetically engineered cells aims at functional repair of induced lesions (Dunnett, 1990). The vast majority of such experiments was carried out in the rat system which is well suited for developmental studies and allows stereotaxic surgical interventions with appropriate accuracy.

However, with the advent of transgenic techniques, it has become possible to study in more detail the role played by single molecules during development and in pathological processes in mice (Aguzzi et al. 1994; Aguzzi et al. 1995). The generation of knockout mice by targeted deletion of genes of interest (Thomas and Capecchi, 1987) has further broadened our insight into molecular mechanisms of neural development and pathogenesis of CNS diseases. A number of transgenic and knockout mice have delivered valuable models for neurodegenerative diseases (Büeler et al. 1993; Games et al. 1995; LaFerla et al. 1995; Aguzzi et al. 1995). Others, however, show early postnatal (Klein et al. 1993; Magyar et al. 1994; Smeyne et al. 1994) or even embryonic (Bladt et al. 1995; Meyer and Birchmeier, 1995) lethal phenotypes which can be difficult to interpret. Although these models provide evidence for a crucial role of the respective gene products during development and hint at an important role of these factors for the determination of cell fates during differentiation (Bladt et al. 1995; Gassmann et al. 1995; Meyer and Birchmeier, 1995), they do not allow to study the role these factors play in secondary pathologic processes such as neurodegeneration. In an effort to overcome this problem we have employed transplantation approaches for neural tissue derived from such mouse embryos. Using grafting techniques, it has been possible to study neural tissue of mice with premature lethal genotypes at time points exceeding by far the life span of the mutant mice (Isenmann et al. 1996; Isenmann et al. 1995).

BIOLOGICAL CHARACTERISTICS OF MOUSE NEUROECTODERMAL GRAFTS

The grafting procedure is relatively simple (Isenmann et al. 1996; Isenmann et al. 1995). Embryos are harvested from timed pregnant dams at defined stages at mid-gestation. Graft tissue can be radiolabelled for later identification by autoradiography (Isenmann et al. 1996; Jaeger and Lund, 1980) and injected into the caudoputamen or lateral ventricles of recipient mice using a stereotaxic frame (Aguzzi et al. 1991; Isenmann et al. 1996). If histocompatible strains of mice are used, signs of graft rejection, such as lymphocytic infiltration and tissue necrosis, remain an exceptional finding and are detected in less than 5% of neural grafts (Isenmann et al. 1996).

In an effort to determine the optimal time point for embryonic tissue preparation and transplantation, we compared the final size of grafts resulting from tissue harvested at various embryonic stages.

We found murine telencephalic tissue from embryonic day (E) 12.5 to reliably differentiate into large neural grafts which are suitable for detailed graft analysis. Tissue harvested at earlier embryonic stages often resulted in grafts containing non-neural tissue portions, since it was difficult to clearly separate mesenchymal tissue from the neural anlage at E 9.5 - E 11.5. Such tissue portions induce permanent BBB leakage after grafting (Isenmann et al., 1996b) and were thus considered unfavorable. In contrast, neural tissue harvested at later embryonic stages (E 13.5 - E16.5) was easily separated from the meninges. However, proliferation and growth potential were markedly reduced, resulting in smaller transplants that were only partially accessible to thorough examination. Moreover, when tissue was harvested and transplanted at E 12.5, the total number of neural grafts was higher than with tissue harvested at other embryonic stages (Isenmann et al., 1996a).

Graft cell proliferation, as determined with immunocytochemical detection of incorporated 5-bromo-2'-deoxyuridine (Gratzner, 1982) showed that proliferation indices of graft cells decreased sharply from initially 35% of grafted cells to around 5% during the second week after transplantation and to less than 1% after more than seven weeks (Isenmann et al. 1996). At the same time, differentiation of grafted cells proceeds to the terminal postmitotic state. Thus, mature neuroepithelial grafts contain neurons with myelinated processes and a dense synaptic network, glia (astrocytes, oligodendrocytes, and microglia), and blood vessels four weeks after grafting (projected age of grafted tissue: approx. P20) (Isenmann et al. 1996; Isenmann et al. 1995). Taken together, these findings indicate that embryonic neuroepithelial tissue grafted into an adult host brain follows a program of maturation and differentiation similar to the in vivo time course (Isenmann et al. 1996).

BLOOD-BRAIN BARRIER (BBB) AND BRAIN GRAFTS

The BBB maintains the homeostatic environment in the brain by preventing blood-borne compounds from free entry into the CNS parenchyma. The barrier is formed by tight junctions in the vascular endothelia which are probably induced by astrocytes (Janzer and Raff, 1987; Risau and Wolburg, 1990). A number of pathological CNS processes, such as inflammation, demyelination, tumor or degeneration can induce breakdown of the BBB. In turn, BBB leakage might induce CNS dysfunction caused by blood-borne neurotoxic compounds normally excluded from the brain parenchyma (Rosenstein and Brightman, 1983; Svendgaard et al. 1975).

Various investigators have reported controversial findings on the post-transplantation status of the BBB in rodents. An early, yet most valuable study has suggested that the type of donor tissue determines the characteristics and BBB properties of graft supplying vessels (Stewart and Wiley, 1981). According to this hypothesis, neural grafts would be expected to induce BBB properties in the supplying blood vessels. In fact, several authors described a complete BBB reconstitution after neural grafting to the CNS. Some even find no residual BBB leakage as early as one week after grafting (Bertram et al. 1994; Broadwell et al. 1990; Broadwell et al. 1989; Swenson et al. 1989). Other studies, however, have claimed the BBB to remain permanently disrupted after neural grafting to the CNS (Rosenstein, 1987; Rosenstein and Brightman, 1986). Our group carried out studies in the model described using four independent marker molecules to detect damage to the BBB (Isenmann et al. 1996). The results obtained with various techniques were surprisingly consistent, with minor variations owing to variable sensitivity of individual techniques rather than differing findings. Magnetic resonance imaging (MRI) using a contrast agent *in vivo* indicates, in agreement with the histological data, that in our paradigm, i.e. grafting of tissue fragments as opposed to single cell suspensions, the BBB is reconstituted in 67% of all grafts after 3 weeks, and in 90% of the grafts seven weeks after grafting (Isenmann et al. 1996). These findings are particularly important with respect to the exploitation of neurografting techniques in neurodegenerative diseases. They indicate that the grafting procedure as such does not induce permanent BBB leakage that might expose the grafted tissue to a non-physiological environment, and suggest that the genotype of the grafted tissue determines the BBB properties of the graft. Thus, a pathologic condition affecting exclusively the graft can result in secondary BBB disruption.

NEUROGRAFTS IN PRION RESEARCH

Having established that a neurograft from a donor which would have undergone early lethality may be kept alive in a non-affected surrounding, we decided to apply this technique to the study of mouse scrapie. $Prnp^{o/o}$ mice, which are devoid of PrP^C, are resistant to scrapie and do not propagate prions (Büeler et al. 1993; Sailer et al. 1994). Because these mice show normal development and behavior (Büeler et al. 1992; Manson et al. 1994), it has been argued that scrapie pathology may come about because PrP^{Sc} deposition is neurotoxic (Forloni et al. 1993), rather than by depletion of cellular PrP^C. In the latter case, one might expect that the lack of PrP^C would result in embryonic or perinatal lethality, especially since PrP^C is encoded by a unique gene for which no related family members have been found.

To address the question of neurotoxicity, we set out to expose brain tissue of $Prnp^{o/o}$ mice to a continuous source of PrP^{Sc}. For the reasons exposed above, we thought that a convenient way to do so would be to graft embryonic telencephalic tissue from transgenic mice overexpressing PrP into various structures (in most experiments we chose the caudoputamen or the lateral ventricles or the forebrain) of $Prnp^{o/o}$ mice and to inoculate it with scrapie prions. The donors, tg20 mice (Fischer et al. 1996), contain two arrays of 30-50 gene copies encoding PrP, over-express PrP^C 5-8 fold and show incubation times of around 60 days as compared to 160 days for CD-1 wild-type mice. In the terminal stage of scrapie, both mouse strains exhibit similar prion titers while PrP^{Sc} levels in tg20 are 25-50% lower than in CD-1 brain.

The recipient mice were monitored clinically for the development of scrapie symptoms. In addition, we performed histological analyses of the grafts and their surroundings (conventional histology, immunohistochemistry, in situ hybridization), and we determined the content of PrP^C and PrP^{SC} by western blotting and histoblotting. Since to date it would not be appropriate to equate the amount of PrP^{SC} with infectivity, we decided to independently determine the amount of infectivity present in the graft and in regions of the brain at various distance from the graft by bioassay titration using tg20 recipient mice.

To our surprise, we observed that all mice remained free of scrapie symptoms for at least 70 weeks; this exceeds at least sevenfold the survival time of scrapie-infected tg20 mice. Therefore, the presence of a continuous source of PrP^{Sc} and of scrapie prions does not exert any clinically detectable adverse effects onto the physiological functions of a mouse devoid of PrP^C.

On the other hand, histological analysis revealed that tg20 and wild-type grafts developed characteristic histopathological features of scrapie 70 and 160 days after inoculation (p.i.), respectively, reflecting the incubation time of scrapie in the respective donor animals (Büeler et al. 1993; Fischer et al. 1996). Uninfected or mock-infected tg20 grafts occasionally showed mild gliosis but never spongiosis. Therefore, high expression of PrP by itself did not induce neurodegeneration in grafts.

Early stages of the disease in the graft (70-140 days p.i.) were characterized by spongiosis and gliosis. In addition, we observed reduced synaptophysin immunoreactivity, which we take as a sign of damage to the neuronal trees and subsequent decrease in the density of synaptic junctions. These changes are similar to those found in terminally sick tg20 mice, and since they occurred at graft ages similar to the life expectancy of scrapie-exposed donor mice, we concluded that neurografted PrP^C-expressing tissue indeed constitutes a realistic model for the scrapie encephalopathy. Intermediate stage grafts (140-280 days p.i.) showed *status spongiosus* with dramatic ballooning and loss of neurons, gliosis and stripping of neuronal processes. At late stages (280-480 days p.i.), cellular density increased from 900 to over 4000 cells/mm^2. Astrocytes constituted the main cell population and synaptophysin immunoreactivity was almost completely absent.

Although grafts had extensive contact with the recipient $Prnp^{o/o}$ brain, histopathology never extended into host tissue, even at the latest stages. Wild-type mice engrafted with *tg20* tissue showed severe histopathology in the graft and milder changes in the recipient brain, in good accordance with the general observation that the level of available PrP^C determines the speed of onset and often also the extent of pathology. Surprisingly, histoblot analysis (Taraboulos et al. 1992) of non-inoculated, engrafted $Prnp^{o/o}$ brain revealed that PrP immunoreactivity extended to the white matter of the recipient brains. In infected grafts, PrP^{Sc} was detected in both grafts and recipient brain, where it formed fine granules along white matter tracts and even in the contralateral hemisphere. Further, immunohistochemistry revealed PrP deposits in the host hippocampus and occasionally in the parietal cortex of all animals harbouring PrP-expressing grafts. Up to 35 clusters of PrP deposits per section appeared late in infection, each consisting of 25-120 globules of 2-4 µm diameter closely associated to astrocytic processes; no deposits were observed in inoculated or mock inoculated non-engrafted $Prnp^{o/o}$ brains or in $Prnp^{o/o}$ brains engrafted with $Prnp^{o/o}$ tissue. It is unlikely that such deposits were produced locally by *tg20* cells emigrating from the graft, since the graft borders were always sharply demarcated. PCR analysis of host brain regions containing PrP deposits failed to reveal PrP-encoding DNA (detection limit <1:10^4) and graft-derived cells were never detected distant from the graft by *in situ* hybridization or autoradiography of brain engrafted with ^{3}H-thymidine-labelled (Isenmann et al. 1995) tissue. We conclude that graft-derived PrP^C and PrP^{Sc} are transferred from the graft to distant areas of the host brain.

It could be argued that pathology does not spread to the surroundings because there may not be sufficient physiological connections between the graft and the surrounding tissue. To address this potential issue, we grafted *tg20* PrP^C overexpressing tissue in wild-type mice and determined the effects of intracerebral prion inoculations. This procedure resulted in histopathologically verifiable scrapie in both graft and in host tissue and in clinical scrapie of the host mouse without any modulation of the time course of the disease. However, the extent of pathology in the graft was more pronounced in the graft than in the host, once again reinforcing the general observation that the availability of PrP^C, rather than the deposition of PrP^{Sc}, is the rate-limiting step and also the major pathogenetic determinant in the development of scrapie.

To determine infectivity, various portions of brains containing grafts with severe histopathology were inoculated into *tg20* indicator mice. Samples of graft led to terminal scrapie after 74 days, indicating a titer of approximately 5.7 logLD$_{50}$ units per ml 10% homogenate. While frontal brain and cerebellum, in which no deposits were detected, did not show infectivity, about 1.5 logLD$_{50}$ units per ml 10% homogenate were detected in the contralateral hemisphere. Infectivity is not due to residual inoculum, because within 4 days after inoculation no infectivity can be detected in recipient brain (Büeler et al. 1993). Thus, infectious prions moved from the grafts to some regions of the PrP-deficient host brain without causing pathological changes or clinical disease. The distribution of PrP^{Sc} in the white matter tracts of the host brain suggests diffusion within the extra-cellular space (Jeffrey et al. 1994) rather than axonal transport (S.B. and A.A., unpublished results).

Why was no scrapie pathology observed in PrP^C-deficient tissue even in regions adjoining the graft which contained high levels of PrP^{Sc} and was clearly leaking such material? Perhaps PrP^{Sc} is inherently non-toxic and PrP^{Sc} plaques found in spongiform encephalopathies are an epiphenomenon rather than a cause of neuronal damage. Indeed, the extent of PrP deposition in the brains of humans succumbing to prion diseases with similar clinical presentation is extremely variable (Collinge et al. 1990; Hayward et al. 1994).

Alternatively, PrP^Sc is only toxic when it is formed and accumulated within the cell, but not when it is presented from without. Finally, it may be that PrP^Sc is pathogenic when presented from without, but only to cells expressing PrP^C, either because it initiates conversion of PrP^C to PrP^Sc at the cell surface and/or because it is internalized by way of association with PrP^C, which is endocytosed efficiently (Shyng et al. 1993). Along with previous transgenetic studies (Büeler et al. 1994; Manson et al. 1994) showing delayed onset of disease in $Prnp^{o/+}$ mice despite massive accumulation of PrP^Sc and reports of typical scrapie histopathology in FFI-inoculated mouse brains devoid of detectable levels of PrP^Sc (Collinge et al. 1995), the present data imply that not deposition of PrP^Sc but rather availability of PrP^C for some intracellular process elicited by the infectious agent is directly linked to spongiosis, gliosis, and neuronal death.

ACKNOWLEDGMENTS

The authors wish to thank Karin Stepinski, Rosmarie von Rotz and Regula Chase for secretarial assistance. The work described is being supported by the Kanton of Zürich, and by grants top A.A. from the Schweizerischer Nationalfonds and Nationales Forschungsprogramm NFP38, the European Union, the Migros Foundation, and the Bundesamt für Gesundheitswesen.

REFERENCES

Aguzzi A, Brandner S, Isenmann S, Steinbach J, Sure U. (1995): Transgenic and gene disruption techniques in the study of neurocarcinogenesis. In: *Glia*. **15**:348-364.

Aguzzi A, Brandner S, Sure U, Ruedi D, Isenmann S. (1994): Transgenic and knock-out mice: models of neurological disease. In: *Brain Pathol*. **4**:3-20.

Aguzzi A, Kleihues P, Heckl K, Wiestler OD. (1991): Cell type-specific tumor induction in neural transplants by retrovirus-mediated oncogene transfer. In: *Oncogene*. **6**:113-118.

Bertram KJ, Shipley MT, Ennis M, Sanberg PR, Norman AB. (1994): Permeability of the blood-brain barrier within rat intrastriatal transplants assessed by simultaneous systemic injection of horseradish peroxidase and Evans blue dye. In: *Exp Neurol*. **127**:245-252.

Bladt F, Riethmacher D, Isenmann S, Aguzzi A, Birchmeier C. (1995): Essential role for the c-met receptor in the migration of myogenic precursor cells into the limb bud [see comments]. In: *Nature*. **376**:768-771.

Broadwell RD, Charlton HM, Ebert P, Hickey WF, Villegas JC, Wolf AL. (1990): Angiogenesis and the blood-brain barrier in solid and dissociated cell grafts within the CNS. In: *Prog Brain Res*. **82**:95-101.

Broadwell RD, Charlton HM, Ganong WF, Salcman M, Sofroniew M. (1989): Allografts of CNS tissue possess a blood-brain barrier. I. Grafts of medial preoptic area in hypogonadal mice. In: *Exp Neurol*. **105**:135-151.

Brüstle O. (1995): Targeted Introduction of Neurons into the Embryonic Brain. In: *Neuron*. **15**:1275-1285.

Brüstle O, Aguzzi A, Talarico D, Basilico C, Kleihues P, Wiestler OD. (1992): Angiogenic activity of the K-fgf/hst oncogene in neural transplants. In: *Oncogene*. **7**:1177-1183.

Büeler H, Raeber A, Sailer A, Fischer M, Aguzzi A, Weissmann C. (1994): High prion and PrP^Sc levels but delayed onset of disease in scrapie-inoculated mice heterozygous for a disrupted PrP gene. In: *Molecular Medicine*. **1**:19-30.

Büeler HR, Fischer M, Lang Y, Bluethmann H, Lipp HP, DeArmond SJ, Prusiner SB, Aguet M, Weissmann C. (1992): Normal development and behaviour of mice lacking the neuronal cell-surface PrP protein. In: *Nature*. **356**:577-582.

Büeler HR, Aguzzi A, Sailer A, Greiner RA, Autenried P, Aguet M, Weissmann C. (1993): Mice devoid of PrP are resistant to scrapie. In: *Cell*. **73**:1339-1347.

Campbell C, Olsson M, Björklund A. (1995): Regional Incorporation and Site-Specific Differentiation of Striatal Precursors Transplanted to the Embryonic Forebrain Ventricle. In: *Neuron.* **15:**1273

Collinge J, Owen F, Poulter M, Leach M, Crow TJ, Rossor MN, Hardy J, Mullan MJ, Janota I, Lantos PL. (1990): Prion dementia without characteristic pathology. In: *Lancet.* **336:**7-9.

Collinge J, Palmer MS, Sidle KCL, Gowland I, Medori R, Ironside JW, Lantos PL. (1995): Transmission of fatal familial insomnia to laboratory animals. In: *Lancet.* **346:**569-570.

Das GD. (1990): Neural transplantation: an historical perspective. In: *Neurosci Biobehav Rev.* **14:**389-401.

Dunnett SB. (1990): Neural transplantation in animal models of dementia. In: *Eur J Neurosci.* **2:**567-587.

Dunnett SB, Björklund A, Stenevi U, Iversen SD. (1981): Grafts of embryonic substantia nigra reinnervating the ventrolateral striatum ameliorate sensorimotor impairments and akinesia in rats with 6-OHDA lesion of the nigrostriatal pathway. In: *Brain Res.* **229:**209-217.

Fischer M, Rülicke T, Raeber A, Sailer A, Moser M, Oesch B, Brandner S, Aguzzi A, Weissmann C. (1996): Prion protein with amino terminal deletions restoring susceptibilty of PrP knockout mice to scrapie. In: *EMBO J.* **15:**1255-1264.

Fisher LJ, Gage FH. (1993): Grafting in the mammalian central nervous system. In: *Physiol Rev.* **73:**583-616.

Forloni G, Angeretti N, Chiesa R, Monzani E, Salmona M, Bugiani O, Tagliavini F. (1993): Neurotoxicity of a prion protein fragment. In: *Nature.* **362:**543-546.

Gage FH, Coates PW, Palmer TD, Kuhn HG, Fisher LJ, Suhonen JO, Peterson DA, Suhr ST, Ray J. (1995): Survival and differentiation of adult neuronal progenitor cells transplanted to the adult brain. In: *Proc Natl Acad Sci U S A.* **92:**11879-11883.

Games D, Adams D, Alessandrini R, Barbour R, Berthelette P, Blackwell C, Carr T, Clemens J, Donaldson T, Gillespie F, et al. (1995): Alzheimer-type neuropathology in transgenic mice overexpressing V717F beta-amyloid precursor protein. In: *Nature.* **373:**523-527.

Gassmann M, Casagranda F, Orioli D, Simon H, Lai C, Klein R, Lemke G. (1995): Aberrant neural and cardiac development in mice lacking the ErbB4 neuregulin receptor. In: *Nature.* **378:**390-394.

Gratzner HG. (1982): Monoclonal antibody to 5-bromo- and 5-iododeoxyuridine: A new reagent for detection of DNA replication. In: *Science.* **218:**474-475.

Hayward PA, Bell JE, Ironside JW. (1994): Prion protein immunocytochemistry: reliable protocols for the investigation of Creutzfeldt-Jakob disease. In: *Neuropathol Appl Neurobiol.* **20:**375-383.

Isenmann S, Molthagen M, Brandner S, Kühne G, Bartsch U, Magyar JP, Sure U, Schachner M, Aguzzi A. (1995): Neural transplants of AMOG/□2-deficient mice survive but do not express the □2-subunit of the Na,K-ATPase. In: *Glia.* 377-388.

Isenmann S, Brandner S, Kühne G, Boner J, Aguzzi A. (1996): Comparative in vivo and pathological analysis of the blood-brain barrier in mouse telencephalic transplants. In: *Neuropathol Appl Neurobiol.* 118-128.

Isenmann S, Brandner S, Sure U, Aguzzi A. (1996): Telencephalic transplants in mice: characterization of growth and differentiation patterns. In: *Neuropathol Appl Neurobiol.* 108-117.

Jaeger CB, Lund RD. (1980): Transplantation of embryonic occipital cortex to the tectal region of newborn rats: a light microscopic study of organization and connectivity of the transplants. In: *J Comp Neurol.* **194:**571-597.

Janzer RC, Raff MC. (1987): Astrocytes induce blood-brain barrier properties in endothelial cells. In: *Nature.* **325:**253-257.

Jeffrey M, Goodsir CM, Bruce ME, McBride PA, Fowler N, Scott JR. (1994): Murine scrapie-infected neurons in vivo release excess prion protein into the extracellular space. In: *Neurosci Lett.* **174:**39-42.

Klein R, Smeyne RJ, Wurst W, Long LK, Auerbach BA, Joyner AL, Barbacid M. (1993): Targeted disruption of the trkB neurotrophin receptor gene results in nervous system lesions and neonatal death. In: *Cell*. **75**:113-122.

Kromer LF, Bjorklund A, Stenevi U. (1983): Intracephalic embryonic neural implants in the adult rat brain. I. Growth and mature organization of brainstem, cerebellar, and hippocampal implants. In: *J Comp Neurol*. **218**:433-459.

LaFerla FM, Tinkle BT, Bieberich CJ, Haudenschild CC, Jay G. (1995): The Alzheimer's A beta peptide induces neurodegeneration and apoptotic cell death in transgenic mice. In: *Nat Genet*. **9**:21-30.

Lindvall O. (1991): Prospects of transplantation in human neurodegenerative diseases. In: *Trends Neurosci*. **14**:376-384.

Lund RD, Hauschka SD. (1976): Transplanted neural tissue develops connections with host rat brain. In: *Science*. **193**:582-584.

Magyar JP, Bartsch U, Wang ZQ, Howells N, Aguzzi A, Wagner EF, Schachner M. (1994): Degeneration of neural cells in the central nervous system of mice deficient in the gene for the adhesion molecule on Glia, the beta 2 subunit of murine Na, K-ATPase. In: *J Cell Biol*. **127**:835-845.

Manson JC, Clarke AR, Hooper ML, Aitchison L, McConnell I, Hope J. (1994): 129/Ola mice carrying a null mutation in PrP that abolishes mRNA production are developmentally normal. In: *Mol Neurobiol*. **8**:121-127.

Manson JC, Clarke AR, McBride PA, McConnell I, Hope J. (1994): PrP gene dosage determines the timing but not the final intensity or distribution of lesions in scrapie pathology. In: *Neurodegeneration*. **3**:331-340.

Meyer D, Birchmeier C. (1995): Multiple essential functions of neuregulin in development [see comments]. In: *Nature*. **378**:386-390.

O'Leary DD, Stanfield BB. (1989): Selective elimination of axons extended by developing cortical neurons is dependent on regional locale: experiments utilizing fetal cortical transplants. In: *J Neurosci*. **9**:2230-2246.

Renfranz PJ, Cunningham MG, McKay RD. (1991): Region-specific differentiation of the hippocampal stem cell line HiB5 upon implantation into the developing mammalian brain. In: *Cell*. **66**:713-729.

Risau W, Wolburg H. (1990): Development of the blood-brain barrier [see comments]. In: *Trends Neurosci*. **13**:174-178.

Rosenstein JM. (1987): Neocortical transplants in the mammalian brain lack a blood-brain barrier to macromolecules. In: *Science*. **235**:772-774.

Rosenstein JM, Brightman MW. (1983): Circumventing the blood-brain barrier with autonomic ganglion transplants. In: *Science*. **221**:879-881.

Rosenstein JM, Brightman MW. (1986): Alterations of the blood-brain barrier after transplantation of autonomic ganglia into the mammalian central nervous system. In: *J Comp Neurol*. **250**:339-351.

Sailer A, Büeler H, Fischer M, Aguzzi A, Weissmann C. (1994): No propagation of prions in mice devoid of PrP. In: *Cell*. **77**:967-968.

Shyng SL, Huber MT, Harris DA. (1993): A prion protein cycles between the cell surface and an endocytic compartment in cultured neuroblastoma cells. In: *J Biol Chem*. **268**:15922-15928.

Smeyne RJ, Klein R, Schnapp A, Long LK, Bryant S, Lewin A, Lira SA, Barbacid M. (1994): Severe sensory and sympathetic neuropathies in mice carrying a disrupted Trk/NGF receptor gene. In: *Nature*. **368**:246-249.

Stewart PA, Wiley MJ. (1981): Developing nervous tissue induces formation of blood-brain barrier characteristics in invading endothelial cells: a study using quail--chick transplantation chimeras. In: *Dev Biol*. **84**:183-192.

Svendgaard NA, Bjorklund A, Hardebo JE, Stenevi U. (1975): Axonal degeneration associated with a defective blood-brain barrier in cerebral implants. In: *Nature*. **255**:334-336.

Swenson RS, Shaw P, Alones V, Kozlowski G, Zimmer J, Castro AJ. (1989): Neocortical transplants grafted into the newborn rat brain demonstrate a blood-brain barrier to macromolecules. In: *Neurosci Lett.* **100:**35-39.

Taraboulos A, Jendroska K, Serban D, Yang SL, DeArmond SJ, Prusiner SB. (1992): Regional mapping of prion proteins in brain. In: *Proc Natl Acad Sci U S A.* **89:**7620-7624.

Thomas KR, Capecchi MR. (1987): Site-directed mutagenesis by gene targeting in mouse embryo-derived stem cells. In: *Cell.* **51:**503-512.

Wiestler OD, Brüstle O, Eibl RH, Radner H, Aguzzi A, Kleihues P. (1992): Retrovirus-mediated oncogene transfer into neural transplants. In: *Brain Pathol.* **2:**47-59.

Woerly S, Marchand R. (1990): [A century of neurotransplantation in mammals]. In: *Neurochirurgie.* **36:**71-95.

STRUCTURAL PROPERTIES OF RECOMBINANT HUMAN PRION PROTEIN

Graham Jackson[1], Andrew Hill[1], Catherine Joseph[1],
Anthony Clarke[2] and John Collinge[1]

[1] Neurogenetics Unit, Imperial College School of Medicine at St Mary's, London, UK
[2] Department of Biochemistry, University of Bristol, Bristol, UK

Prion diseases are a group of fatal, neurodegenerative conditions affecting both humans and animals. Previously called transmissible spongiform encephalopathies or slow virus diseases they are unique in that they may have a sporadic, inherited or transmissible origin. The transmissible prion diseases have become an area of increasing public concern because of the epidemic of a novel bovine prion disease, bovine spongiform encephalopathy, and despite recent advances, many uncertainties exist in understanding of the nature of the infectious agent.

Over seventeen different mutant human prion proteins are associated with the aetiology of the human inherited prion diseases, most of which are known also to be transmissible to laboratory animals (see Collinge, 1997 for review). Prions appear to consist principally or entirely of an abnormal isoform of a host encoded glycoprotein, prion protein (PrP). This abnormal isoform (PrP^{Sc}) has different physiochemical properties from cellular PrP (PrP^{C}) and is derived from PrP^{C} by a post-translational process. The precise nature of this change is unknown, although indications are that this involves a conformational rather than covalent modification of the protein (Pan et al., 1993). Essential to the study and understanding of prion diseases are the structures of both PrP^{C} and PrP^{Sc}, and the mechanism of conversion that leads to generation of infectivity.

PrP^{Sc} is extremely insoluble and exists as an aggregated polymer limiting the effectiveness of study using conventional biophysical techniques. The study object of choice must therefore be the soluble PrP^{C}, the low abundance and localisation of which makes it difficult to purify from endogenous sources.

We chose the human PrP gene for heterologous expression as the normal primary structure will adopt the PrP^{C} conformation whilst the introduction of one of more of the known human pathogenic mutations would be anticipated to disturb this conformation and predispose to conversion to the PrP^{Sc} isoform.

Human PrP is encoded within a single exon and expressed as a 253 amino acid polypeptide (Basler et al., 1986), the N-terminal 22 amino acids of which encode a signal sequence (Turk et al., 1988) and the C-terminal 23 amino acids are cleaved on addition of a glycosylphosphatidylinositol (GPI) anchor (Stahl et al., 1987).

One characteristic of the post-translational modification of PrP^C is the acquisition of partial protease resistance. After digestion with proteinase K the protease sensitive N-terminal is cleaved to produce a fragment in the range 27-30 kDa (PrP27-30) as determined by SDS-PAGE. Around 89 residues are cleaved to produce a fragment of approximately 164 amino acids. This region of the open reading frame of HuPrP (residues 89-230) was chosen for heterologous expression as purified PrP27-30 retains infectivity and contains many of the sites of known human pathogenic mutations. This region has been shown to contain a significant proportion of secondary structure (Hornemann, et al., 1996). Indeed the structure of a smaller fragment within this region of the equivalent mouse sequence has been determined by NMR spectroscopy (Riek, et al., 1996).

In order to provide large quantities of protein for biophysical study and facilitate protein purification we selected an E. coli expression vector, pTrcHisA (Invitrogen), which produces a fusion protein with an N-terminal poly-histidine tag. The recombinant PrP (rPrP) was expressed at a level of 30% total cell protein in an insoluble form, localised to inclusion bodies, which aided protein purification. Following solubilisation in 6M guanidine hydrochloride the rPrP was isolated by column chromatography and cleaved to remove the N-terminal fusion. Refolding of the recombinant prion protein by rapid dilution yielded soluble rPrP for characterisation and biophysical study.

We have begun a biophysical characterisation and full structural determination of human PrP^C utilising various techniques. Circular dichroism indicates this protein contains mainly α-helix with a small proportion of β-sheet. This secondary structure can be reversibly denatured by titration with the chaotropic agent guanidine hydrochloride (GuHCl). The helical structure observed at 222nm is lost as a single co-operative transition with a mid-point of 2.3M GuHCl (see Figure 1). This data set was fitted to the equation : $-RT.\ln K = \Delta G_w + n.[\Delta G_{sm}.[D]/(K_{den} + [D])]$ using values for ΔG_{sm} and K_{den} of 0.465 cal/mol and 3.07M respectively, according to the method described by Staniforth (1993). This demonstrated the polypeptide was folded with 19 residues protected from bulk solvent in a hydrophobic core of a predominantly (43% of

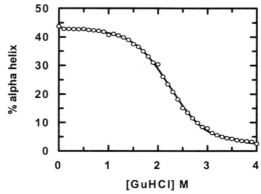

[FIGURE 1] Circular dichroism was measured at 222nm in a Jan Yobin CD6 spectrometer thermostated to 25°C. Samples of recombinant human PrP were made up in cuvettes of 0.1mm pathlength in PBS at a concentration of 1mg/ml. Varying concentrations of GuHCl were added and left for 12 hrs to equilibrate before the CD signal was recorded. Percentage helix was calculated using the algorithm of Chen, et al., (1972) : % α-helix = $100.[-\theta_{222} - 2340]/30300$. The solid line through the data represents a fit to the equation : $-RT.\ln K = \Delta G_w + n.[\Delta G_{sm}.[D]/(K_{den} + [D])]$.

amides) α-helix containing structure. The small number of residues protected from bulk solvent was reflected in a relatively low free energy of folding of only -3.5 kcal/mol which suggests a high level of conformational plasticity.

Intrinsic fluorescence from the native tryptophan at residue 99 was used to monitor conformational changes during denaturation as were the ten tyrosine residues. The signal changes observed with tryptophan fluorescence alone were poor due to a high level of solvent exposure of this residue which suggests this local area of structure may be unstructured or highly mobile. However, by monitoring the ratio of fluorescence emission at tryptophan (350nm) and tyrosine (303nm) wavelengths resonance energy transfer and internal quenching effects between tyrosine residues and the tryptophan indicate there are two distinct transitions during the unfolding of PrP^C (see Figure 2). The major transition is the second with a mid-point of approximately 2.3M GuHCl which corresponds to the loss of secondary structure observed by CD at 222nm. Interestingly an additional earlier transition with a mid-point of 0.5M GuHCl is observed in these fluorescence experiments. This transition may represent a conformational change in an area of structure local to the tryptophan residue. The absence of a change in CD signal over this range of GuHCl concentrations suggests this change may involve a loss of tertiary structure at the N terminus which does not disrupt secondary structure.

Similar experiments performed on sonicated PrP27-30 show a distinctly different fluorescence profile, with the tryptophan residues buried and protected from solvent in the native structure. Two unfolding transitions are observed with mid-points of 1.5M and 3M GuHCl respectively (see Figure 3). The early transition may again represent structural change local to the environment of tryptophan 99 with the second transition corresponding to loss of secondary structure and global unfolding of the polypeptide chain.

In order to define the 3D solution structure of human PrP^C we started NMR data collection. Preliminary 1-D NMR spectra of the recombinant PrP^C show a large dispersion and narrow bandwidths confirming the protein exists as a folded monomer

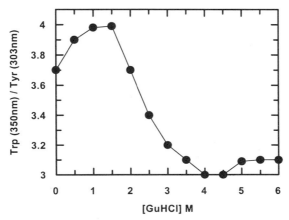

[FIGURE 2] Fluorescence measurements were carried out on a Schimadzu RF 530X spectrofluorimeter thermostatted to 25°C. Recombinant PrP was made up to 5μM in 20mM BisTrisPropane pH 8.0 in a 1cm pathlength cuvette. Fluorescence excitation was at 275nm and emision monitored at 303nm and 350nm, with excitation and emision bandwidths of 1.5nm. Measurements were repeated with varying concentrations of denaturant after allowing the sample to equilibrate for 12 hrs.

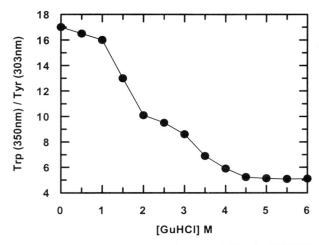

[FIGURE 3] Fluorescence measurements were carried out on a Schimadzu RF 530X spectrofluorimeter thermostatted to 25°C. Sonicated PrP27-30 was made up to 5μM in 20mM BisTrisPropane pH 8.0 in a 1cm pathlength cuvette. Fluorescence excitation was at 275nm and emision monitored at 303nm and 350nm, with excitation and emision bandwidths of 1.5nm. Measurements were repeated with varying concentrations of denaturant after allowing the sample to equilibrate for 12 hrs.

(see Figure 4). There are resonances visible at -2 ppm which originate from methyl groups held close to aromatic rings. This is a strong indication of highly ordered tertiary structure in the molecule. This coupled to the large number of NOEs observed in the NOESY spectrum (see Figure 5) is an indication that recombinant human PrP^C represents a good study object for NMR.

Estimates of secondary structure content from a proton NOESY spectrum match those obtained from CD, ie approximately 45% helix and 15% sheet. Here the number of NOEs exceeds 1500 (see Figure 5) and will provide sufficient distance constraints to define the 3-D structure of the polypeptide. With a molecular weight of 16kDa the

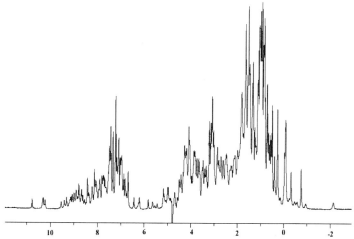

[FIGURE 4] One dimensional homonuclear proton spectrum collected on a Bruker 600mHz NMR spectrometer at 25°C referenced to water at 4.87ppm. Recombinant PrP was concentrated to 800μM in 90% H$_2$O/10% D$_2$O pH 5.5 and allowed to equilibrate to temperature in the spectrometer for 1 hr before data collection.

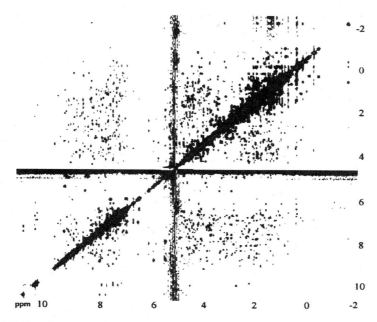

[FIGURE 5] Two dimensional [^{1}H - ^{1}H] NOESY spectrum collected on a Bruker 600mHz NMR spectrometer at 25°C referenced to water at 4.87ppm. Recombinant PrP was concentrated to 800μM in 90% H_2O/10% D_2O pH 5.5 and allowed to equilibrate to temperature in the spectrometer for 1 hr before data collection.

recombinant human prion protein is too large for full resonance assignment in homonuclear form. We are currently producing prion protein uniformly labelled with ^{15}N and double labelled with ^{15}N/^{13}C to overcome this restriction.

Discussion

Heterologous expression of amino acids 89 to 231 of the human prion protein yields a polypeptide that adopts a defined three dimensional conformation containing a large proportion of secondary structure and a defined tertiary structure. Reversible denaturation proceeds via a folding intermediate that is populated at 1.5M GuHCl. This intermediate appears to contain most or all of the secondary structure present in the native state and differs only in the tertiary structure around tryptophan 99. The overall stability of the folded conformation is low (-3.5kcal/mol) which could permit conformational switching between states that are commonly derived from the same folding intermediate or denatured state. Indeed a folding energy of only -3.5kcal/mol implies the ratio of folded to unfolded molecules is only 1600 to 1. This represents a high proportion of PrP molecules that exist in an unfolded state at equilibrium and are competent for conversion to the PrP^{Sc} isoform.

The area of tertiary structure around trp-99 is of even lower stability and may represent the initiation site for conversion. The precise conformation of the PrP^{Sc} isoform is unknown although it would appear to contain a high proportion of β-sheet type secondary structure as determined by circular dichroism (Safar et al., 1990). Equilibrium denaturation of PrP^{Sc} reveals two transitions the second of which centred at 3M GuHCl is likely to represent the loss of secondary structure. So far insufficient purified material has been available to monitor the loss of CD with increasing denaturant concentration to confirm this. Should this be the case then the earlier transition is likely to result from local rearrangement of the N-terminus around trp-99. The stability of the structure in this region is significantly greater in the PrP^{Sc} isoform than in PrP^C as reflected in the higher concentration of GuHCl for the mid-point of the transition (1.5M as opposed to 0.5M GuHCl).

This implies the N-terminus of our expressed fragment may be a crucial area of sequence involved in early folding events in the conversion of PrP^C to PrP^{Sc}. To further investigate the folding of PrP^C and conversion to PrP^{Sc} we intend to create site-directed mutations in the PrP gene that are known to be implicated in prion disease. The identification of differences in the folding pathways of these mutants should be very revealing as to the primary event involved in the misfolding of PrP^C leading to the formation of PrP^{Sc}.

Bibliography

Basler, K., Oesch, B., Scott, M., Westaway, D., Walchli, M., Groth, D.F., McKinley, M.P., Prusiner, S.B., and Weissmann, C. (1986). Scrapie and cellular PrP isoforms are encoded by the same chromosomal gene. Cell *46*, 417-428.

Collinge, J. Human prion diseases and bovine spongiform encephalopathy (BSE). *Hum Mol Genetics* 6(10):1699-1705, 1997.

Pan, K., Baldwin, M.A., Nguyen, J., Gasset, M., Serban, A., Groth, D., Mehlhorn, I., Huang, Z., Fletterick, R.J., Cohen, F.E., and Prusiner, S.B. (1993). Conversion of α-helices into β-sheets features in the formation of the scrapie prion proteins. Proc Natl Acad Sci USA *90*, 10962-10966.

Hornemann, S. and Glockshuber, R. (1996). Autonomous and reversible folding of a soluble amino-terminally truncated segment of the mouse prion protein. J. Mol. Biol. *262*, 614-619.

Riek, R., Hornemann, S., Wider, G., Billeter, M., Glockshuber, R. and Wuthrich, K. (1996). NMR structure of the mouse prion protein domain PrP(121-231). Nature *382*, 180-182.

Safar, J., Wang, W., Padgett, M.P., Ceroni, M., Piccardo, P., Zopf, D., Gajdusek, D.C., and Gibbs, C.J.J. (1990). Molecular mass, biochemical composition, and physicochemical behavior of the infectious form of the scrapie precursor protein monomer. Proc. Natl. Acad. Sci. U. S. A. *87*, 6373-6377.

Stahl, N., Borchelt, D.R., Hsiao, K., and Prusiner, S.B. (1987). Scrapie prion protein contains a phosphatidylinositol glycolipid. Cell *51*, 229-240.

Staniforth, R.A., Burston, S.G., Smith, C.J., Jackson, G.S., Badcoe, I.G., Atkinson, T., Holbrook, J.J. and Clarke, A.R. (1993). The energetics and cooperativity of protein folding. A simple experimental analysis based upon the solvation of internal residues. Biochemistry *32*, 3842-3851.

Turk, E., Teplow, D.B., Hood, L.E., and Prusiner, S.B. (1988). Purification and properties of the cellular and scrapie hamster prion proteins. Eur. J Biochem. *176*, 21-30.

AUTONOMOUS FOLDING AND THREE-DIMENSIONAL STRUCTURE OF THE CARBOXY-TERMINAL DOMAIN OF THE MOUSE PRION PROTEIN, PrP(121- 231)

Rudi Glockshuber, Simone Hornemann, Roland Riek, Gerhard Wider, Martin Billeter, and Kurt Wüthrich

Institut für Molekularbiologie und Biophysik
Eidgenössische Technische Hochschule Hönggerberg
CH-8093 Zürich
Switzerland

INTRODUCTION

The "protein-only" hypothesis (Alper et al., 1967; Griffith, 1967; Prusiner, 1982) states that the infectious agent of mammalian transmissible spongiform encephalopathies, such as *Scrapie* in sheep, bovine spongiform encephalopathy (BSE), and the Creutzfeldt-Jakob disease (CJD), the Gerstmann-Sträussler-Scheincker syndrome (GSS), fatal familial insomnia (FFI) and Kuru in man, is mainly, if not entirely, composed of a single protein, the "scrapie" form of the prion protein, PrP^{Sc} (Prusiner, 1982; Prusiner, 1991; Weissmann, 1994; Weissmann, 1996). This infectious form of the prion prtotein, PrP, is believed to be chemically identical to its monomeric, cellular form, PrP^C (Stahl & Prusiner, 1991), but PrP^{Sc} appears to possess a different three-dimensional structure with increased ß-sheet content (Pan et al., 1993) and forms oligomeric, protease-resistant aggregates (Prusiner, 1982; Bolton et al., 1982).

Mammalian PrP^C is a cell surface protein which occurs in neurons and most other tissues. PrP^C has been reported, from observations with certain strains of PrP knockout mice, to be necessary for normal synaptic function (Collinge *et al.*, 1994), long-term survival of Purkinje neurons (Sakaguchi et al., 1996), and the regulation of circadian activity rhythms and sleep (Tobler et al., 1996). However, knockout mice derived from other strains do not show neurological disorders (Ch. Weissmann, private communication; Lledo et al., 1996). The known amino acid sequences of mammalian prion proteins are highly conserved, and pairs of sequences are generally more than 90 percent identical (Schätzl *et al.*, 1995). PrP^C is a secretory protein and undergoes several posttranslational modifications. The mature form of murine PrP^C consists of 208 amino acids (residues 23 to

231, with deletion of codon 55) (Westaway *et al.*, 1987). The protein has a single disulfide bond between residues 179 and 214, two N-glycosylation sites at residues 181 and 197 and is attached to the cell surface via a glycosyl-phosphatidyl-inositol (GPI) anchor at the carboxy-terminal serine 231 (Stahl & Prusiner, 1991) (amino acid numbering according to hamster PrP, see Schätzl et al., 1995).

The findings that mice devoid of PrP are resistant to infection by prions (Büeler et al., 1993), that inherited prion diseases in humans are associated with mutations in the prion protein gene (Prusiner, 1993), and that a spontaneous prion disease occurs in uninoculated, transgenic mice expressing PrP with a substitution analogous to that observed in an inherited human GSS case (Hsiao et al., 1989) are presumably the most convincing *in vivo* data supporting the protein-only hypothesis. Despite these results, the protein-only hypothesis cannot be considered as proven, since extensive attempts to generate infectious prion material *in vitro* have failed so far. However, excess PrP^{Sc} has recently been shown to convert PrP^C into an insoluble form *in vitro* which exhibits biochemical properties such as protease resistance similar to PrP^{Sc} (Kocisko et al., 1994; Bessen et al., 1995). The failure to produce infectious prions *in vitro* may also be explained by the absence of an essential host factor, for which indirect evidence was obtained recently (Telling et al., 1995).

Two different mechanisms for the generation of an infectious prion particle from PrP^C are presently discussed: A first model proposes that initial transformation of PrP^C to PrP^{Sc} is rate-limiting and that subsequent autocatalytic formation and polymerization of PrP^{Sc} is fast (Prusiner, 1991). In a second model, a rapid equilibrium between monomeric PrP^C and monomeric PrP^{Sc} is proposed, and the rate-limiting step would be the formation of a stable, ordered aggregate of PrP^{Sc} molecules, which would act as a nucleus for growth of the infectious prion (Jarret and Lansbury, 1993).

A prerequisite for understanding the processes underlying the conversion of PrP^C to PrP^{Sc} is the knowledge of folding, thermodynamic stablility and the three-dimensional structure of PrP^C. Thermodynamic and structural studies of PrP^C in solution have however been hampered in the past by its low solubility in the absence of detergents and the low yields of PrP^C obtained after purification from its natural source.

In this article, we describe a recombinant carboxy-terminal fragment of residues 121-231 in mouse PrP which represents an intrinsically stable domain of PrP. PrP(121-231) is monomeric, refolds spontaneously and quantitatively after chemical denaturation with guanidinium chloride (GdmCl) and has a free energy of folding of - 22 kJ/mol (Hornemann & Glockshuber, 1996). Its high solubility in water allowed the determination of its three-dimensional structure in solution by nuclear magnetic resonance (NMR) spectroscopy (Wüthrich, 1986). We show that PrP(121-231) possesses a so far unknown protein fold and, in contrast to model predictions of an all-helical structure of PrP^C, contains an antiparallel ß-sheet. Implications of the three-dimensional structure of PrP(121-231) will be discussed in conjunction with the formation of PrP^{Sc} from PrP^C, the location of mutated residues associated with inherited human prion diseases, and the species barrier of prion disease transmission.

A THEORETICAL THREE-DIMENSIONAL MODEL OF PrP^C AND THE IDENTIFICATION OF THE CARBOXY-TERMINAL DOMAIN OF MOUSE PrP, PrP(121-231)

The primary structures of mature, mammalian prion proteins share a rather unusual sequence motif in the amino-terminal part of the protein. It consists of an octapeptide sequence which is repeated five times in most of the cases and has a high glycine content of about 50 percent (Schätzl et al., 1995). The amino-terminal segment of PrP^C may therefore lack a well-defined three-dimensional fold. In any case, attempts failed to predict regular secondary structures within the amino-terminal part of the protein (residues 23-108) including this characteristic, five-fold octapeptide repeat (Huang *et al.*, 1994). However, plausible three-dimensional models for the carboxy-terminal segment of PrP comprising residues 108 to 218 were presented (Huang *et al.*, 1994; Huang *et al.*, 1996). These models predict that the segment 108-218 in PrP^C forms a four-helix bundle.

To study folding and stability of the predicted four-helix bundle domain PrP(108-218), we recombinantly expressed two segments of mouse PrP, comprising residues 95-231 and 107-231, respectively, in *Escherichia coli*. Both segments were fused to the bacterial OmpA sequence for secretory expression to allow formation of the single disulfide bond in the oxidizing environment of the periplasm of *Escherichia coli* under the control of the T7 promoter (Studier & Moffat, 1986). Large amounts of soluble protein were obtained in the periplasmic fraction. However, Edman sequencing of the proteins revealed that both fragments 95-231 and 107-231 were amino-terminally degraded *in vivo*. All cleavage sites were found within the predicted first helix of PrP^C (amino acids 109-122), after residues 112, 118, and 120 (Hornemann & Glockshuber, 1996). This finding was surprising since proteolytic cleavage of folded proteins generally occurs at domain borders or within exposed loop regions, but not within regular secondary structures (Price & Johnson, 1993). We thus assumed that the segment 121-231, which is resistant to degradation in the periplasm of *E. coli*, represents an intrinsically stable domain and that residues 108-120 are not part of this domain.

In order to obtain the presumed carboxy-terminal domain PrP(121-231) with homogeneous amino-terminus, the segment 121-231 was directly fused to the OmpA signal sequence. To allow efficient cleavage of the signal sequence *in vivo* and to avoid a negative charge at the carboxylate to Ser231, recombinant PrP(121-231) was elongated by an additional serine residue at both the amino- and carboxy-terminus.

Periplasmic expression yielded a soluble 13.3 kDa protein. PrP(121-231) could be purified by conventional chromatographic techniques in the absence of any detergents by anion exchange chromatography, hydrophobic chromatography and gel filtration (Hornemann & Glockshuber, 1996). HPLC-analysis before and after reduction of purified

PrP(121-231) with dithiothreitol revealed that the single disulfide bond was formed quantitatively. The mass of the 113-residue protein was verified by electrospray mass spectrometry (error: 1 Da). PrP(121-231) proved to be soluble at concentrations up to 1 mM in distilled water without aggregation between pH 4 and 8.5 (Hornemann & Glockshuber, 1996). Analytical gel filtration on a Superdex 75 HR column revealed that the protein was a monomer in solution, even at high protein concentrations (Figure 1).

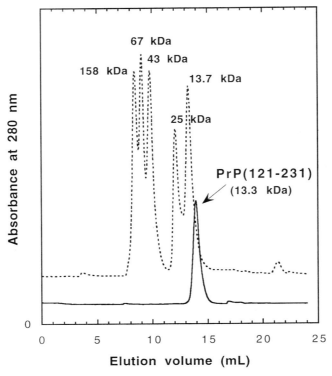

Figure 1. Analytical gel filtration of PrP(121-231) on a Superdex 75 HR column (30 x 1 cm). Gel filtration was performed at room temperature at a flow rate of 0.2 ml/min in 50 mM sodium phosphate, pH 7.0. The protein concentration of the applied sample (50 μl) was 0.1 mM. The solid line corresponds to the elution profile of purified PrP(121-231). The dashed line represents the elution profile of a mixture of standard proteins with known molecular mass.

Far-UV circular dichroism (CD) spectra of purified PrP(121-231) revealed two minima at 208 and 222 nm and a mean residue ellipticity of - 15600 deg cm^2/dmol at 222 nm (Hornemann & Glockshuber, 1996), which demonstrate a relatively high content of α-helices (Johnson Jr., 1994) (Figure 2). Comparison with the far-UV CD spectrum of intact PrP^C from hamster (Pan *et al.*, 1993) indicates that the relative content of regular secondary structure is significantly higher in PrP(121-231). In addition, a characteristic near-UV CD spectrum was obtained for PrP(121-231), which is diagnostic for the presence of tertiary structure (Figure 2, inset).

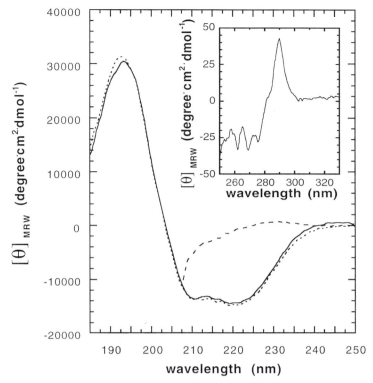

Figure 2. Far-UV circular dichroism spectra of native (solid line), unfolded (dashed line) and refolded (dotted line) PrP(121-231) at pH 7.0 and 22 °C (Inset: near-UV CD spectrum of native PrP(121-231)). Spectra were measured at protein concentrations of 0.5 mg/ml in 20 mM sodium phosphate pH 7.0. The sample of unfolded PrP(121-231) additionally contained 6 M GdmCl. Refolding of PrP(121-231) was achieved by dialysis against 20 mM sodium phosphate pH 7.0. Spectra were corrected for the buffer. The concentration of PrP(121-231) was determined by its absorbance at 280 nm ($A_{280nm, 1mg/ml, 1 cm}$ = 1.55) (reproduced from Hornemann & Glockshuber (1996) with permission).

In order to prove that PrP(121-231) represents an autonomous folding unit, its guanidinium chloride-dependent unfolding and refolding was measured at pH 7.0 using its far-UV CD signal at 222 nm. PrP(121-231) exhibits a cooperative and completely reversible unfolding transition (Figure 3). The spectroscopic properties of the refolded protein are identical to those of native PrP(121-231) (Figure 2). Evaluation of the data according to a two-state mechanism of folding yielded a free energy of folding of -21.8 ± 1.4 kJ/mol and a midpoint of unfolding at 2.53 M GdmCl (Hornemann & Glockshuber, 1996). The cooperativity of folding (m-value), which is genreally proportional to the mass of the protein and the difference in accessible surface area between the unfolded and folded state, has a value of 8.6 ± 0.5 kJ mol^{-1} M^{-1} GdmCl and is in the range expected for a 13.3 kDa protein (Myers *et al.*, 1995).

The biochemical analysis of PrP(121-231) thus gave strong indications that this amino-terminally truncated segment of the mouse prion protein is an isolated domain with defined tertiary structure and high intrinsic stability. Since there is no protein glycosylation in *E.coli*, folding and solubility do not require N-glycosylation at residues 181 and 198. Most importantly, refolding of chemically denatured PrP(121-231) is cooperative and reversible and yields a molecule that is indistinguishable from the native, recombinant protein. It is thus most likely that the three-dimensional structure of recombinant PrP(121-231) expressed in *E.coli* is identical to the structure of the corresponding segment in the cellular prion protein.

Within the framework of the protein-only hypothesis, the reversible unfolding of PrP(121-231) by GdmCl (Figure 3) has an intriguing implication if one assumes that the GdmCl-unfolded, intact cellular prion protein with all its posttranslational modifications also folds reversibly. If PrP^C and PrP^{Sc} indeed have identical covalent structres (Stahl & Prusiner, 1996), the two froms of PrP will yield identical unfolded forms in GdmCl (and probably in other denaturants), so that after reconstitution *in vitro* one would obtain folded PrP^C,in experiments strted either with PrP^C or PrP^{Sc}. This would explain why all attempts to reconstitute infectivity after solubilization of infectious PrP^{Sc} with high concentrations of GdmCl or urea have failed (Prusiner et al., 1993).

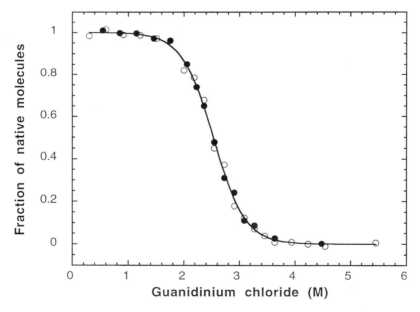

Figure 3. Reversible unfolding (o) and refolding (•) of PrP(121-231) in the presence of guanidinium chloride at pH 7.0 and 22 °C. Native or GdmCl-denatured PrP(121-231) was diluted 1:11 with 20 mM sodium phosphate pH 7.0 containing different GdmCl concentrations (final concentration of PrP(121-231): 38 μM). After incubation for 36 h, the mean residue ellipticity at 222 nm was recorded. The original data were evaluated according to a two-state transition using a six-parameter fit (solid line; Santoro & Bolen, 1988) and normalized (reproduced from Hornemann & Glockshuber (1996) with permission).

The intrinsically stable folding unit PrP(121-231) is not in accordance with the proposed three-dimensional structure of the PrP^C segment 108-218 (Huang *et al.*, 1994), as it lacks the first helix of residues 109-122 in the predicted four-helix bundle. In the following section we describe the three-dimensional structure of PrP(121-231) in solution and will show that it possesses a novel protein fold. We believe that the high stability and solubility of PrP(121-231) over a wide range of conditions will provide a basis for investigation of its kinetics of folding and unfolding, and for the identification of possible folding intermediates of PrP^C (Baldwin, 1996) that may be involved in the mechanism of transition from PrP^C to PrP^{Sc}.

THE NMR SOLUTION STRUCTURE OF PrP(121-231)

We decided to determine the three-dimensional structure of mouse PrP(121-231) for the following reasons: PrP(121-231) contains all residues which are posttranslationally modified in mature PrP^C, i.e. both glycosylation sites, the single disulfide bond and the residue bearing the GPI anchor. In addition, it contains most of the point-mutation sites in PrP which have been associated with inherited human prion diseases (Prusiner, 1993). Finally, experiments with transgenic mice exclusively expressing the segment 81–231 of mouse PrP proved that this segment is sufficient for generation and propagation of the prion disease *in vivo* (Fischer et al., 1996), which indicated that the carboxy-terminal part of PrP is of special functional importance.

The NMR structure determination of PrP(121-231) was performed at pH 4.5, 20 °C and protein concentrations of 0.8 mM using uniformly ^{15}N-labeled and ^{15}N/^{13}C doubly labeled samples. The ^{1}H, ^{13}C and ^{15}N resonances of the backbone were assigned by establishing intraresidual and sequential correlations of the amide ^{1}H and ^{15}N resonances with C$^\alpha$, C$^\beta$ and H$^\alpha$ signals using three-dimensional (3D) triple resonance experiments and 3D ^{15}N-resolved [^{1}H,^{1}H]-NOESY (nuclear Overhauser enhancement spectroscopy) experiments. The side-chain signals were assigned from 3D through-bond correlation NMR experiments. The initial NMR structure determination was based on sequence-specific resonance assignments at 93 % completeness. It was calculated with the program DIANA (Güntert et al., 1991) using 1368 NOE distance constraints and 227 dihedral angle constraints. The 20 conformers with the lowest DIANA target function values were energy-minimized using the program OPAL (Luginbühl et al., 1996) with the AMBER force field (Cornell et al., 1995). For residues 125–166 and 177–219 of these 20 conformers, the RMDS (root mean square distance) to the mean structure was 1.4 Å for the N, C$^\alpha$ and C' atoms, and 2.0 Å for all heavy atoms (Riek et al., 1996).

The NMR structure of PrP(121–231) revealed a so far unknown protein fold which differs clearly and extensively from the proposed three-dimensional model of PrP^C (Figure 4). Importantly, the relative orientation of the three helices in PrP(121–231) is clearly different from the relative orientation of any three helices in the proposed four-helix-bundle

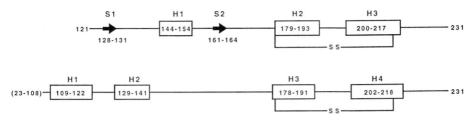

Figure 4. Comparison of the location of regular secondary structures in the NMR solution structure of PrP(121-231) and in the proposed four-helix bundle model by Huang et al. (1994).

model (Huang et al., 1994). A systematic search of the Brookhaven data bank with the program DALI (Holm & Sander, 1994) did not lead to the identification of other proteins with similar folds (Riek et al., 1996).

PrP(121-231) contains three α-helices and a two-stranded antiparallel β-sheet (Figures 5, 6a). The approximate lengths of the helices are from residues 144−154, 179−193 and 200−217, and of the β-strands from residues 128−131 and 161−164. The first turn of the second helix and the last turn of the third helix are linked by the single disulfide bond in the protein. The twisted V-shaped arrangement of these two longest helices forms a scaffold onto which the short ß-sheet and the first helix are anchored. At the present stage of refinement all regular secondary structure elements and the connecting loops are well defined with the sole exception of residues 167−176, which apprear to lie in a rather flexible loop region. 18 residues contribute to the hydrophobic core of PrP(121-231), which

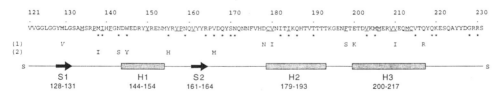

Figure 5. Location in the three-dimensional structure of PrP(121−231) of residues involved in sequence variations among mammalian prion proteins and of residues that have been associated with the species barrier of prion disease transmission and with inherited prion diseases. The primary structure of mouse PrP(121-231) and the location of regular secondary structures is shown. Residues contributing to the hydrophobic core of the domain are underlined, variable residues among mammalian prion proteins (Schätzl et al., 1995) are marked with asterisks. Line (1), mutations in human PrP which have been associated with inherited prion diseases (Prusiner, 1993; a stop codon at residue 145, which has been reported in addition to these point mutations, is not considered here, nor is the Met232Arg mutation, which is not contained in mature PrP^C). All of these residues are identical in wild type human PrP and mouse PrP. The polymorphism at codon 129 in human PrP, where homozygosity appears to increase susceptibility to sporadic CJD, is marked by italics. Line (2), residues in PrP(121−231) for which experimental evidence has been presented that they contribute to the species barrier of prion disease transmission between mice and humans (Telling et al., 1995). (Reproduced with permission from Riek et al., 1996)

includes side chains of the second helix (residues 179, 180, and 184), the third helix (residues 203, 206, 209, 210, 213, and 214), the second strand of the ß-sheet (Val161), the first, mostly hydrophilic helix (Tyr150), and three loop regions (residues 134, 137, 139, 141, 157, 158, and 198). The majority of the residues forming the hydrophobic core are invariant in the known mammalian prion protein sequences, with the exceptions of Ile139, Ile184 and Val203 (Figure 5). In contrast, most residues which are variable in mammalian prion proteins are solvent accessible (Riek et al., 1996).

The calculation of the electrostatic surface potential (Honig et al., 1994) of PrP(121-231) revealed that the protein possesses a pronouncedly uneven spatial distribution of acidic and basic residues, which might contribute to direct its orientation relative to the cell membrane (Riek et al., 1996). The positively charged side of PrP, which includes the ß-sheet (Figure 6a), would then probably bind to the membrane surface, while the negatively charged side of the protein, which contains both glycosylation sites and the single tryptophan residue, would be exposed to the solvent. The single, invariant disulfide bond of the prion protein is highly shielded from solvent contact and forms an important part of the hydrophobic core. This is in accordance with the observation that the disulfide bond is essential for folding of hamster PrP(90-231) (Mehlhorn et al., 1996). Another remarkable feature of the structure of PrP(121-231) is the fact that the first α-helix

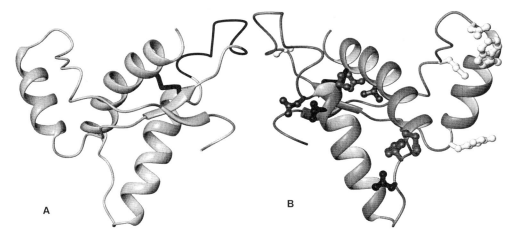

Figure 6. Globular fold and locations of selected residues in the three-dimensional structure of PrP(121−231). **a.** Ribbon diagram of the structure of PrP(121−231), indicating the positions of the three helices and the antiparallel two-stranded β-sheet. The connecting loops are displayed in grey if their structure is well-defined, and in black otherwise. The disulfide bond between Cys179 and Cys214 is shown in black. The N-terminal segment of residues 121−124 and the C-terminal segment 220−231 are disordered and not displayed. **b.** Locations of selected residues in the three-dimensional structure of PrP(121−231). The backbone is shown in grey and compared to **a** the molecule has been rotated by 180° around a vertical axis. The side chains of residues with mutations that have been associated with inherited prion diseases are highlighted in dark grey (cf. Figure 5, line (1)). The solvent-accessible glycosylation sites at Asn181 and Asn197 and the disulfide bond are shown in black. Five residues which may be involved in the mouse/human species barrier (cf. Figure 5, line (2)), are displayed in light grey (reproduced from Riek et al. (1996) with permission). The Figure was generated with the program MOLMOL (Koradi et al., 1996).

(residues 144-154) is relatively isolated from the rest of the structure. This helix is unusually hydrophilic, contains more that 50 % charged residues and does not show amphipathic character. It only contibutes a single residue, Tyr 150, to the hydrophobic core of the carboxy-terminal PrP domain (Riek et al., 1996).

FUNCTIONAL IMPLICATIONS OF THE THREE-DIMENSIONAL STRUCTURE OF PRP(121-231) FOR THE GENERATION OF PRION DISEASES

Regarding previous Fourier-transform infrared data (Pan et al., 1993) and structure predictions for PrP^C (Huang et al., 1994, 1996), the presence of a β-sheet in the NMR structure of PrP(121–231) is an unexpected finding. Since the fraction of β-sheet structure in PrP may increase upon transition from PrP^C to PrP^{Sc} (Pan et al., 1993), it is tempting to speculate that the short antiparallel β-sheet might be a "nucleation site" in PrP^C for a conformational transition to PrP^{Sc}. Such a transition could also include the loops connecting both strands of the ß-sheet to the first helix. In this context, it is interesting to note that the polymorphism at codon 129 in human PrP (methionine or valine), which has been linked to predisposition to sporadic CJD (Palmer et al., 1991), lies within the first strand of the ß-sheet.

The three-dimensional solution structure of PrP(121–231) allows the spatial localization of residues which have been proposed to be important for the species barrier of prion disease transmission (Scott et al., 1993; Telling et al., 1995) and which have been associated with inherited human prion diseases (Prusiner, 1993) (Figure 6b). The species barrier of prion disease transmission could principally result from an altered PrP^{Sc} binding site in PrP^C. The species barrier between mice and humans appears to be caused by one or more sequence differences between human and mouse PrP within the segment comprising residues 96–167 (Telling et al., 1995). There are a total of eight sequence differences between mouse and human PrP within this segment, of which five are contained in PrP(121–231). Four of these differences are located within or adjacent to the first helix, which might thus be part of a single binding site in PrP^C for PrP^{Sc} (Figure 6b). Regarding the aforementionned possible orientation of PrP^C relative to the cell membrane, the potential binding site for PrP^{Sc} would then be located on the accessible, solvent-exposed side of the cellular prion protein. Importantly, two of the three residues in PrP(121-231) which may determine the species barrier between mouse and hamster (Scott et al., 1993) are also located in the region of the first helix.

PrP(121-231) contains six of the point-mutation sites in mature, human PrP which have been linked with familial human prion diseases (Figure 5, Figure 6b). These six residues are exclusively located within or immediately adjacent to the last two helices which form the scaffold of the structure of PrP(121-231). Three of these six mutations (Val180Ile, Phe198Ser, Val210Ile) affect the hydrophobic core of the protein and three cause changes of charges at exposed residues (Asp178Asn, Glu200Lys, and Gln207Arg). It may be that some of these point mutations detabilize the three-dimensional structure of

PrP^C, which might facilitate its conversion to PrP^{Sc}. Other mutations might affect the self-aggregation of the prion protein and in this way contribute to easier transition to PrP^{Sc}. Replacement of the three exposed residues might however also influence the ligand binding properties of PrP^C. Involvement of a natural ligand that competes with PrP^{Sc} for binding to PrP^C would thus be another explanation why these mutations might favour the generation and propagation of infectious prions.

Besides the fact that homozygosity at codon 129 in human PrP leads to a predisposition to sporadic CJD (Palmer et al., 1991), the Met/Val 129 polymorphism has another striking consequence. In conjunction with the inherited Asp178Asn mutation, it determines the phenotype of this familial prion disease: While the Met129/Asn178 allele segregates with fatal familial insomnia, the Val129/Asn178 allele segregates with inherited CJD (Goldfarb et al., 1992). Analysis of the three-dimensional structure of PrP(121-231) shows that, although there is no close approach of the polypeptide backbone near the residues 129 and Asn 178, side chain/side chain interactions would be sterically possible. More precise information is expected from the ongoing structure refinement.

In conclusion, we believe that the present three-dimensional NMR structure of PrP(121−231) will provide a basis for rational design of *in vitro* and *in vivo* experiments to test the above-mentionned hypotheses on the generation and propagation of prion diseases. For example, if PrP^C did indeed undergo a substantial conformational change during the transition to PrP^{Sc}, it might be possible to generate PrP^{Sc}-specific antibodies using buried segments in the structure of PrP(121-231) as haptens.

ACKNOWLEDGEMENTS

We thank M. Fischer, A. Raeber, C. Mumenthaler, and C. Weissmann for discussions. This project was supported by the Schweizerischer Nationalfonds (Project 31.32035.91) and the ETH Zurich. S. H. was supported by the Boehringer-Ingelheim-Fonds.

REFERENCES

Alper, T., Cramp, W. A., Haig, D. A., and Clarke, M. C., 1967, Does the agent of scrapie replicate without nucleic acid?, *Nature* 14:764-766.

Baldwin, R. L., 1996, On-pathway versus off-pathway folding intermediates, *Folding & Design* 1:R1-R8.

Bessen, R. A., Kocisko, D. A., Raymond, G. J., Nandan, S., Lansbury, P. T., and Caughey, B., 1995, Non-genetic propagation of strain-specific properties of scrapie prion protein, *Nature* 375:698-700.

Bolton, D. C., McKinley, M. P., and Prusiner, S. B., 1982, Identification of a protein that purifies with the scrapie prion, *Science* 218:1309-1311.

Büeler, H., Aguzzi, A., Sailer, A., Greiner, R. A., Autenried, P., Aguet, M., and Weissmann, C., 1993, Mice devoid of PrP are resistant to scrapie, *Cell* 73:1339-1347.

Collinge, J., Whittington, M. A., Sidle, K. C., Smith, C. J., Palmer, M. S., Clarke, A. R., and Jefferys, J. G. R., 1994, Prion protein is necessary for normal synaptic function, *Nature* 370:295-297.

Cornell, W. D., Cieplak, P., Bayly, C. I., Gould, I. R., Merz Jr., K. M., Ferguson, D. M., Spellmeyer, D. C., Fox, T., Caldwell, J. W., and Kollmam, P. A., 1995, A second generation force field for the simulation of proteins, nucleic acids, and organic molecules, *J. Am. Chem. Soc.* 117:5179-5197.

Fischer, M., Rülicke, T., Raeber, A., Sailer, A., Moser, M., Oesch, B., Brandner, S., Aguzzi, A., and Weissmann, C. , 1996, Prion protein (PrP) with amino-proximal deletions restoring susceptibility of PrP knockout mice to scrapie, *EMBO J.* 15:1255-1264.

Goldfarb, L. G., Petersen, R. B., Tabaton, M., Brown, P., LeBlanc, A. C., Montagna, P., Cortelli, P., Julien, J., Vital., C., Pendelbury, W. W., Haltia, M., Wills, P. R., Hauw, J. J., McKeever, P. E., Monari, L., Schrank, B., Swergold, G. D., Autilio-Gambetti, L., Gajdusek, D. C., Lugaresi, E., and Gambetti, P., 1992, Fatal familial insomnia and familial Creutzfeldt-Jakob disease: Disease phenotype determined by a DNA polymorphism, *Science* 258: 806-808.

Griffith, J. S., 1967, Self-replication and scrapie, *Nature* 215:1043-1044.

Güntert, P., Braun, W., and Wüthrich, K., 1991, Efficient computation of three-dimensional protein structures in solution from nuclear magnetic resonance data using the program DIANA and the supporting programs CALIBA, HABAS and GLOMSA, *J. Mol. Biol.* 217:517-530.

Holm, L. and Sander, C., 1994, Protein structure comparison by alignment of distance matrices, *J. Mol. Biol.* 233:123-138.

Honig, B. and Nicholls, A., 1994, Classical electrostatics in biology and chemistry, *Science* 268:1144-1149.

Hornemann, S. and Glockshuber, R., 1996, Autonomous and reversible folding of a soluble amino-terminally truncated segment of the mouse prion protein, *J. Mol. Biol.* 261:614-619.

Hsiao, K., Baker, H. F., Crow, T. J., Poulter, M., Owen, F., Terwilliger, J. D., Westaway, D., Ott, J., and Prusiner, S. B., 1989, Linkage of a prion protein missense variant to Gerstmann-Sträussler syndrome, *Nature* 338:342-345.

Huang, Z., Gabriel, J.-M., Baldwin, M. A., Fletterick, R. J., Prusiner, S. B., and Cohen, F. E., 1994, Proposed three-dimensional structure for the cellular prion protein, *Proc. Natl. Acad. Sci., U.S.A.* 91:7139-7143.

Huang, Z., Prusiner S. B. & Cohen, F. E., 1995, Scrapie prions: a three-dimensional model of an infectious fragment, *Folding & Design* 1:13-19.

Jarrett, J. T. and Lansbury Jr., P. T., 1993, Seeding "one-dimensional crystallization" of amyloid: a pathogenic mechanism in Alzheimer's disease and scrapie? *Cell* 73:1055-1058.

Johnson Jr., W. C., 1990, Protein secondary structure and circular dichroism: A practical guide, *Proteins: Struct. Funct. Genet.* 7:205-214.

Koradi, R., Billeter, M., and Wüthrich, K., 1996, MOLMOL: a program for display and analysis of macromolecular structures, *J. Mol. Graph.* 14: 51-55 (1996).

Lledo, P.-M., Tremblay, P., DeArmond, S.J., Prusiner, S. B., and Nicoll, R. A., 1996, Mice deficient for prion protein exhibit normal neuronal excitability and synaptic transmission in the hippocampus, *Proc. Natl. Acad. Sci. U.S.A.* 93:2403-2407.

Luginbühl, P., Güntert, P., Billeter, M., and Wüthrich, K., 1996, The new program OPAL for molecular dynamics simulations and energy refinements of biological macromolecules, *J. Biomol. NMR* 8:136-146.

Mehlhorn, I., Groth, D., Stöckel, J., Moffat, B., Reilly, D., Yansura, D., Willett, W.S., Baldwin, M., Fletterick, R., Cohen, F. E., Vandlen, R., Henner, D., and Prusiner, S. B., 1996, High-level expression and characterization of a purified 142 residue polypeptide of the prion protein, *Biochemistry* 35:5528-5537.

Myers, J. K., Pace, C. N. & Scholtz, J. M., 1995, Denaturant m values and heat capacity changes: Relation to changes in accessible surface areas of protein unfolding, *Protein Sci.* 4:2138-2148.

Palmer, M. S., Dryden, A. J., Hughes, J. T., and Collinge, J., 1991, Homozygous prion protein genotype predisposes to sporadic Creutzfeldt-Jakob disease, *Nature* 352:340-342.

Pan, K.-M., Baldwin, M., Nguyen, J., Gasset, M., Serban, A., Groth, D., Mehlhorn, I., Huang, Z., Fletterick, R. J., Cohen, F. E., and Prusiner, S. B., 1993, Conversion of α-helices into ß-sheets features in the formation of the scrapie prion proteins, *Proc. Natl. Acad. Sci., U.S.A.* 90:10962-10966.

Price, N. C. and Johnson, C. M., 1993, Proteinases as probes of conformation of soluble proteins. In *Proteolytic Enzymes, A Practical Approach* (Beynon, R. J. & Bond, J. S., eds.), pp. 163-180, IRL Press, Oxford.

Prusiner, S. B., 1982, Novel proteinaceous infectious particles cause scrapie, *Science* 216:136-144.

Prusiner, S. B., 1991, Molecular biology of prion diseases, *Science* 252:1515-1522.

Prusiner, S. B., 1993, Genetic and infectious prion diseases, *Arch. Neurol.* 50:1129-1153.

Prusiner, S. B., Groth, D., Serban, A., Stahl, N., and Gabizon, R., 1993, Attempts to restore scrapie infectivity after exposure to protein denaturants, *Proc. Natl. Acad. Sci. U.S.A.* 90:2793-2797.

Riek, R., Hornemann, S., Wider, G., Billeter, M., Glockshuber, R., and Wüthrich, K., 1996, NMR structure of the mouse prion protein domain PrP(121-231), *Nature* 382:180-182.

Santoro, M. M. and Bolen, D. W., 1988, Unfolding free energy changes determined by the linear extrapolation method. 1. Unfolding of phenylmethanesulfonyl α-chymotrypsin using different denaturants, *Biochemistry* 27:8063-8068.

Sakaguchi, S., Katamine, S., Nishida, N., Moriuchi, R., Shigematsu, K., Sugimoto, T., Nakatani, A., Katoaka, Y., Houtani, T., Shirabe, S., Okada, H., Hasegawa, S., Miyamoto, T., and Noda, T., 1996, Loss of cerebellar Purkinje cells in aged mice homozygous for a disrupted *PrP* gene, *Nature* 380:528-531.

Schätzl, H. M., Da Costa, M., Taylor, L., Cohen F. E., and Prusiner, S. B., 1995, Prion protein gene variation among primates, *J. Mol. Biol.* 245:362-374.

Scott, M., Groth, D., Foster, D., Torchia, M., Yang, S-L., DeArmond, S. J., and Prusiner, S. B., 1993, Propagation of prions with artificial properties in transgenic mice expressing chimeric PrP genes, *Cell* 73:979-988.

Stahl, N. and Prusiner S. B., 1991, Prions and prion proteins. *FASEB J.* 5:2799-2807.

Studier, F. W. and Moffatt, B. A., 1986, Use of bacteriophage T7 RNA polymerase to direct selective high-level expression of cloned genes, *J. Mol. Biol.* 189:113-130.

Telling, G. C., Scott, M., Mastrianni, J., Gabizon, R., Torchia, M., Cohen, F. E., DeArmond, S. J., and Prusiner, S. B., 1995, Prion propagation in mice expressing human and chimeric PrP transgenes implicates the interaction of cellular PrP with another protein, *Cell* 83:79-90.

Tobler, I., Gaus, S. E., Deboer, T., Achermann, P., Fischer, M., Rülicke, T., Moser, M., Oesch, B., McBride, P. A., and Manson, J. C., 1996, Altered circadian activity rhythms and sleep in mice devoid of prion protein, *Nature* 380:639-642.

Weissmann, C., 1994, Molecular biology of prion diseases, *Trends Cell Biol.* 4:10-14.

Weissmann, C., 1996, Molecular biology of transmissible spongiform encephalopathies, *FEBS Lett.* 398:3-11.

Westaway, D., Goodman, P. A., Mirenda, C. A., McKinely, M.P., Carlson, G. A., and Prusiner, S. B., 1987, Distinct prion proteins in short and long scrapie incubation period mice, *Cell* 51:651-662.

Wüthrich, K., 1986, NMR of Proteins and Nucleic Acids, Wiley, New York

PROTEASE-RESISTANT PRION PROTEIN FORMATION

Byron Caughey and Gregory J. Raymond

Laboratory of Persistent Viral Diseases
Rocky Mountain Laboratories
National Institute of Allergy and Infectious Diseases
National Institutes of Health
Hamilton, MT 59840, USA

INTRODUCTION

In 1981, unique fibrillar structures called scrapie associated fibrils (SAF) were identified in in scrapie-infected, but not normal, brain tissue [1]. Biochemical analysis of these fibrils and of other infectious tissue preparations from hosts with different transmissible spongiform encephalopathies (TSE) led to the discovery that they contained an abnormal form of a host protein which is now most widely known as prion protein (PrP) [2,3]. The abnormal TSE-associated PrP forms insoluble aggregates (e.g. amyloid plaques) and is partially resistant to proteinase K (PK) [3,4], which removes approximately 67 amino acid residues (6-7 kDa) from the N-terminus of each molecule in the polymer [5,6]. Although the PK-resistance of various abnormal TSE-associated PrPs can differ somewhat, we refer to them generically as protease-resistant PrP (PrP-res), but the forms associated with specific TSE diseases are often labelled with a superscript designating that disease, e.g. PrP^{Sc} for scrapie PrP and PrP^{BSE} for bovine spongiform encephalopathy-associated PrP. In contrast, the normal, non-pathogenic form, PrP-sen or PrP^C, is usually soluble in mild detergents and fully degraded by PK [7,8].

There are many indications that PrP plays a critical role in TSEs. Thus, it is of interest to understand how PrP-res is made and how it relates to the infectious agents of TSE diseases. One popular idea is that PrP-res might be the self-inducing infectious protein modelled by Griffith[9]. In TSE-infected hosts, PrP-res usually accumulates in the central nervous system and lymphoreticular tissues [2,3,8,10-12]. Both the normal and abnormal PrP isoforms are encoded by the same host gene [13] and no TSE associated differences have been observed in either the PrP mRNA levels [6,14] or the chemical structures of the PrP isoforms [5,15]. Pulse-chase metabolic labelling studies have shown that PrP-res is derived from mature PrP-sen posttranslationally [16,17]. These observations led to suggestions that the primary difference between PrP-res and PrP-sen might be conformational or due to interactions with cofactors[5]. Indeed, conformational analyses have provided evidence that PrP-res has a much higher beta sheet secondary structure composition than PrP-sen[18-20].

In scrapie-infected cells, the conversion of PrP to the protease-resistant state likely occurs at the plasma membrane or along an endocytic pathway to the lysosomes [17,21,22]. Once formed, PrP-res accumulates in secondary lysosomes[23], on the cell surface[24] or in extracellular spaces in the form of amorphous deposits, diffuse fibrils[25,26] or dense amyloid plaques[27].

A major unresolved question is how, at the molecular level, the posttranslational conversion of PrP-sen to PrP-res occurs. Most protein-only models for the TSE agent predict that the putative infectious protein, e.g. PrP-res, directly interacts with the endogenous PrP-sen to induce the conversion of the latter to more PrP-res [9,28-31]. Passage of PrP-res from one host to another might then constitute "infection" which could initiate pathogenic PrP-res formation in the new host. In the case of familial TSE diseases, mutations in the host PrP gene might facilitate spontaneous PrP-res formation without the need for infection with preformed PrP-res. Interestingly, PrP molecules containing mutations that are associated with humans familial TSE diseases exhibit some biochemical properties of PrP-res in when expressed in uninfected tissue culture cells [32,33].

The importance of direct homologous PrP-sen -PrP-res interactions in PrP-res formation was implied by a number of indirect lines of evidence. For example, in heterozygous humans with GSS (Indiana and Swedish kindreds) who express both the normal and mutant PrP alleles only the mutant PrP is incorporated into amyloid fibrils[34]. Studies in scrapie-infected transgenic mice and cell cultures have also shown that specific sequence homologies between PrP-sen and PrP-res molecules are required for efficient PrP-res formation[35-37]. Conversely, expression of incompatible PrP-sen molecules differing by as little as one amino acid residue from the endogenous PrP can completely block PrP-res formation in scrapie-infected cells[37,38]. Analogous interference effects have been observed in the growth of synthetic amyloid fibrils by PrP peptide fragments when peptides of slightly different sequence are mixed together[39].

SELF-INDUCED PrP-RES FORMATION

Direct evidence that PrP-res can induce its own formation has come from recent *in vitro* studies in a cell-free system showing that PrP-sen can convert to the protease-resistant form in the presence of PrP-res isolated from scrapie-infected brain tissue[40]. Striking specificity was observed in the cell-free conversion reaction which were consistent with, and might account for, biological aspects of TSE infections such as the existence of TSE strains and barriers in transmitting TSE strains between species (species barrier effects) [18,41].

Species specificity in the conversion of PrP-res to PrP-sen: a basis for TSE species barriers

The molecular basis for species barrier effects has been investigated by combining PrP-res and PrP-sen from several different species (mice, hamsters, sheep, cattle and human) in the cell-free conversion reaction [[18,42] and Raymond et al., submitted]. To a remarkable extent, the efficiency of the cell-free conversion reactions correlated with in vivo transmissibilities of the corresponding agents within or between species. The homologous conversion reactions between PrP-res and PrP-sen molecules of the same sequence were most efficient, just as TSE agents are most efficiently transmitted to hosts of the same PrP genotype. In contrast, little or no conversion was observed with PrP-res-PrP-sen combinations associated with a lack of transmissibility *in vivo*. For example, PrP^{BSE} readily converted bovine, murine, and two different ovine PrP-sen molecules to PrP-res and this correlates with the fact that BSE is transmissible to hosts expressing these types of PrP-sen. However, PrP^{BSE} failed to convert the PrP-sen of two BSE-resistant hosts, Syrian hamsters and sheep of the A^{136} R^{171} PrP genotype. Sheep PrP-res with the $V^{136}Q^{171}$ sequence efficiently converted the PrP-sen of

$V^{136}Q^{171}$ and $A^{136}Q^{171}$ sheep, which are susceptible to scrapie agent from $V^{136}Q^{171}$ sheep, but not the $A^{136}R^{171}$ PrP-sen of a scrapie-resistant strain of sheep [42]. Conversion results from these and other PrP-res-PrP-sen combinations have provided strong support for the concept that the sequence specificity in the conversion of PrP-sen to protease-resistant forms strongly modulates the interspecies or intraspecies transmissibilities of TSE agents *in vivo*.

Strain specificity in PrP-res formation: a role for conformational variations

Various strains of TSE agents can be differentiated on the basis of species tropism, incubation period, clinical disease, neuropathological manifestations and PrP-res distribution in brain tissue [for review [43]]. Multiple TSE strains have been documented even within individual host species with invariant PrP genotype. This fact poses an interesting challenge for the protein-only hypothesis for the infectious agent: it requires that the "inheritance" or propagation of the agent strain differences must be mediated by stable variations in PrP-res structure rather than mutations in an agent-specific nucleic acid.

Strain differences in PrP-res structure have now been described in different host species [44-47]. Most remarkable are the types of PrP-res associated with the hyper (HY) and drowsy (DY) strains of hamster-adapted transmissible mink encephalopathy (TME) because they have identical amino acid sequence and seem to differ solely in conformation [45]. Furthermore, using the cell-free conversion reaction, we have demonstrated that HY and DY PrP-res molecules impart their strain-specific proteinase K resistance on hamster PrP-sen during formation of new PrP-res [41]. This represents the in vitro propagation of these strain specific biochemical phenotypes of PrP-res by a nongenetic mechanism. These data are consistent with the concept of the self-propagation of PrP-res polymers with distinct 3-D structures as a molecular basis for scrapie strains.

Correlation between scrapie infectivity and the converting activity, protease-resistance and aggregation of PrP-res

Considering the possibility that PrP-res might be the infectious agent which depends on the converting activity for its self-propagation, we tested whether the effects of GdnHCl on the converting activity, PK-resistance and aggregation of PrP^{Sc} also affect scrapie infectivity [48]. Large GdnHCl-induced reductions in infectivity were associated with the irreversible elimination of both the proteinase K-resistance and self-propagating converting activity of PrP^{Sc}. In intermediate GdnHCl concentrations that stimulate converting activity and partially disaggregate PrP^{Sc}, both scrapie infectivity and converting activity were associated with residual partially protease-resistant multimers of PrP^{Sc}.

In situ conversion of PrP-sen to PrP-res in TSE-infected brain tissue

To determine if the induced conversion of PrP-sen to PrP-res could be observed in TSE-infected brain tissue as well as in the cell-free system, an *in situ* conversion assay was developed [49]. Infected brain tissue slices were bound to microscope slide, incubated with radiolabelled PrP-sen and then treated with PK to digest any PrP-sen that was not converted to the PK-resistant form. Biochemical analysis indicated that the expected labelled PrP-res conversion products were formed in, and bound to, infected, but not uninfected brain slices. Autoradiographic analysis of the pattern of newly formed PrP-res in the infected brain slice closely matched the distribution of the preexisting PrP-res. Punctate *in situ* PrP conversion was observed in brain regions containing PrP-res amyloid plaques and a more dispersed conversion product was detected in areas containing diffuse PrP-res deposits. These studies provide direct evidence that PrP-res formation involves the incorporation of soluble PrP-sen into both non-fibrillar and fibrillar PrP-res deposits in TSE-infected brain.

Mechanism of PrP-res formation

The insolublility, protease resistance, birefringent staining with Congo red and high beta sheet content make at least some PrP-res deposits similar to the abnormal amyloids deposits found in a number of other diseases [4,29]. Amyloids can be composed of a number of proteins, depending upon the disease[50]. Amyloid deposits are composed of linear fibrils that result from the polymerization of a ususally soluble precursor protein or peptide. Seeding solutions of the precursor with preexisting amyloid fibril fragments can greatly accelerate the polymerization of amyloidogenic proteins [29,31]. Amyloid polymerization often involves an increase in the beta sheet content of the constituent protein. The similarities between PrP-res and other amyloids suggested that the mechanism of PrP-res formation is like that of other amyloids[29,31,51]. Additional support for this notion is the fact that small synthetic peptide fragments of the PrP sequence can form amyloid fibrils by a seeded polymerization mechanism[52].

Our recent studies using the full PrP-res protein have provided evidence that only ordered aggregates of PrP-res, albeit widely variable in size, can induce the conversion of PrP-sen to the protease-resistant form in the cell-free system[53]. The polymerized state of PrP-res also correlates with its PK-resistance, its ability to renature to full proteinase K resistance after partial denaturation, and with the presence of scrapie infectivity [48,54]. Furthermore, the in situ conversion reaction in brain slices shows that the conversion product is bound to those deposits and not released into the medium [49]. These observations are also consistent with a seeded polymerization mechanism for PrP-res formation.

However, some investigators have argued that polymerization of PrP is not required for PrP-res formation [30,55] because not all deposits of PrP-res in vivo show birefringent staining with Congo red or have readily visible amyloid fibril structures by electron microscopy [e.g. [7]]. One proposed mechanism, often called the heterodimer model, holds that a PrP-res monomer binds to a monomer of PrP-sen to form a heterodimer[9,28,30]. The heterodimer then spontaneously converts to a PrP-res homodimer and then splits into two PrP-res monomers. The aggregation of PrP-res would then be a secondary phenomenon that results in vivo or in vitro only from partial proteolysis and/or extraction from the membrane environment [56]. The fact that no proteinase K-resistant monomer of PrP has been identified runs counter to this argument. Nonetheless, there have been reports of scrapie infectivity that cofractionates with monomeric forms of PrP [57,58], but these studies not been confirmed[59,60]. Our studies showing that ordered aggregates of PrP-res are active in converting PrP-sen to PrP-res demonstrate that there is, at least, no obligate requirement for a free PrP-res monomer, if one should exist, in the conversion mechanism[53].The lack of visible fibrils in some tissue and membrane fractions containing PrP-res might be explained by a prevalence of short PrP-res polymers or the association of the PrP-res polymer with other fibrils, PrP-sen or other factors that obscure its underlying fibrillar ultrastructure and affect its birefringent staining with Congo red. Consistent with this idea is the fact that, in the in situ conversion reaction in brain slices, the diffuse and amyloid plaque PrP-res deposits are functionally similar in that they are both capable of inducing PrP conversion [49]. The nucleated polymerization model suggests that PrP-res polymers ranging in size from huge amyloid plaques down to stable oligomers containing only several PrP monomers could seed the polymerization reaction. Theoretical consideration of the likely volume of a PrP monomer compared to the dimensions of SAF suggests that fibrils containing 60 PrP molecules might be no longer than they are wide and, therefore, would not be visible as a fibril ultrastructurally. Furthermore, unless the PrP-res polymers are long and aligned in large oriented bundles or radiating amyloid plaques that are visible by light microscopy ($>\sim0.5$ µm), staining by Congo red would not appear birefringent under polarized light. Thus, the absence of readily visible, congophilic fibrils containing hundreds or thousands of PrP-res molecules is *not* persuasive evidence that PrP-res is usually monomeric.

REFERENCES

1. P.A. Merz, R.A. Somerville, H.M. Wisniewski, and K. Iqbal, Abnormal fibrils from scrapie-infected brain, *Acta Neuropathol.* 54:63 (1981).
2. D.C. Bolton, M.P. McKinley, and S.B. Prusiner, Identification of a protein that purifies with the scrapie prion, *Science* 218:1309 (1982).
3. H. Diringer, H. Gelderblom, H. Hilmert, M. Ozel, C. Edelbluth, and R.H. Kimberlin, Scrapie infectivity, fibrils and low molecular weight protein, *Nature* 306:476 (1983).
4. S.B. Prusiner, M.P. McKinley, K.A. Bowman, P.E. Bendheim, D.C. Bolton, D.F. Groth, and G.G. Glenner, Scrapie prions aggregate to form amyloid-like birefringent rods, *Cell* 35:349 (1983).
5. J. Hope, L.J.D. Morton, C.F. Farquhar, G. Multhaup, K. Beyreuther, and R.H. Kimberlin, The major polypeptide of scrapie-associated fibrils (SAF) has the same size, charge distribution and N-terminal protein sequence as predicted for the normal brain protein (PrP), *EMBO J.* 5:2591 (1986).
6. B. Oesch, D. Westaway, M. Walchli, M.P. McKinley, S.B.H. Kent, R. Aebersold, R.A. Barry, P. Tempst, D.B. Teplow, L.E. Hood, et al. A cellular gene encodes scrapie PrP 27-30 protein, *Cell* 40:735 (1985).
7. R.K. Meyer, M.P. McKinley, K.A. Bowman, M.B. Braunfeld, R.A. Barry, and S.B. Prusiner, Separation and properties of cellular and scrapie prion protein, *Proc. Natl. Acad. Sci. USA* 83:2310 (1986).
8. R. Rubenstein, R.J. Kascsak, P.A. Merz, M.C. Papini, R.I. Carp, N.K. Robakis, and H.M. Wisniewski, Detection of scrapie-associated fibril (SAF) proteins using anti-SAF antibody in non-purified tissue preparations, *J. Gen. Virol.* 67:671 (1986).
9. J.S. Griffith, Self-replication and scrapie, *Nature* 215:1043 (1967).
10. M. Shinagawa, E. Munekata, S. Doi, K. Takahashi, H. Goto, and G. Gato, Immunoreactivity of a synthetic pentadecapeptide corresponding to the N-terminal region of the scrapie prion protein, *J. Gen. Virol.* 67:1745 (1986).
11. R. Rubenstein, P.A. Merz, R.J. Kascsak, C.L. Scalici, M.C. Papini, R.I. Carp, and R.H. Kimberlin, Scrapie-infected spleens: analysis of infectivity, scrapie-associated fibrils, and protease-resistant proteins, *J. Infect. Dis.* 164:29 (1991).
12. R.E. Race and D. Ernst, Detection of proteinase K-resistant prion protein and infectivity in mouse spleen by 2 weeks after scrapie agent inoculation, *J. Gen. Virol.* 73:3319 (1992).
13. K. Basler, B. Oesch, M. Scott, D. Westaway, M. Walchli, D.F. Groth, M.P. McKinley, S.B. Prusiner, and C. Weissmann, Scrapie and cellular PrP isoforms are encoded by the same chromosomal gene, *Cell* 46:417 (1986).
14. B. Chesebro, R. Race, K. Wehrly, J. Nishio, M. Bloom, D. Lechner, S. Bergstrom, K. Robbins, L. Mayer, J.M. Keith, et al. Identification of scrapie prion protein-specific mRNA in scrapie-infected and uninfected brain, *Nature* 315:331 (1985).
15. N. Stahl, M.A. Baldwin, D.B. Teplow, L. Hood, B.W. Gibson, A.L. Burlingame, and S.B. Prusiner, Structural studies of the scrapie prion protein using mass spectrometry and amino acid sequencing, *Biochemistry* 32:1991 (1993).
16. D.R. Borchelt, M. Scott, A. Taraboulos, N. Stahl, and S.B. Prusiner, Scrapie and cellular prion proteins differ in the kinetics of synthesis and topology in cultured cells, *J. Cell Biol.* 110:743 (1990).
17. B. Caughey and G.J. Raymond, The scrapie-associated form of PrP is made from a cell surface precursor that is both protease- and phospholipase-sensitive, *J. Biol. Chem.*

266:18217 (1991).

18. D.A. Kocisko, S.A. Priola, G.J. Raymond, B. Chesebro, P.T. Lansbury, Jr., and B. Caughey, Species specificity in the cell-free conversion of prion protein to protease-resistant forms: a model for the scrapie species barrier, *Proc. Natl. Acad. Sci. USA* 92:3923 (1995).

19. K.-M. Pan, M. Baldwin, J. Nguyen, M. Gasset, A. Serban, D. Groth, I. Mehlhorn, Z. Huang, R.J. Fletterick, F.E. Cohen, et al. Conversion of alpha-helices into beta-sheets features in the formation of the scrapie prion protein, *Proc. Natl. Acad. Sci. USA* 90:10962 (1993).

20. J. Safar, P.P. Roller, D.C. Gajdusek, and C.J. Gibbs, Jr. Conformational transitions, dissociation, and unfolding of scrapie amyloid (prion) protein, *J. Biol. Chem.* 268:20276 (1993).

21. B. Caughey, G.J. Raymond, D. Ernst, and R.E. Race, N-terminal truncation of the scrapie-associated form of PrP by lysosomal protease(s): implications regarding the site of conversion of PrP to the protease-resistant state, *J. Virol.* 65:6597 (1991).

22. D.R. Borchelt, A. Taraboulos, and S.B. Prusiner, Evidence for synthesis of scrapie prion protein in the endocytic pathway, *J. Biol. Chem.* 267:16188 (1992).

23. M.P. McKinley, A. Taraboulos, L. Kenaga, D. Serban, A. Stieber, S.J. DeArmond, S.B. Prusiner, and N. Gonatas, Ultrastructural localization of scrapie prion proteins in cytoplasmic vesicles of infected cultured cells, *Lab. Invest.* 65:622 (1991).

24. M. Jeffrey, C.M. Goodsir, M.E. Bruce, P.A. McBride, J.R. Scott, and W.G. Halliday, Infection specific prion protein (PrP) accumulates on neuronal plasmalemma in scrapie infected mice, *Neurosci. Lett.* 147:106 (1992).

25. C.A. Wiley, P.G. Burrola, M.J. Buchmeier, M.K. Wooddell, R.A. Barry, S.B. Prusiner, and P.W. Lampert, Immuno-gold localization of prion filaments in scrapie-infected hamster brains, *Lab. Invest.* 57:646 (1987).

26. M. Jeffrey, C.M. Goodsir, M.E. Bruce, P.A. McBride, N. Fowler, and J.R. Scott, Murine scrapie-infected neurons in vivo release excess prion protein into the extracellular space, *Neurosci. Lett.* 174:39 (1994).

27. P.E. Bendheim, R.A. Barry, S.J. DeArmond, D.P. Stites, and S.B. Prusiner, Antibodies to a scrapie prion protein, *Nature* 310:418 (1984).

28. D.C. Bolton and P.E. Bendheim, A modified host protein model of scrapie, in: *Novel Infectious Agents and the Central Nervous System,* G. Bock, J. Marsh, eds., John Wiley & Sons, Chichester (1988).

29. D.C. Gadjusek, Transmissible and nontransmissible amyloidoses: Autocatalytic post-translational conversion of host precursor proteins to beta-pleated configurations, *J. Neuroimmunol.* 20:95 (1988).

30. S.B. Prusiner, Molecular biology of prion diseases, *Science* 252:1515 (1991).

31. J.T. Jarrett and P.T. Lansbury, Jr. Seeding "One-Dimensional Crystallization" of Amyloid: A Pathogenic Mechanism in Alzheimer's Disease and Scrapie? *Cell* 73:1055 (1993).

32. S. Lehmann and D.A. Harris, A mutant prion protein displays an aberrant membrane association when expressed in cultured cells, *J. Biol. Chem.* 270:24589 (1995).

33. S. Lehmann and D.A. Harris, Mutant and infectious prion proteins display common biochemical properties in cultured cells, *J. Biol. Chem.* 271:1633 (1996).

34. F. Tagliavini, F. Prelli, G. Porro, G. Rossi, G. Giaccone, M.R. Farlow, S.R. Dlouhy, B. Ghetti, O. Bugiani, and B. Frangione, Amyloid fibrils in Gerstmann-Straussler-Scheinker disease (Indiana and Swedish kindreds) express only PrP peptides encoded by the mutant allele, *Cell* 79:695 (1994).

35. S.B. Prusiner, M. Scott, D. Foster, K.M. Pan, D. Groth, C. Mirenda, M. Torchia, S.L. Yang, D. Serban, G.A. Carlson, et al. Transgenetic studies implicate interactions between homologous PrP isoforms in scrapie prion replication, *Cell* 63:673 (1990).

36. M.R. Scott, R. Kohler, D. Foster, and S.B. Prusiner, Chimeric prion protein expression in cultured cells and transgenic mice, *Protein Sci.* 1:986 (1992).

37. S.A. Priola, B. Caughey, R.E. Race, and B. Chesebro, Heterologous PrP molecules interfere with accumulation of protease-resistant PrP in scrapie-infected murine neuroblastoma cells. *J. Virol.* 68:4873 (1994).

38. S.A. Priola and B. Chesebro, A single hamster amino acid blocks conversion to protease-resistant PrP in scrapie-infected mouse neuroblastoma cells, *J. Virol.* 69:7754 (1995).

39. J.H. Come and P.T. Lansbury, Jr. Predisposition of prion protein homozygotes to Creutzfeldt-Jakob disease can be explained by a nucleation-dependent polymerization mechanism, *J. Am. Chem. Soc.* 116:4109 (1994).

40. D.A. Kocisko, J.H. Come, S.A. Priola, B. Chesebro, G.J. Raymond, P.T. Lansbury, and B. Caughey, Cell-free formation of protease-resistant prion protein. *Nature* 370:471 (1994).

41. R.A. Bessen, D.A. Kocisko, G.J. Raymond, S. Nandan, P.T. Lansbury, Jr., and B. Caughey, Nongenetic propagation of strain-specific phenotypes of scrapie prion protein, *Nature* 375:698 (1995).

42. A. Bossers, P.B.G.M. Belt, G.J. Raymond, B. Caughey, R. de Vries, and M.A. Smits, Scrapie susceptibility-linked polymorphisms modulate the *in vitro* conversion of sheep prion protein to protease-resistant forms, *Proc. Natl. Acad. Sci. USA* 94:4931 (1997).

43. M.E. Bruce, Strain typing studies of scrapie and BSE, in: *Methods in Molecular Medicine: Prion Diseases,* H. Baker, R.M. Ridley, eds., Humana Press, Totowa (1996).

44. R.J. Kascsak, R. Rubenstein, P.A. Merz, R.I. Carp, N.K. Robakis, H.M. Wisniewski, and H. Diringer, Immunological comparison of scrapie-associated fibrils isolated from animals infected with four different scrapie strains, *J. Virol.* 59:676 (1986).

45. R.A. Bessen and R.F. Marsh, Distinct PrP properties suggest the molecular basis of strain variation in transmissible mink encephalopathy, *J. Virol.* 68:7859 (1994).

46. L. Monari, S.G. Chen, P. Brown, P. Parchi, R.B. Petersen, J. Mikol, F. Gray, P. Cortelli, P. Montagna, B. Ghetti, et al. Fatal familial insomnia and familial Creutzfeldt-Jakob disease: different prion proteins determined by a DNA polymorphism, *Proc. Natl. Acad. Sci. U. S. A.* 91:2939 (1994).

47. P. Parchi, R. Castellani, S. Capellari, B. Ghetti, K. Young, S.G. Chen, M. Farlow, D.W. Dickson, A.A.F. Sima, J.Q. Trojanowski, et al. Molecular basis of phenotypic variability in sporadic Creutzfeldt-Jakob disease, *Ann. Neurol.* 39:767 (1996).

48. B. Caughey, G.J. Raymond, D.A. Kocisko, and P.T. Lansbury, Jr. Scrapie infectivity correlates with converting activity, protease resistance, and aggregation of scrapie-associated prion protein in guanidine denaturation studies, *J. Virol.* 71:4107 (1997).

49. R.A. Bessen, G.J. Raymond, and B. Caughey, *In situ* formation of protease-resistant prion protein in TSE-infected brain slices, [In Press] *J. Biol. Chem.* (1997).

50. G.G. Glenner, Amyloid deposits and amyloidosis: the beta-fibrillosa (second of two parts), *N. Engl. J. Med.* 302:1333 (1980).

51. P.T. Lansbury, Jr. and B. Caughey, The chemistry of scrapie infection: implications of the 'ice 9' metaphor, *Chem. & Biol.* 2:1 (1995).

52. J.H. Come, P.E. Fraser, and P.T. Lansbury, Jr. A kinetic model for amyloid formation in the prion diseases: importance of seeding. *Proc. Natl. Acad. Sci. USA* 90:5959 (1993).

53. B. Caughey, D.A. Kocisko, G.J. Raymond, and P.T. Lansbury, Aggregates of scrapie associated prion protein induce the cell-free conversion of protease-sensitive prion protein to the protease-resistant state, *Chem. & Biol.* 2:807 (1995).

54. D.A. Kocisko, P.T. Lansbury, Jr., and B. Caughey, Partial unfolding and refolding of scrapie-associated prion protein: evidence for a critical 16-kDa C-terminal domain,

Biochemistry 35:13434 (1996).

55. H. Wille, G. Zhang, M.A. Baldwin, F.E. Cohen, and S.B. Prusiner, Separation of scrapie prion infectivity from PrP amyloid polymers, *J. Mol. Biol.* 259:608 (1996).

56. M.P. McKinley, R.K. Meyer, L. Kenaga, F. Rahbar, R. Cotter, A. Servan, and S.B. Prusiner, Scrapie prion rod formation in vitro requires both detergent extraction and limited proteolysis, *J. Virol.* 65:1340 (1991).

57. J. Safar, W. Wang, M.P. Padgett, M. Ceroni, P. Piccardo, D. Zopf, D.C. Gajdusek, and C.J. Gibbs, Jr. Molecular mass, biochemical composition, and physicochemical behavior of the infectious form of the scrapie precursor protein monomer, *Proc. Natl. Sci. USA* 87:6373 (1990).

58. P. Brown, P. Liberski, A. Wolff, and D.C. Gajdusek, Conservation of infectivity in purified fibrillary extracts of scrapie-infected hamster brain after sequential enzymatic digestion or polyacrylamide gel electrophoresis, *Proc. Natl. Acad. Sci. USA* 87:7240 (1990).

59. R. Rubenstein, R.I. Carp, W. Ju, C. Scalici, M.C. Papini, A. Rubenstein, and R.J. Kascsak, Concentration and distribution of infectivity and PrP^{Sc} following partial denaturation of a mouse-adapted and a hamster-adapted scrapie strain, *Arch. Virol.* 139:301 (1994).

60. J. Hope, The nature of the scrapie agent: the evolution of the virino, *Ann. N. Y. Acad. Sci.* 724:282 (1994).

BIOPHYSICAL STUDIES ON STRUCTURE STRUCTURAL TRANSITIONS AND INFECTIVITY OF THE PRION PROTEIN

Detlev Riesner[1], Klaus Kellings[1], Karin Post[1],
Martin Pitschke[1], Holger Wille[2], Hana Serban[2],
Darlene Groth[2], Michael A. Baldwin[2], and Stanley B. Prusiner[2,3]

[1]Institut für Physikalische Biologie und Biologisch-Medizinisches Forschungszentrum,
Heinrich-Heine-Universität Düsseldorf, 40225 Düsseldorf, Germany

[2]Departments of Neurology and [3]Biochemistry and Biophysics, University of California,
San Francisco, CA 94143-0518, U.S.A.

INTRODUCTION

Prions are composed largely, if not entirely, of an abnormal isoform of the prion protein (PrP) designated PrPSc. A protease resistant polypeptide, PrP 27-30, can be derived from PrPSc by limited proteolysis with retention of infectivity. Both PrPSc and the cellular isoform PrPC are encoded by a chromosomal gene; PrPSc is produced from the cellular isoform by a posttranslational process (for review see Prusiner, 1991).

Several mechanisms been considered which might explain the transformation of the cellular isoform PrPC to PrPSc. Among these are a second component, a covalent posttranslational modification or a conformational change. Despite an intensive search, no second component such as a scrapie-specific nucleic acid has been found; quantitative analyses of residual nucleic acids has eliminated oligonucleotides greater than about 80 nucleotides (Kellings et al., 1992; Riesner, 1991). Proteolysis and peptide mapping failed to identify any covalent posttranslational modification that could account for the conversion (Stahl et al., 1993). However, conformational differences were identified by infrared spectroscopy (IR) and circular dichroism (CD). PrP[C] is predominantly an α-helical protein; whereas, PrP[Sc] exhibits substantial ß-sheet and reduced amounts of α-helix (Pan et al., 1993; Safar et al., 1993a). Treatment of PrP 27-30 or PrP[Sc] with solvents such as hexafluoroisopropanol (HFIP) that increased the α-helix content of PrP[Sc] resulted in a substantial reduction in scrapie infectivity (Gasset et al., 1993; Safar et al., 1993b).

Conformational studies of the PrP isoforms are complicated by the insolubility of some conformations. For example, PrP[C] is soluble in non-denaturing detergents whereas PrP[Sc] is not. To obtain high resolution structural data, conditions for solubilization of PrP[Sc] under which scrapie prion infectivity is retained need to be identified. Early attempts to solubilize scrapie infectivity were unsuccessful (Millson et al., 1979), and the insolubility of prions retarded the development of effective purification protocols (Prusiner et al., 1980; Siakotos et al., 1976). Once the insolubility of scrapie infectivity was appreciated, it was used to enrich fractions for infectivity which led to the discovery of PrP 27-30 (Prusiner et al., 1982). Some studies were also misled by the assumption that scrapie is caused by a virus and that the smallest scrapie prion particles were likely to be quite large despite ionizing radiation data indicating the contrary (Alper et al., 1966). The insolubility of PrP[Sc] has prevented also to determine its molecular homogeneity or heterogeneity, respectively. Since as much as 10^5-10^6 PrP molecules correlate with one infectious unit, it could be that much less PrP molecules represent the essential portion of the infectious agent but are hidden among many more proteinase resistant PrP molecules. A homogeneous population of 10^5-10^6 PrP[Sc] molecules is the other alternative.

Solubilization of PrP[Sc] and PrP 27-30 was effected when they were mixed with phospholipids to form liposomes or detergent-lipid-protein complexes (DLPCs) (Gabizon et al., 1987). The use of phospholipids to solubilize scrapie prions has not been useful in structural studies of the prion proteins because such high concentrations of lipid are required. In some studies, scrapie infectivity was reported to have observed sedimentation coefficients as low as 2 - 3 S (Malone et al., 1978; Prusiner et al., 1978b). Recent claims that PrP 27-30 can be recovered after SDS-PAGE with retention of prion infectivity (Brown et al., 1990) could not be confirmed (Prusiner et al., 1993).

To identify conditions for solubilization of PrP 27-30, we examined a variety of ionic and non-ionic detergents under conditions that were likely to retain scrapie prion infectivity. Sonication of the prion rods in the presence of 0.2 - 0.3% SDS produced a soluble fraction that did not sediment during centrifugation at 100,000 x g for 1h. This soluble fraction had a high α-helical content and relatively little ß-sheet in contrast to PrP 27-30 amyloid polymers and it contained spherical particles of ~10 nm in diameter and an observed sedimentation coefficient of about 6S. The spheres were composed of four to six PrP 27-30 molecules; they contain little, if any, scrapie infectivity. A lipid-rich, infectious fraction stayed insoluble at the meniscus after ultracentrifugation. Although our studies did not yield infectious preparations of soluble prions, they provided different fractions of PrP 27-30 which could be characterized by infectivity, protein conformation and conformational transitions.

MATERIALS AND METHODS

Chemicals and Enzymes. Hexafluoroisopropanol (HFIP) and acetonitrile (AcN) were of spectroscopic grade from Aldrich (Milwaukee, Wisc., USA). Lipids were from Avanti Polar Lipids (Alabaster, Alabama, USA). Dodecylmaltoside (dodecyl-b-D-maltopyranoside), lauryldimethylamineoxide (LDAO) and thesit were kindly provided by Dr. W. Welte (Freiburg, Germany); the other detergents were from Calbiochem (San Diego, Calif., USA). BenzonaseTM and proteinase K were from Merck (Darmstadt, Germany).

Preparation of PrP 27-30. PrP 27-30 was purified from the brains of scrapie infected Syrian hamsters (Lak:LVG from Charles River Laboratories) as described previously (Prusiner *et al.*, 1982b). The final purification step was either a sucrose gradient centrifugation or an ultrafiltration in which proteins of less than 300 kDa were removed and PrP was obtained in the retentate (Hecker *et al.*, 1992). Prions from both sources gave similar results.

Bioassays. Bioassays of scrapie prions were performed in Syrian hamsters by an incubation time interval procedure (Prusiner *et al.*, 1982b). The same correlation between inoculated dose and incubation time was found for homogenates, purified prion rods from sucrose gradient centrifugation and DLPCs, indicating that a variety of chemical treatments and fractionation procedures did not alter the relationship between inoculated dose and incubation time.

Buffers. Sonication buffer A (20 mM MOPS, 20 mM Tris, pH 7.2); sonication buffer B (10 mM Na phosphate, pH 7.0); TBST (10 mM Tris-HCl, pH 8.0, 0.15 M NaCl, 0.01% Tween-20).

Sonication and centrifugation. Sonication was carried out with a cup horn sonicator (Branson) at 40 W for 5 min at temperatures not exceeding 17°C. Details are described elsewhere (Riesner *et al.*, 1996). After sonication, the sample was centrifuged in a 1.5 ml Eppendorf spin tube (Beckman Instruments, Palo Alto, CA, USA) in a tabletop TL-100 ultracentrifuge (Beckman Instruments, Palo Alto, CA, USA). Two protocols were applied. The "soluble fraction" protocol: Sonication was carried out in 440 μl buffer followed by centrifugation in thick wall polycarbonate tubes in a TLS-55 swinging bucket rotor at 40,000 rpm for

115 min at 4°C (100,000 x g) and a 220 µl supernatant fraction was collected from immediately below the meniscus. The "total supernatant" protocol: Sonication was carried out in 250 µl buffer and either the whole sample was loaded onto the sucrose gradient or the supernatant of 220 - 230 µl was separated from the pellet by careful pipetting after centrifugation using a fixed angle rotor TLA-100.3 at 40,000 rpm (60,000 x g) or at 50,000 rpm (100,000 x g) for 1 hour at 4°C.

Sucrose gradient centrifugation. Sucrose gradient centrifugation (5%-20% sucrose in 10 mM Na phosphate buffer, pH 7)was performed in a SW-60 rotor (4 ml polyallomer tubes) at 60,000 rpm for 5.5 hours. It was estimated that the largest particles in the supernatant from the solubilization centrifugation (~25 S) would just reach the bottom of the gradient under these conditions. The details are given elsewhere (Riesner et al., 1996).

PrP analysis by PAGE, western blots and quantitation. PrP-samples were analyzed by SDS-PAGE (12% acrylamide) from which the proteins were transferred onto an Immobilon P membrane (Millipore), incubated with primary anti-PrP antibody then with alkaline phosphatase-conjugated secondary antibody. In most cases PrP was determined from dot blots visualized with the chemiluminescent system (ECL) (Amersham) and scanned with a densitometer (for details see Riesner et al., 1996).

Digestion by proteinase K. Samples were incubated with either 50 µg proteinase K/ml at 37°C for one hour or 10 µg proteinase K/ml at 37°C for times between 1 and 30 minutes.

Circular dichroism (CD) spectroscopy. CD spectra were recorded with a Jasco model 720 spectropolarimeter using 1 or 5 mm path length cylindrical cells at room temperature. All measurements were carried out in buffer B (10 mM Na phosphate, pH 7.0) containing 0.2% SDS, which allowed readings to be taken down to 190 nm. Signal averaging allowed satisfactory CD spectra to be obtained with PrP 27-30 concentrations down to ~10 µg/ml. Quantitative secondary structure assignments were carried out with a computer program kindly provided by Dr. W. Curtis Johnson, University of Oregon (Johnson, 1990). Further details were described elsewhere (Riesner et al., 1996).

Electron microscopy (EM). The samples were adsorbed for about 20 s onto glow discharged, carbon coated grids. Excess material and sucrose were washed away with three drops each of 0.1 M and 0.01 M ammonium acetate. The grids were then stained with 2% uranyl acetate and viewed in a JEOL 100CX II electron microscope at 80 kV. The diameter of the particles was measured parallel to one side of each print, independent of the orientation of the particles.

RESULTS

Sonication of PrP 27-30 with Detergents

Prion rods (~300 μg PrP 27-30/ml) were sonicated under controlled conditions in buffer A (20 mM MOPS, 20 mM Tris, pH 7.2) with detergents that were

Table 1. Efficiency of solubilization of PrP 27-30 by sonication in buffer A containing various detergents[a]

Detergent in sonication buffer	Supernatant		
	PrP content (%)	Infectivity (%)	Log ID_{50}/ml
None	1 - 2	2	6.4
SDS (0.05 - 10%)	20 - 30*	8 - 50*	7.4 - 8.4*
Sarkosyl (2%)	20	50	9.1
Zwittergent 3-12 (1-3%)	10	2	7.2 - 7.6
LDAO (1%)	5	<0.1	4.9
Thesit (3%)	10	7	7.4
Dodecylmaltoside (3%)	10	6	7.6
DLPC	1 - 2	4	7.1

[a]The PrP 27-30 was determined with serial dilutions by western blots and the infectivity by bioassays in Syrian hamsters. After solubilization centrifugation according to the total supernatant protocol the PrP 27-30 content and the infectivity of supernatant and pellet were added; the percentage of the soluble fraction is listed.
*The values are not related to the SDS concentrations in a monotonic way. They refer to SDS concentration >0.2%.

representative of ionic, non-ionic, zwitterionic and glycosidic classes. After sonication the samples were centrifuged in a fixed angle rotor at 60,000 x g or at 100,000 x g at 4°C for 1 hour and the supernatant was sampled according to the total supernatant protocol. The effectiveness of the detergent was evaluated based on the fraction of the PrP 27-30 and the scrapie infectivity remaining in the supernatant, as determined by western blots with anti-PrP antibodies and bioassays in Syrian hamsters, respectively (Table 1). The most effective detergents were SDS and Sarkosyl, both of which resulted in 20 - 30% of PrP 27-30 and a similar fraction of the infectivity in the supernatant fraction. SDS concentrations of 0.05% up to 10% were investigated, and the effectiveness of solubilization saturated around 0.3% in good agreement with the critical micelle concentration (CMC, ~0.24%). The finding that SDS and Sarkosyl did not inactivate scrapie infectivity at low concentrations is in agreement with earlier findings (Prusiner et al., 1980). For all subsequent experiments the sonication was carried out with 0.2% or 0.3% SDS.

Effectiveness of solubilization did not change, when entrifugation was changed from 60,000 x g to 100,000 x g under otherwise identical parameters. Also three different sonication time protocols (5 minutes, three periods of 5 minutes with 5 minute breaks, and 30 minutes continuously) yielded very similar results. The initial temperature was set either to 7°C, which increased during a five minute sonication period to a maximum of 17°C, or was set to 30°C, increasing to 37°C; the efficiency did not change. In an attempt to minimize denaturation, we employed 5 minutes of sonication starting at 7°C with an increase to no more than 17°C.

Chaotropic agents known to denature PrP 27-30 and destory infectivity were used at low concentrations, e.g., 1 M urea or 0.5 M Gdn-HCl (Prusiner, et al., 1993). The effect of HFIP on prion infectivity was evaluated at a concentration of 10% (vol/vol). Sucrose was also added up to 25% (wt/vol), either alone or in combination with the other reagents. After sonication in 0.3% SDS, the fraction of PrP 27-30 remaining in the supernatant was considerably more sensitive to digestion with proteinase K than that in the prion rods found in the pellet. Bioassays of these digested supernatant fractions showed reductions in prion titers by factors 5, 16, 3, 50, and 5 in five independent experiments. While such changes are not significant, there was always a small reduction of titer. In contrast, HFIP diminished the infectivity drastically; other reagents did not appear to be promising, and so further studies were not performed.

Sucrose Gradient Centrifugation Analysis

Sucrose gradient centrifugation was performed to determine whether PrP 27-30 from the supernatant fraction after sonication and infectivity cosedimented. To minimize contamination of the supernatant fraction by any particles floating in the meniscus or particles loosely adherent to the pellet, centrifugation of the SDS sonicated prion rods was performed in a swinging bucket rotor and only 50% of the supernatant was collected with a pipette from the upper portion of the liquid column (soluble fraction protocol). The gradients were preformed with either no detergent or 0.2% SDS to test for reaggregation in the absence of SDS. Gradients were fractionated from either the top or the bottom to assess possible artifacts in the profile arising from the fractionation protocol. Each fraction from the sucrose gradient was analyzed for PrP 27-30 content and scrapie infectivity. Profiles in the presence and absence of 0.2% SDS are similar, each profile represents average values from four or five independent experiments (Fig. 1A).

The majority of PrP 27-30 was found in fractions 2 to 5 and sedimented as a symmetrical peak of ~6S, but only ~0.02% of the infectivity was recovered in these fractions. It is noteworthy that the supernatant collected by the soluble fraction protocol contained only 6% of the total infectivity, of which 50% was loaded onto the gradient. Because the specific infectivity in fractions 2 to 5 was nearly 3 orders of magnitude lower than that of the prion rods, we conclude that soluble PrP 27-30 prepared under these conditions is devoid of prion infectivity.

The low recovery of the infectivity in the supernatant loaded onto the gradients in Fig. 1 compared with the nearly 50% infectivity in the supernatant in the initial SDS sonication experiments (Table 1) appeared contradictory. Therefore prion rods were sonicated in 0.2% SDS and loaded directly onto the sucrose gradient without prefractionation by differential centrifugation. Again, most of the PrP 27-30 was found in fraction 2 to 5 (Fig. 1B), but some PrP 27-30 was also found in the meniscus fraction. Most notably, $\sim 10^7$ 50% infectious doses (ID_{50}) of prion infectivity per ml were found in the meniscus and the adjacent fraction as well as at the bottom of the gradient, in contrast to fractions 3 to 5, containing most of the PrP 27-30, in which prion titers were lower by nearly a factor of 1,000.

The prions in the meniscus fractions would seem to be of either small size or low density. Since the supernatant fractions prepared with the soluble fraction protocol would have eliminated the prions of low density but not those of small size, we conclude that the absence of infectivity in the meniscus fraction in Fig. 1A indicates a low density population of prions. Most likely the low density of the prions is due to lipids and possibly detergents bound to PrP 27-30. The origin of the lipids has to be the prion rods.

Ultrastructural Analysis

Fractions 2 to 5 of the sucrose gradients (Fig. 1) contained numerous sperical particles (Fig. 2A and 3E). Since these fractions contain only PrP 27-30 as judged by SDS-PAGE, and no particles other than spheres were found by electron microcopy, we conclude that the spheres are composed of PrP 27-30. Most of the spheres ranged from 8 to 12 nm in diameter; a substantial portion had diameters of ~ 10 nm (Fig. 2B). We were unable to identivy any consistent substructure by using negative staining.

Since the particles in fractions 2 to 5 were of spherical shape we estimated the molecular weight from the average sedimentation coefficient (~ 6 S) determined from the position of the peak fraction. The partial specific volume (v) for a sedimenting particle was calculated according to the method of Durchschlag (1986) to be 0.71 cm^3 g^{-1}, based on the following composition of PrP 27-30: 54% protein (v = 0.71 cm^3 g^{-1}), 18% carbohydrate (v = 0.65 cm^3 g^{-1}), 7% glycolipid anchor (v = 0.59 cm^3 g^{-1}) and 21% SDS, i.e. 0.4 g SDS per gram protein (v = 0.82 cm^3 g^{-1}) according to Tanford (1980) and assuming hydration as 0.5 g water per gram PrP 27-30, (the values for SDS binding and hydration being

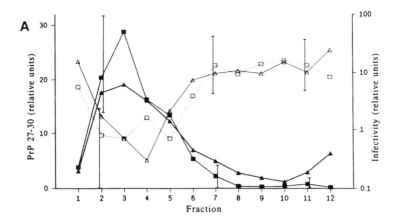

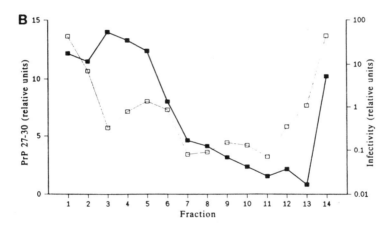

Fig. 1 Analysis by 5 - 20% sucrose gradient centrifugation of the soluble fraction (A) or of the total sample (B) after sonication in 0.2% SDS. The sedimentation profiles show PrP 27-30 (■) and infectivity (☐) in the presence of 0.2% SDS and PrP27-30 (▲) and infectivity (△) without SDS in the gradient. The numbers in (A) are the mean of four independent experiments. Relative units are depicted (in percent) for PrP 27-30 and infectivity; the sum of all fractions is set to 100%. Error bars are defined by the highest and the lowest values respectively; they are schown for three fractions only, but are characteristic for all fractions. The larger error bars in fraction 2 are due to the flank position in the PrP peak or due to the very low and therefore inaccurate infectivity data, respectively. Figure is modified from Riesner *et al.*, 1996.

rough estimates only). Applying the standard method for evaluation of a spherical molecule of 6 S (Cantor and Schimmel, 1982), a molecular weight of 117 kDa was obtained, which is interpreted best as pentameric PrP 27-30. Allowing for the errors and uncertainties in this calculation, we conclude that each of these spheres contains between four and six PrP 27-30 molecules. On the other hand, the observed sedimentation values for PrP 27-30 and the spheres suggest that PrP 27-30 is unlikely to form particle other than the spheres visualized by electron microscopy. Sample before and after sonication with 0.2% SDS and from selected fractions of the sucrose gradient centrifugation (Fig. 1) were examined by EM. We confirmed that the samples before sonication consisted of prion rods which were 100 - 300 nm in length (Fig. 3A). Rods were still present after sonication in SDS but they were smaller (50 - 150 nm); in addition, the PrP 27-30 was dispersed into small spherical particles and elongated aggregates (Fig. 3B).

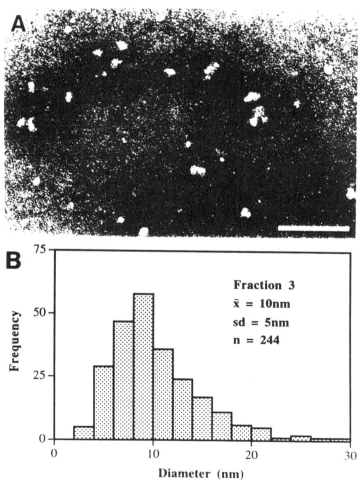

Fig. 2 (A) Electron microscopy of the peak fraction of sucrose gradient centrifugation as shown in Fig. 1A (bar = 100 nm); (B) size distribution of the spherical particles in (A). Figure is modified from Riesner et al., 1996.

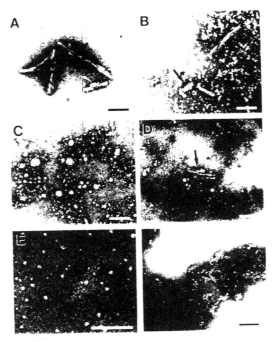

Fig. 3 Electron micrographs of different fractions of PrP 27-30. Prion rods before sonication (A) and after sonication (B) in 0.2% SDS; (C) and (D) meniscus fractions from the sucrose gradient illustrated in Fig. 1B; (E) fraction 3 from the same gradient; (F) after addition of 25% acetonitrile to the supernatant prepared according to the total supernatant protocol. The arrows point to residual rod fragments; the bar corresponds to 100 nm. Figure is modified from Riesner *et al.*, 1996.

The meniscus fraction of the sucrose gradient contained a heterogeneous array of particles (Fig. 3C) and aggregates of small rods (Fig. 3D). The finding of rod fragments in the meniscus fractions implies that the density of these fragments is below 1.018 g/cm^3 (5% sucrose) and that the rod fragments are associated with substantial amounts of lipid. Fraction 3 which had the highes concentration of PrP 27-30, contained only small spherical particles of ~10 nm in diameter (Fig. 3E). These spheres were indistinguishable from those shown in fig. 2A). The pellet fraction contained rod fragments which withstood the sonication procedure (data not shown).

Conformational Transitions

Conformational analyses were performed on the supernatant and the resuspended pellet fractions. After the supernatant sample was removed from the 100,000 x g centrifugation for the sucrose gradient (soluble fraction protocol, Fig. 1A), the remaining supernatant was collected and examined by CD spectroscopy. The pellet was resuspended in an equal volume of the same buffer. The supernatant exhibited a spectrum characteristic of a protein of high α-helical content; whereas, the spectrum of the pellet was characteristic for structures containing high ß-sheet content. These spectra were similar to those obtained on fractions prepared using the total supernatant protocol where the supernatant (Fig. 4A) showed minima at 208 and 222 nm, characteristic for α-helical structures; whereas, the pellet fraction gave a maximum at 195 nm and a minimum at 218 nm, representative of ß-sheets. The spectrum of the supernatant showed no changes over a 24 hour period suggesting that the PrP 27-30 remained soluble over this time interval. In contrast, the ellipticity of the resuspended pellet decreased measurably over one hour in accord with the insolubility of the rods. When the prion rods were sonicated for up to one hour without SDS, the supernatant fraction gave a lower ellipticity spectrum with a characteristic ß-sheet signal suggesting that the conformation of PrP 27-30 in the "broken rods" was unaltered by sonication alone. However, the amplitudes in the CD-spectrum cannot be correlated with molar ellipticities.

Addition of acetonitrile is known to induce ß sheets in peptides. Increasing the acetonitrile concentration from 0 to 50% in increments of 10% induced a progressive shift from a- helix to ß-sheet (Fig. 4B). The proportions of α-helical and ß-sheet structures determined from the CD spectra were tabulated (Table II): The α-helical content of the supernatant fractions ranged from 40 to 60%; this range of values may be due to variations in the SDS sonication procedure as well as the separation protocol used to separate the supernatant from the pellet after centrifugation in a fixed angle rotor. The addition of 10 or 20% acetonitrile effected a small reduction in the α-helical content and a concomitant increase in ß-sheet; 30% acetonitrile produced approximately equal amounts of α-helix and ß-sheet. The spectra of the resuspended pellet resembled that of the supernatant to which 30% acetonitrile had been added (Table 2). A concentration of 40 - 50% acetonitrile seems to denature PrP 27-30. The addition of 30 - 50% acetonitrile to the supernatant precipitated PrP 27-30 after exposure overnight as assessed by 100,000 x g centrifugation. The increase in the ß-sheet content correlated with the formation of insoluble aggregates in additional experiments (data not shown).

Since infectious prion rods containing PrP 27-30 are rich in ß-sheet, we investigated whether the increase in ß-sheet due to acetonitrile was accompanied by an increase in infectivity. Addition of 10 - 20% acetonitrile had no effect on prion infectivity while 30% acetonitrile reduced infectivity ~75%. Thus, the conformational change induced by acetonitrile is not equivalent to that which occurs during the formation of PrPSc. The bioassays were started before the

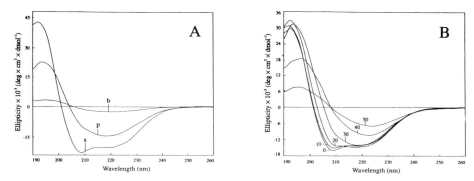

Fig. 4 Circular dichroism (CD) spectra. (A) supernatant (s) and pellet (p) prepared according to the total supernatant protocol in 10 mM phosphate buffer, pH 7.0, containing 0.2% SDS; "broken rods" (b) were obtained by sonication for 1 hour in the absence of SDS and without centrifugation. (B) the supernatant after addition of 0 to 50% (in steps of 10%) acetonitrile. Ellipticities were calibrated according to an amino acid analysis of the sample and were corrected for dilution with acetonitrile. Figure from Riesner et al., 1996.

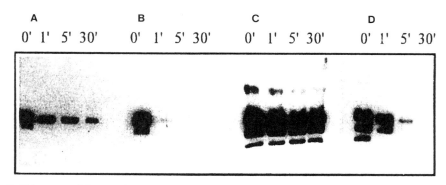

Fig. 5 Western blots showing variable proteinase K resistance of PrP 27-30 fractions from solubilization experiments. (A) from the meniscus of the gradient in Fig. 1B, (B) from fraction 3 of the gradient in Fig. 1A, (C) pellet from soluble fraction protocol and (D) fraction 3 (cf. 5B) with addition of 30% acetonitrile. The incubation times were 0, 1, 5 and 30 minutes at 37°C at a final concentration of 10 µg proteinase K per ml. Figure from Riesner et al., 1996.

Table 2. Effects of acetonitrile (AcN) on the infectivity of the supernatant after solubilization and centrifugation and on the secondary structure as determined by CD spectroscopy

Sample	AcN (%)	log ID$_{50}$/ ml	Infectivity[a] (%)	Secondary structure (%)			
				α-helix	ß-sheet	turns	other
Pellet fraction	0	9.6	83	29	31	10	30
Supernatant fraction	0	8.9	17	61	5	16	18
	10	9.1	26	57	9	15	19
	20	8.9	17	53	18	9	20
	30	8.4	5	37	33	5	25
	40	5.9	0.02	26	36	6	32
	50	5.0	0.002	22	29	13	36

[a] The total infectivity of pellet and supernatant (100%) after solubilization and centrifugation was 9.7 log ID$_{50}$/ml

separation of the SDS sonicated rods in infectious lipid-rich fraction and non-infectious spherical particles was known. With that knowledge the infectivity tests would have been performed on isolated spherical particles, after reaggregated to ß-sheet by 30% acetonitrile. Thus, we cannot exclude completely at present that a portion of reaggregated PrP 27-30 regained infectivity. Addition of 40 - 50% acetonitrile diminished prion titers by a factor of 10^3 to 10^4. It seems likely that acetonitrile at concentrations above 30% acts as a denaturant for PrP.

Electron microscopy was also used to monitor the effect of acetonitrile on the supernatant following SDS sonication and centrifugation. No rods were found in the presence of 25% acetonitrile; only amorphous aggregates were observed (Fig. 3F).

Characteristic fractions containing PrP 27-30 were tested for proteinase K resistance. The majority of the PrP 27-30 in the meniscus fraction was digested within the first minute; whereas, the remaining PrP 27-30 was stable for 30 minutes (Fig. 5A). This behavior may reflect the array of structures found by EM (Fig. 4 C and D) with aggregates of rods being responsible for the residual proteinase K resistance. In contrast, the soluble PrP 27-30 from the peak fractions was digested nearly completely within the first minute (Fig. 5B). The pellet obtained after SDS sonication of prion rods followed by centrifugation in a fixed angle rotor exhibited substantial proteinase K resistance (Fig. 5C). When the peak fraction was treated with 30% acetonitrile and precipitable PrP 27-30 recovered by a 100,000 x g centrifugation, measurable but modest proteinase K resistance was regained (Fig. 5D).

DISCUSSION

Considerable evidence supports the hypothesis that a conformational change in PrP underlies both the propagation of prions and the pathogenesis of prion diseases (Pan et al., 1993; Cohen et al., 1994). Although no experimental studies have demonstrated the refolding of PrPC into PrPSc in cell free systems, the studies reported here address this issue from the perspective of modifying the structure of PrP 27-30 under relatively gentle conditions. The conformational changes of PrP 27-30 were characterized by the chemical parameters like solubility, proteinase K resistance, and secondary structure and by the biological property of infectivity.

Disruption of Prion Rods

After sonication of prion rods in the presence of 0.2% SDS, ~25% of the PrP 27-30 was found in the supernatant after centrifugation at 100,000 x g for 1 h. The pellet contained the majority of PrP 27-30, consisting of prion rods which withstood the SDS-sonication procedure, but the rods were smaller as measured by electron microscopy (data not shown).The CD spectrum of this pellet fraction showed a more pronounced ß-sheet signal than that of prion rods subjected to sonication in the absence of SDS, which is most probably a consequence of the larger particles size if sonicated in the absence of SDS.

The data of Table 1 had suggested that a major part of scrapie infectivity as well as PrP 27-30 content was transferred into the supernatant containing PrP 27-30 in a soluble state, but the sucrose gradient centrifugations (Fig. 1) demonstrated that the supernatant was heterogeneous. Soluble PrP 27-30 formed a fairly homogeneous population of spherical particles sedimenting with 6S and lacking significant infectivity, whereas lipid-rich, infectious prions were identified in the meniscus fraction of sucrose gradients when the total supernatant procedure was used to generate the sample loaded onto the gradient (Fig. 1B). It should be emphasized that the lipids or other compounds which are responsible for the low density of the prions were not added during sonication but are constituents of the prion rods. A heterogeneous array of particles including spheres of different sizes, aggregates of spheres, and even small prion rod-like structures, were found by electron microscopy of lipid-rich fractions (Fig. 3). Also a portion of the PrP 27-30 (Fig. 5) as well as the prion infectivity was destroyed by proteinase K digestion, whereas residual PrP 27-30 and infectivity resisted limited digestion. Possibly the rod-shaped structures in the lipid-rich fraction represent a proteinase K resistant subpopulation of PrP 27-30 molecules that are associated with prion infectivity. A possible explanation of our results would be, that lipids may stabilize the conformation of PrP 27-30 and thus preserve prion infectivity. One would expect that PrP 27-30 in the lipid-rich fraction assumes the ß-sheet structure, but this could not yet be measured. The stabilizing effect of lipids on the preservation of infectivity might be similar to that reported previously for the detergent-lipid-protein-complexes (DLPCs) (Gabizon et al. 1987). In DLPSs, the ß-sheet content of PrP 27-30 was similar to that measured for the prions rods by FTIR spectroscopy (Gasset et al., 1993).

Formation of Spherical Particles

The PrP 27-30 molecules that form the spherical oligomers resemble PrPC in several aspects: like PrPC the PrP 27-30 of the spheres is soluble, sensitive to proteinase K, α-helical and non-infectious; both molecules have a low ß-sheet content (Gasset et al., 1993; Pan et al., 1993; Safar et al., 1993a). The transformation of prion rods to the spherical particles has to be differentiated from a denaturation of PrP 27-30 which had been studied several times before, most recently by Safar et al. (1993a) and Oesch et al. (1994). Why such spheres form is unclear but elucidating the mechanism of their formation may provide insight into the molecular forces that participate in the polymerization of PrP 27-30 into prion rods. The spheres of PrP 27-30 reported here resemble morphologically spheres generated by sonication of prion rods in the absence of SDS; a "non-sedimenting" fraction contained spherical particles of 8 - 20 nm diameter as judged by electron microscopy but the yield was so low that the particles could not be characterized in more detail (McKinley et al., 1986).

Some investigators attempted to determine the size of infectious prions using sucrose gradient centrifugation after extraction CJD-infected mouse brain preparations with Sarkosyl and sonication (Slaviadis et al., 1989). A non-infectious peak of ~7 S containing PrP was found which was interpreted as composed of "monomeric or dimeric" PrP. From these findings, they concluded that PrP can be separated from infectivity and therefore, cannot be the major or essential component of the infectious prion. Indeed, the 7 S PrP peak in those studies is likely to be composed of the same spherical oligomers of PrP 27-30 reported here. In other studies, spherical particles were described with an average diameter of 12 nm in brain homogenates from scrapie-infected Syrian hamsters and not from uninfected controls (Ozel and Diringer, 1994). Similar spheres have also been found in homogenates prepared from the brains of patients who died of CJD (Ozel et al., 1994). Based on these ultrastructural findings, the investigators suggest that those "virus-like" particles are the causative agents of scrapie and CJD. Although these 12 nm spheres look very similar as the spherical particles shown in Fig. 2A of this work, it cannot be stated that they are identical. The particles of this work, however, are oligomers of PrP 27-30 and definitely not "virus-like".

Conformational Transitions of PrP 27-30

A portion of PrP 27-30 from prion rods could be converted into PrP spheres with high α-helical content and converted back into ß-sheet by addition of 25 - 30% acetonitrile. The sedimentation behavior demonstrated that the ß-sheet conformation is connected with aggregation and thus insolubility, and only the α-helical state of PrP spheres is soluble. Our experiemnts showed, however, that ß-sheet formation is faster than forming larger aggregates, since ß-sheet spectra could be measured within minutes whereas aggregates formed over night. Also proteinase K resistance is connected with aggregation, whereas the soluble PrP spheres have lost any proteinase K resistance (Fig. 5). The aggregates reformed by 25% acetonitrile are morphologically different from prion rods, and did not posses scrapie infectivity as measured by bioassay in hamster. It should be mentioned, however, that both re-aggregation and denaturation are superimposed during AcN addition.Furthermore, the infectious lipid-rich fraction was present in the sample; therefore, if small titers would have been re-generated, they might have escaped our notice.

Our findings may be relevant to a recent study in which PrPC was reported to acquire proteinase K resistance in the presence of a 50-fold excess of PrP 27-30 (Kocisko et al., 1994). The protocol involves exposure of PrPC to 3 M GdnHCl prior to mixing with PrPSc which was either exposed to 3 M GdnHCl or not. Whether the conformation of PrP 27-30 in 0.2% SDS after sonication is equivalent to that of PrPC or PrPSc in 3 M GdnHCl remains to be established, but seems to be unlikely, if the analysis by Safar et al. (1994) is taken into account. Safar et al. have characterized by spectroscopic methods "molten globules" structure, which was induced either by 2 M Gdn HCL or at a pH < 2. Our results are similar to their report in that respect that a structure of higher α-helix content could be induced. The marked difference is, however, in the infectivity, since they described their intermediates as infectious state, whereas we seperated infectivity from the soluble α-helical state and could not restore infectivity. Since our studies show that reaggregation can be accompanied by the acquisition of proteinase K resistance without restoration of prion infectivity, it is possible that some of the results of the mixing studies noted above can be explained by protein aggregation in which PrPC is not converted into PrPSc. It should be noted that subsequent investigation failed to confirm the renaturation from 3M Gdn-HCl (Kaneko et al., 1995). The relatively uniform size of the spheres suggests that the PrP 27-30 molecules comprising them possess a unique metastable conformation which produces this distinct quaternary structure. Whether conditions which renature the PrP 27-30 in the spheres and restore scrapie infectivity can be identified remains unknown, but is certainly worthy of further investigation.

ACKNOWLEDGEMENTS

This work was supported by research grants from the National Institutes of Health (NIH), U.S.A. (NS22786), the Ministerium für Wissenschaft und Forschung (MWF) of NRW, Germany and the Bundesministerium für Forschung und Technologie (BMFT), Germany.

REFERENCES

Alper, T., D. A. Haig and M. C. Clarke. 1966. The exceptionally small size of the scrapie agent. Biochem. Biophys. Res. Commun., 22, 278.

Brown, P., P. P. Liberski, A. Wolff and D. C. Gajdusek. 1990. Conservation of infectivity in purified fibrillary extracts of scrapie-infected hamster brain after sequential enzymatic digestion or polyacrylamide gel electrophoresis. Proc. Natl. Acad. Sci. USA, 87, 7240-7244.

Cantor, C.R. and Schimmel, P.R. 1980. Biophysical Chemistry. W.H. Freeman and Company, San Francisco.

Cohen, F.E., K.M. Pan, Z. Huang, M.A. Baldwin, R.J. Fletterick, and S.B. Prusiner. 1994. Structural clues to prion replication. Science, 264, 530-531.

Durchschlag, H. 1986. Specific Volumes of Biological Macromolecules and some Other Molecules of Biological Interest. In Thermodynamic Data for Biochemistry and Biotechnology (Hinz, H.-J., ed.), Springer-Verlag, Berlin-Heidelberg-New York-Tokio.

Gabizon, R., M. P. McKinley and S. B. Prusiner. 1987. Purified prion proteins and scrapie infectivity copartition into liposomes. Proc. Natl. Acad. Sci. USA, 84, 4017-4021.

Gasset, M., M. A. Baldwin, R. J. Fletterick and S. B. Prusiner. 1993. Perturbation of the secondary structure of the scrapie prion protein under conditions associated with changes in infectivity. Proc. Natl. Acad. Sci. USA, 90, 1-5.

Hecker, R., A. Taraboulos, M. Scott, K.-M. Pan, M. Torchia, K. Jendroska, S. J. DeArmond and S. B. Prusiner. 1992. Replication of distinct prion isolates is region specific in brains of transgenic mice and hamsters. Genes Dev., 6, 1213-1228.

Johnson, W. C., Jr. 1990. Protein secondary structure and circular dichroism: a practical guide. Proteins, 7, 205-214.

Kaneko, K., Paretz, D., Pan, K.M., Blochberger, T., Gabizon, R., Griffith, O.H., Cohen, F.E., Baldwin, M.A. and S.B. Prusiner. 1995. Prion Protein (PrP) synthetic peptides induce cellular PrP to acquire propeties of the scrapie isoform. Proc. Natl. Acad. Sci. USA 92, 11160-11164.

Kellings, K., N. Meyer, C. Mirenda, S. B. Prusiner and D. Riesner. 1992. Further analysis of nucleic acids in purified scrapie prion preparations by improved return refocussing gel electrophoresis (RRGE). J. Gen. Virol., 73, 1025-1029.

Kocisko, D. A., J. H. Come, S. A. Priola, B. Chesebro, G. J. Raymond, P. T. Lansbury Jr. and B. Caughey. 1994. Cell-free formation of protease-resistant prion protein. Nature, 370, 471-474.

Malone, T. G., R. F. Marsh, R. P. Hanson and J. S. Semancik. 1978. Membrane-free scrapie activity. J. Virol., 25, 933-935.

McMinlex, M.P., Braunfeld, M.B., Bellinger, C.G., and Prusiner, S.B. 1986. Molecular characteristics of prion rods purified from scrapie-infected hamster brains. J. Infect. Dis. 154, 110-120.

Millson, G. C. and E. J. Manning. 1979. The effect of selected detergents on scrapie infectivity. In Slow Transmissible Diseases of the Nervous System, Vol. 2. New York, Academic Press. 409.

Oesch, B., M. Jensen, P. Nilsson and J. Fogh. 1994. Properties of the scrapie prion protein: quantitative analysis of protease resistance. Biochemistry, 33, 5926-5931.

Ozel, M. and Diringer, H. 1994. small virus-like strucutrue in fractions from scrapie hamster brain, Lancet 343, 894.

Ozel, M., Xi, Y.-G., Baldauf, E., Diringer, H., and Pocchiari, M. 1994. Small virus-like structure in brains from cases of sporadic and familial Creutzfeldt-Jakob disease. Lancet 344, 923.

Pan, K.-M., M. Baldwin, J. Nguyen, M. Gasset, A. Serban, D. Groth, I. Mehlhorn, Z. Huang, R. J. Fletterick, F. E. Cohen and S. B. Prusiner. 1993. Conversion of a-helices into ß-sheets features in the formation of the scrapie prion proteins. Proc. Natl. Acad. Sci. USA, 90, 10962-10966.

Prusiner, S.B. 1991. Molecular biology of prion diseases. Science, 252, 1515-1522.

Prusiner, S. B., W. J. Hadlow, C. M. Eklund, R. E. Race and S. P. Cochran. 1978b. Sedimentation characteristics of the scrapie agent from murine spleen and brain. Biochemistry, 17, 4987.

Prusiner, S. B., D. F. Groth, S. P. Cochran, F. R. Masiarz, M. P. McKinley and H. M. Martinez. 1980a. Molecular properties, partial purification, and assay by incubation period measurements of the hamster scrapie agent. Biochemistry, 19, 4883.

Prusiner, S. B., D. C. Bolton, D. F. Groth, K. A. Bowman, S. P. Cochran and M. P. McKinley. 1982a. Further purification and characterization of scrapie prions. Biochemistry, 21, 6942-6950.

Prusiner, S. B., S. P. Cochran, D. F. Groth, D. E. Downey, K. A. Bowman and H. M. Martinez. 1982b. Measurement of the scrapie agent using an incubation time interval assay. Ann. Neurol., 11, 353-358.

Prusiner, S. B., D. Groth, A. Serban, N. Stahl and R. Gabizon. 1993. Attempts to restore scrapie prion infectivity after exposure to protein denaturants. Proc. Natl. Acad. Sci. USA, 90, 2793-2797.

Riesner, D.. 1991. The search for a nucleic acid component to scrapie infectivity. Sem. Vir., 2, 215-226.

Riesner, D., K. Kellings, K. Post, H. Wille, H. Serban, D. Groth, M.A. Baldwin, and S.B. Prusiner. 1996. Disruption of Prion Rods Generates 10-nm Spherical Particles Having High α-Helical Content and Lacking Scrapie Infectivity. J. Virol 70, 1714-1722.

Safar, J., P. P. Roller, D. C. Gajdusek and C. J. Gibbs Jr. 1993a. Conformational transitions, dissociation, and unfolding of scrapie amyloid (prion) protein. J. Biol. Chem., 268, 20276-20284.

Safar, J., P. P. Roller, D. C. Gajdusek and C. J. J. Gibbs. 1993b. Thermal-stability and conformational transitions of scrapie amyloid (prion) protein correlate with infectivity. Protein Sci., 2, 2206-2216.

Siakotos, A. N., D. C. Gajdusek, C. J. Gibbs Jr., R. D. Traub and C. Bucana. 1976. Partial purification of the scrapie agent from mouse brain by pressure disruption and zonal centrifugation in sucrose-sodium chloride gradients. Virology, 70, 230.

Sklaviadis, T.K., Manuelidis, L., and Manuelidis, E.E. 1989. Physical properties of the Creutzfeldt-Jakob disease agent. J. Virol. 63, 1212.

Stahl, N., M. A. Baldwin, D. B. Teplow, L. Hood, B. W. Gibson, A. L. Burlingame and S. B. Prusiner. 1993. Structural analysis of the scrapie prion protein using mass spectrometry and amino acid sequencing. Biochemistry, 32, 1991.

Tanford, C. 1980. The Hydrophobic Effect: formation of micelles and biological membranes (2nd edition). John Wiley & Sons Inc., New York-London-Sydney.

AMYLOIDOGENESIS IN TRANSMISSIBLE SPONGIFORM ENCEPHALOPATHIES

Franco Cardone and Maurizio Pocchiari

Laboratory of Virology
Istituto Superiore di Sanità
Viale Regina Elena 299
00161 Roma, Italy

Transmissible spongiform encephalopathies (TSE) or prion diseases are degenerative pathologies of the central nervous system characterised by a long incubation period and a fatal clinical course. They comprehend Creutzfeldt-Jakob disease (CJD), Gerstmann-Sträussler-Scheinker syndrome (GSS), fatal familial insomnia (FFI) and kuru in humans, scrapie of sheep and goats, bovine spongiform encephalopathy and other rare pathologies in animals[1].

The vast majority of human TSE cases occurs in a sporadic form. GSS, FFI and about 10-15% of CJD cases occur in families with an autosomal dominant pattern of inheritance associated with mutations of the PrP gene (*PRNP*)[1]. At present, 13 different point mutations have been described and, although a correlation between type of mutation and clinical presentation has been reported, a great clinical heterogeneity is often seen within affected members of the same family, i.e., bearing the same mutation[2,3]. Insertions of different length, located around the octapeptide repeat coding region, are also associated with familial CJD or GSS[4]. On the contrary, a single repeat deletion appears as a low-frequency polymorphism with an uncertain role, if any, in the manifestation of CJD[5,6]. Familial cases are also transmissible to laboratory animals[7,8], though at a lower rate than sporadic CJD, presenting the striking biological enigma of a disease which is simultaneously genetically inherited and infectious. No point or insert mutations have ever been found in the *PRNP* of sporadic CJD cases[6]. However, about 80-85% of sporadic CJD cases are homozygous at the polymorphic codon 129 (Met or Val) compared to only about 50-60% of the control population[6,9]. Moreover, all patients affected by the new variant CJD (nvCJD)[10], thought to be caused by the oral infection with the BSE agent[11], are 129Met homozygous[12] suggesting that these individuals have an increase susceptibility to the infectious agent[13]. The residue at position 129 also seems to influence, to a certain extent, the clinical course of the disease in some familial cases, but not in others[3,14]. In addition, the association on the same allele of Met or Val at position 129 and the Asp178Asn mutation plays an important role in determining FFI or CJD, respectively[15].

The formation and accumulation of amyloid in the brain of affected, subjects is considered the key pathogenetic event in these disorders since its formation precedes the histological lesions and the clinical appearance of the disease. Similarly to other amyloidosis, the deposit of amyloid is formed by a host-encoded protein which, in TSE, is represented by PrP. PrP is a highly conserved membrane-associated protein whose physiological function is still unclear. It consists of 253 residues in humans, with five tandem repeats of 8/9 amino acids near the amino-terminus, two N-glycosylation sites (residues 181 and 197) linked with heterogeneous oligosaccharides, a disulphide bond and a glycosylphosphatidylinositol (GPI) membrane anchor linked to the C-terminal region. PrP assumes two different conformational states: one in which the secondary structure is substantially α-helix (PrPα), whereas the other has a high β-sheet content (PrPβ)[16]. Recently, magnetic resonance studies of mouse PrP indicate that the large C-terminal fragment (residues 121-231) contains three α-helices and two stranded antiparallel β-sheet which may represent the initiation site for the transition from PrPα to PrPβ[17]. On the other hand, the N-terminal segment (residues 23-120) is flexibly disordered[18,19].

Only in TSE-infected cells, PrP aggregates in amyloid fibrils which are insoluble and partially resistant to protease digestion, hence the term PrPres. PrPres is composed of PrPβ. In non infected cells, on the other hand, PrP is soluble and completed hydrolysed by protease treatment (PrPsen). It is still unclear whether PrPsen is always in α-helix conformation or, under physiological conditions, it may also assume a β-sheet structure.

In the first hypothesis, the rate-determining step for the PrPsen to PrPres conversion consists in the conformational change of the first molecule of PrPα into PrPβ (the "refolding" theory[20], Figure 1). PrPβ would promote the transformation of other PrPα molecules, likely through an autocatalytic mechanism[21], and then accumulates as PrPres. The presence of mutant residues may lead to a significant decrease of the activation barrier between the PrPα and PrPβ conformers. This theory predicts that PrPβ preferentially interacts with homologous (i.e. of the same allotype) rather than heterologous PrPα molecules (Figure 1). In the second hypothesis, the two isoforms (PrPα and PrPβ) coexist in the cell, though the concentration of PrPα may largely exceed that of PrPβ. According to this hypothesis, the rate-limiting step for the formation of PrPres is the formation of an ordered nucleus[22,23] (the "nucleation" theory[24], Figure 1), which may depend, among

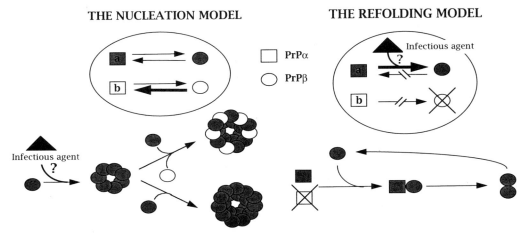

THE NUCLEATION MODEL   THE REFOLDING MODEL

PrPα
PrPβ
Infectious agent
Infectious agent
?

Figure 1. The nucleation and refolding models of amyloidogenesis in transmissible spongiform encephalopathies. In heterozygous subjects, the grey and white symbols represent the **a** and **b** allotypes, respectively. In familial TSE, **a** is the mutant PrP allotype. The infectious agent (black triangles) may initiate the reaction in both models, though at different level of the amyloidogenesis process.

other factors, on the increased concentration of PrPβ in the cell. In familial TSE the initial nucleus may be composed only of the mutant allotype. This occurs because mutations may either increase the self-aggregation of PrPβ molecules or shift the PrPα/PrPβ equilibrium leading to an increased concentration of PrPβ and the formation of the nucleus. In sporadic CJD, the 129Met or the 129Val PrPβ allotype may start the formation of the nucleus. Once the nucleus has formed, however, further addition of only mutant or mutant and wt PrPβ occurs at a high rate (Figure 1).

In the brain of subjects bearing *PRNP* mutations, protein analyses have shown that the amyloid deposits contain either the mutant PrP allotype alone or a mixture of mutant and wt PrPs (Table 1)[3, 25-31].

In the brain of familial CJD patients heterozygous for the Glu200Lys mutation only mutant PrP is converted into PrPres during disease[25]. However, the wt PrP is converted into an insoluble form as well, although it remains sensitive to protease digestion.

Table 1. Amyloid PrP allotypes in familial and sporadic transmissible spongiform encephalopathies

DISEASE TYPE	PrP MUTATIONS OR POLYMORPHISM	AMYLOID PROTEIN	REF.
GSS	Pro102Leu	Mutant only	3, 26
	Ala117Val	Mutant only	27
	Phe198Ser	Mutant only	28
	Gln217Arg	Mutant only	28
	Tyr145stop	Both mutant and wild type°	29
FFI	Asp178Asn/129Met	Mutant only	30
Familial CJD	Asp178Asn/129Val	Mutant only	30
	5 or 6 octarepeat insertions	Both mutant and wild type	30
	Glu200Lys	Both mutant and wild type*	25
	Val210Ile	Both mutant and wild type	31
Sporadic CJD	Met129Val	Both 129Met and 129Val	31

°Proteinase K resistance of wt PrP is unknown

*wt PrP is insoluble in detergents but sensitive to proteinase K

Along with these data is the observation that in the brain of the single Japanese patient affected by PrP cerebral amyloid angiopathy and bearing the Tyr145stop mutation, both wt and mutant PrPs concur to the formation of amyloid deposits around the blood vessels[29]. It remains, however, unsettled whether in this patient the wt PrP deposit is also protease resistant.

Protein sequencing and mass spectrometry analysis have shown that in the brain of one sporadic CJD patient heterozygous at the polymorphic position 129 (Met/Val)[31] PrPres consists of both the wt and mutant allotypes. Heterogeneous solutions of synthetic peptides spanning residue 118-133 (containing Met or Val at position 129) form fibrils more slowly than homogenous solutions[32], indicating that heterogeneous fibrils are less stable than homogeneous ones. These findings may provide an explanation for the low incidence of sporadic CJD in heterozygous individuals[6,9].

The finding that both wt and mutant allotypes of PrP or mutant PrP allotype only are involved in amylogenesis nears familial spongiform encephalopathies with other familial amyloidosis. A brief account of the molecular mechanisms of amyloidogenesis in Alzheimer's disease and transthyretin amyloidosis, which have been regarded as paradigmatic for TSE[33,34], is given.

Alzheimer's disease is the most common form of brain amyloidosis, affecting 3 to 10% of individuals over 65 years[35]. Pathogenetic mechanisms of Alzheimer's disease consist of metabolic and functional alterations of brain neurones leading to severe cytoskeletal alterations, formation of neurofibrillary tangles, axo-dendritic dystrophy and amyloid deposition. Amyloid deposits are commonly found in brain parenchyma and in cerebral vessel walls; they contain predominantly a 4-kDa peptide, composed of 39-43 residues, called β-protein (or A4 or Aβ) which derives from a 770-residues cellular precursor called amyloid precursor protein (APP)[35]. About 1% of Alzheimer's disease cases are linked to mutations in the APP gene[36]. Most of these mutations affect the proteolytic pathway of APP, leading to an overproduction of the 43-residues long β-protein which is the species with the highest fibrillogenicity[36]. A mutation comprised in the sequence of the β-protein that causes a Glu to Gln substitution in position 693, is associated with the deposition of amyloid in small brain vessels leading to fatal stroke or multi-infarct dementia in the fifth or sixth decade[37]. Sequencing studies have shown that in this form of the disease (called hereditary cerebral haemorrhage with amyloidosis Dutch type, HCHWA-D) both wt and mutant β-proteins form amyloid[38]. Studies with synthetic peptides encompassing residues 1-28 of β-protein (which is 39 residues long in HCHWA-D) demonstrated that the mutant 693Gln peptide has an accelerated fibril formation, thus providing a mechanism for the early onset of the disease[39]. The presence of the mutant peptide could then induce the wt β-protein to form amyloid by a nucleation mechanism[40].

Table 2. Composition of amyloid proteins in transthyretin amyloidosis

DISEASE TYPE	MUTATION	AMYLOID PROTEIN	REF.
SSA	No	Wild type only	43
FAP Portuguese and Japanese types	Val30Met	Mutant only	44, 45
FAP Swedish-American type		Both mutant and wild type	46
FAP Swedish type		Both mutant and wild type or wild type only	47
FAP Jewish type	Phe33Ile	Both mutant and wild type or wild type only	48, 49
FAP Appalachian type	Thr60Ala	Both mutant and wild type	50
FAP Indiana-Swiss type	Ile84Ser	Both mutant and wild type	51
FAC Danish type	Leu111Met	Both mutant and wild type	52
FAC Scandinavian type	Val122Ile	Both mutant and wild type or mutant only	53, 54

Transthyretin (TTR) amyloidosis include familial amyloidotic polyneuropathy (FAP) type I and II, senile systemic amyloidosis (SSA), familial amyloidotic cardiomiopathy (FAC) and some other clinical syndromes. All of them are caused by amyloid deposit of the serum protein transthyretin[41]. Familial variants of the disease are associated with more than 40 different point mutations of TTR gene, which, similarly to familial TSE, do not correlate well with clinical symptoms[42]. Interestingly, protein sequencing and mass spectrometry analysis have shown that also in TTR amyloidosis different mutations cause amyloid deposition made of mutant or wt allotype alone, or a mixture of mutant and wt TTRs (Table 2)[43-54]. The reason for this great heterogeneity remains unclear, as it is for TSE[55].

TSE, Alzheimer's disease, TTR and other amyloidoses have been recently defined as "conformational diseases"[56]. In all of them, the common pathogenic mechanism is the structural shift of a cell-encoded protein toward a pathogenic conformation which favours self-aggregation and tissue deposition. The formation and deposition of the misfolded protein represents either the main pathogenetic event (such as in Alzheimer's disease, in sickle cell anaemia or in other congenital conditions) or a secondary phenomenon, linked to a pre-existing disorder (e.g. amyloid A deposition associated with inflammatory diseases or calcitonin amyloidosis in cases of thyroid carcinoma).

Among them, PrP amyloidoses have the unique feature to be also transmissible to a great range of natural and experimental animals[7]. We, and other scientists as well[57-60], therefore think that an infectious agent must either play an important role for the initial conformational change from PrPα to PrPβ (refolding model) or be the responsible "seed" for the formation of the amyloid nucleus (nucleation model).

REFERENCES

1. M. Pocchiari, Prions and related neurological diseases, *Molec. Aspects Med.* 15:195 (1994).
2. P. Brown, L.G. Goldfarb, C.J. Gibbs Jr., and D.C. Gajdusek, The phenotypic expression of different mutations in transmissible familial Creutzfeldt-Jakob disease, *Eur. J. Epidemiol.* 7:469 (1991).
3. P. Barbanti, G. Fabbrini, M. Salvatore, R. Petraroli, F. Cardone, B. Maras, M. Equestre, G. Macchi, G.L. Lenzi, and M. Pocchiari, Polymorphism at codon 129 or codon 219 of PRNP and clinical heterogeneity in a previously unreported family with Gerstmann-Sträussler-Scheinker disease (PrP-P102L mutation), *Neurology* 47:734 (1996).
4. L.G. Goldfarb, and P. Brown, The transmissible spongiform encephalopathies, *Annu. Rev. Med.* 46:57 (1995).
5. C. Masullo, M. Salvatore, G. Macchi, M. Genuardi, and M. Pocchiari, Progressive dementia in a young patient with a homozygous deletion of the PrP gene, *Ann. N. Y. Acad. Sci.* 724:358 (1994).
6. M. Salvatore, M. Genuardi, R. Petraroli, C. Masullo, M. D'Alessandro, and M. Pocchiari, Polymorphisms of the prion protein gene in Italian patients with Creutzfeldt-Jakob disease, *Hum. Genet.* 94:375 (1994).
7. P. Brown, C.J. Gibbs Jr., P. Rodgers Johnson, D.M. Asher, M.P. Sulima, A. Bacote, L.G. Goldfarb, and D.C. Gajdusek, Human spongiform encephalopathy: the National Institutes of Health series of 300 cases of experimentally transmitted disease, *Ann. Neurol.* 35:513 (1994).
8. J. Tateishi, P. Brown, T. Kitamoto, Z.M. Hoque, R. Roos, R. Wollmann, and D.C. Gajdusek, First experimental transmission of fatal familial insomnia. *Nature* 376:434 (1995).
9. O. Windl, M. Dempster, J.P. Estibeiro, R. Lathe, R. DaSilva, T. Esmonde, R. Will, A. Springbett, T.A. Campbell, K.C.L. Sidle, M.S. Palmer, and J. Collinge, Genetic basis of Creutzfeldt-Jakob disease in the United Kingdom: a systematic analysis of predisposing mutations and allelic variation in the *PRNP* gene, *Hum. Genet.* 98:259 (1996).
10. R.G. Will, J.W. Ironside, M. Ziedler, S.N. Cousens, K. Estibeiro, A. Alperovitch, S. Poser, M. Pocchiari, A. Hofman, and Smith, A new variant of Creutzfeldt-Jakob disease in the UK, *Lancet* 347:921 (1996).
11. M. Bruce, R.G. Will, J.W. Ironside, I. McConnell, D. Drummond, A. Suttle, L. McCardle, A. Chree, J. Hope, C. Birkett, S. Cousens, H. Fraser, and C.J. Bostock, Transmissions to mice indicate that 'new variant' CJD is caused by the BSE agent, *Nature* 389:498 (1997).

12. M. Zeidler, G.E. Stewart, C.R. Barraclough, D.E. Bateman, D. Bates, D.J. Burn, A.C. Colchester, W. Durward, N.A. Fletcher, S.A. Hawkins, J.M. Mackenzie, and R.G. Will, New variant Creutzfeldt-Jakob disease: neurological features and diagnostic tests, *Lancet* 350:903 (1997).

13. A.F. Hill, M. Desbruslais, S. Joiner, K.C.L. Sidle, I. Gowland, J. Collinge, L.J. Doey, and P. Lantos , The same prion strain causes vCJD and BSE, *Nature* 389: 448 (1997).

14. M. Pocchiari, M. Salvatore, F. Cutruzzolà, M. Genuardi, C. Travaglini-Allocatelli, C. Masullo, G. Macchi, G. Alemà, S. Galgani, Y.G. Xi, R. Petraroli, M.C. Silvestrini, and M. Brunori, A new point mutation of the prion protein gene in Creutzfeldt-Jakob disease, *Ann. Neurol.* 34:802 (1993).

15. L.G. Goldfarb, R.B. Petersen, M. Tabaton, P. Brown, A.C. LeBlanc, P. Montagna, P. Cortelli, J. Julien, C. Vital, W.W. Pendelbury, M. Haltia, P.R. Wills, J.J. Hauw, P.E. McKeever, L. Monari, B. Schrank, G.D. Swergold, L. Autilio-Gambetti, D.C. Gajdusek, E. Lugaresi, and P. Gambetti, Fatal familial insomnia and familial Creutzfeldt-Jakob disease: disease phenotype determined by a DNA polymorphism, *Science* 258:806 (1992).

16. K.M. Pan, M.A. Baldwin, J. Nguyen, M. Gasset, A. Serban, D. Groth, I. Mehlhorn, Z. Huang, R.J. Fletterick, F.E. Cohen, and S.B. Prusiner, Conversion of α-helices into β-sheets features in the formation of the scrapie prion proteins *Proc. Natl. Acad. Sci. USA* 90:10962 (1993).

17. R. Riek, S. Hornemann, G. Wider, M. Billeter, R. Glockshuber, and K. Wüthrich, NMR structure of the mouse prion protein domain PrP(121-231), *Nature* 382:180 (1996).

18. S. Hornemann, C. Korth, B. Oesch, R. Riek, G. Wider, K. Wütrich, and R. Glockshuber, Recombinant full-length murine prion protein, mPrP(23-231): purification and spectroscopic characterization, *FEBS Lett.* 413:277 (1997).

19. R. Riek, S. Hornemann, G. Wider, R. Glockshuber, and K. Wüthrich, NMR characterization of the full-length murine prion protein, mPrP(23-231), *FEBS Lett.* 413:282 (1997).

20. S.B. Prusiner, Inherited prion diseases, *Proc. Natl. Acad. Sci. USA* 91:4611 (1994).

21. M. Brunori, and B. Talbot, Mechanism for prion replication, *Nature* 314:676 (1985).

22. J. Hofrichter, P.D. Ross, and W.A. Eaton, Kinetics and mechanism of deoxyhemoglobin S gelation: a new approach to understanding sickle cell disease, *Proc. Natl. Acad. Sci. USA* 71:4864 (1974).

23. W.A. Eaton, and J. Hofrichter, Sickle cell hemoglobin polymerization, *Adv. Protein Chem.* 40:63 (1990).

24. D.C. Gajdusek, Nucleation of amyloidogenesis in infectious and noninfectious amyloidoses of brain, *Ann. N. Y. Acad. Sci.* 724:173 (1994).

25. R. Gabizon, G. Telling, Z. Meiner, M. Halimi, I. Kahana, and S.B. Prusiner, Insoluble wild-type and protease-resistant mutant prion protein in brains of patients with inherited prion disease, *Nature Med.* 2:59 (1996).

26. T. Kitamoto, K. Yamaguchi, K. Doh-ura, and J. Tateishi, A prion protein missense variant is integrated in kuru plaque cores in patients with Gerstmann-Straussler syndrome, *Neurology* 41:306 (1991).

27. F. Tagliavini, F. Prelli, M. Porro, G. Rossi, G. Giaccone, T.D. Bird, S.R. Dlouhy, K. Young, P. Piccardo, B. Ghetti, O. Bugiani, and B. Frangione, Only mutant PrP participates in amyloid formation in Gerstmann-Sträussler-Scheinker disease with Ala>Val substitution at codon 117, *J. Neuropathol. Exp. Neurol.* 54:416 (1995).

28. F. Tagliavini, F. Prelli, M. Porro, G. Rossi, G. Giaccone, M.R. Farlow, S.R. Dlouhy, B. Ghetti, O. Bugiani, and B. Frangione, Amyloid fibrils in Gerstmann-Sträussler-Scheinker disease (Indiana and Swedish kindreds) express only PrP peptides encoded by the mutant allele, *Cell* 79:695 (1994).

29. B. Ghetti, P. Piccardo, M.G. Spillantini, Y. Ichimiya, M. Porro, F. Perini, T. Kitamoto, J. Tateishi, C. Seiler, B. Frangione, O. Bugiani, G. Giaccone, F. Prelli, M. Goedert, S.R. Dlouhy, and F. Tagliavini, Vascular variant of prion protein cerebral amyloidosis with τ-positive neurofibrillary tangles: the phenotype of the stop codon 145 mutation in PRNP, *Proc. Natl. Acad. Sci. USA* 93:744 (1996).

30. S.G. Chen, P. Parchi, P. Brown, S. Capellari, W. Zou, E.J. Cochran, C.L. Vnencak-Jones, J. Julien, C. Vital, J. Mikol, E. Lugaresi, L. Autilio-Gambetti and P. Gambetti, Allelic origin of the abnormal prion protein isoform in familial prion diseases, *Nature Med.* 3: 1009 (1997).

31. M.C. Silvestrini, F. Cardone, B. Maras, P. Pucci, D. Barra, M. Brunori, and M. Pocchiari, Identification of the prion protein allotypes which accumulate in the brain of sporadic and familial Creutzfeldt-Jakob disease patients, *Nature Med.* 3:521 (1997).

32. J.H. Come, and P.T. Lansbury Jr., Predisposition of prion protein homozygotes to Creutzfeldt-Jakob disease can be explained by a nucleation-dependent polymerization mechanism, *J. Am. Chem. Soc.* 116:4109 (1994).

33. D.C. Gajdusek, Spontaneous generation of infectious nucleating amyloids in the transmissible and nontransmissible cerebral amyloidoses, *Mol. Neurobiol.* 8:1 (1994).

34. P. Brown, Central nervous system amyloidoses: a comparison of Alzheimer's disease and Creutzfeldt-Jakob disease, *Neurology*, 39:1103(1989).

35. C.P.J. Maury, Biology of disease: molecular pathogenesis of β-amyloidosis in Alzheimer's disease and other cerebral amyloidosis, *Lab. Invest.* 72:4 (1995).

36. J. Hardy, Amyloid, the presenilins and Alzheimer's disease, *Trends Neurol. Sci.* 20:154 (1997).

37. E. Levy, M.D. Carman, I.J. Fernandez-Madrid, M.D. Power, I. Lieberburg, S.G. Van Duinen,G.T.A.M. Bots, W. Luyendijk, and B. Frangione, Mutation of the Alzheimer's disease amyloid gene in hereditary cerebral haemorrhage, Dutch type, *Science* 248:1124 (1990).

38. F. Prelli, E. Levy, S.G. Van Duinen, G.T.A.M. Bots, W. Luyendijk, and B. Frangione, Expression of a normal and variant Alzheimer's -protein gene in amyloid of hereditary cerebral haemorrhage, Dutch type: DNA and protein diagnostic assays, *Biochem. Biophys. Res. Commun.* 170:301 (1990).

39. T. Wisniewski, J. Ghiso, B. Frangione, Peptides homologous to the amyloid protein of Alzheimer's disease containing a glutamine for glutamic acid substitution have accelerated amyloid fibril formation, *Biochem. Biophys. Res. Commun.* 179:1247 (1991).

40. J.T. Jarrett, and P.T.Lansbury Jr., Seeding "one-dimensional crystallization" of amyloid: a pathogenic mechanism in Alzheimer's disease and scrapie? *Cell* 73:1055 (1993).

41. M.D. Benson, Familial amyloidotic polyneuropathy, *Trends Neurol. Sci.* 12:88 (1989).

42. M.J.M. Saraiva, Transthyretin mutations in health and disease, *Hum. Mutat.* 5:191 (1995).

43. A. Gustavsson, H. Jahr, R. Tobiassen, D.R. Jacobson, K. Sletten, and P. Westermark, Amyloid fibril composition and transthyretin gene structure in senile systemic amyloidosis, *Lab. Invest.* 73:703 (1995).

44. M.J.M. Saraiva, S.Birken, P.P. Costa, and D.S. Goodman, Amyloid fibril protein in familial amyloidotic polyneuropathy, Portuguese type, *J. Clin. Invest.* 74:104 (1984).

45. S. Tawara, M. Nakazato, K. Kangawa, H. Matsuo, and S. Araki, Identification of amyloid prealbumin in familial amyloidotic polyneuropathy (Japanese type), *Biochem. Biophys. Res. Commun.* 116:880 (1983).

46. F.E. Dwulet, and M.D. Benson, Primary structure of an amyloid prealbumin and its plasma precursor in a heredofamilial polyneuropathy of Swedish origin, *Proc. Natl. Acad. Sci. USA* 81:694 (1984).

47. P. Westermark, K. Sletten, and B.O. Olofsson, Prealbumin variants in the amyloid fibrils of Swedish familial amyloidotic polyneuropathy, *Clin. Exp. Immunol.* 69:695 (1987).

48. M. Nakazato, K. Kangawa, N. Minamino, S. Tawara, H. Matsuo, and S. Araki, Revised analysis of amino acid replacement in a prealbumin variant (SKO-III) associated with familial amyloidotic polyneuropathy of Jewish origin, *Biochem. Biophys. Res. Commun.* 12:921 (1984).

49. M.Pras, F. Prelli, J. Gafni, and B. Frangione, Genetic heterogeneity of familial amyloid polyneuropathy of Jewish type, in: *Amyloidosis,* G.G. Glenner, E.F. Osserman, E.P. Benditt, E. Calkins, A.S. Cohen and D. Zucker-Franklin, eds., Plenum Publishing Corp., New York (1986).

50. M.R. Wallace, F.E. Dwulet, P.M. Conneally, and M.D. Benson, Biochemical and molecular genetic characterization of a new variant prealbumin associated with hereditary amyloidosis, *J. Clin. Invest.* 78:6 (1986).

51. F.E. Dwulet, and M.D. Benson, Characterization of a transthyretin (prealbumin) variant associated with familial amyloidotic polyneuropathy type II (Indiana/Swiss), *J. Clin. Invest.* 78:880 (1986).

52. L.F. Hermansen, T. Bergman, H. Jörnvall, G. Husby, I. Ranlov, and K. Sletten, Purification and characterization of amyloid-related transthyretin associated with familial amyloidotic cardiomiopathy, *Eur. J. Biochem.* 227:772 (1995).

53. M.J.M. Saraiva, W. Sherman, C. Marboe, A. Figueira, P. Costa, A.F. De Freitas, and A. Gawinowicz, Cardiac amyloidosis: report of a patient heterozygous for thetransthyretin isoleucine 122 variant, *Scand. J. Immunol.* 32:341 (1990).

54. P.D. Gorevic, F. Prelli, J. Wright, M. Pras, and B. Frangione, Systemic senile amyloidosis: Identification of a new prealbumin (transthyretin) variant in cardiac tissue: immunologic and biochemical similarity to one form of familial amyloidotic polyneuropathy, *J. Clin. Invest.* 83:836 (1989).

55. J.W. Kelly, Alternative conformations of amyloidogenic proteins govern their behavior, *Curr. Opin. Struct. Biol.* 6:11 (1996).
56. R.W. Carrell, and D.A. Lomas, Conformational disease, *Lancet*, 350:134 (1997).
57. M.E. Bruce, Scrapie strain variation and mutation, *Br. Med. Bull.*, 49:822 (1993).
58. H. Diringer, M. Beekes, and U. Oberdieck, The nature of the scrapie agent: the virus theory, *Ann. N. Y. Acad. Sci.*, 724:246 (1994).
59. L. Manuelidis, T. Sklaviadis, A. Akowitz, and W. Fritch, Viral particles are required for infection in neurodegenerative Creutzfeldt-Jakob disease, *Proc. Natl. Acad. Sci. USA*, 92:5124 (1995).
60. B. Caughey, and B. Chesebro, Prion protein and the transmissible spongiform encephalopathies, *Trends Cell. Biol.*, 7:56 (1997).

NEURONAL DEGENERATION AND CELL DEATH IN PRION DISEASE

Hans A. Kretzschmar, Armin Giese, Jochen W. Herms, and David R. Brown

Department of Neuropathology
University of Göttingen
Robert-Koch-Str. 40
37075 Göttingen, Germany

INTRODUCTION

The classical pathological features of spongiform encephalopathies include vacuolation of the neuropil, astrocytic gliosis and neuronal degeneration (Masters and Richardson, 1978). Neuronal degeneration is widespread in many cases of CJD and may lead to almost complete loss of nerve cells (Fig. 1A,B). Alfons Jakob described a variety of pathological changes including chromatolysis of Betz cells in the motor cortex as well as satellitosis and neuronophagia ("glial rosettes"). However, there is only limited data available on the extent and pattern of neuronal loss (Jeffrey et al., 1992; Jeffrey et al., 1995; Masters and Richardson, 1978; Hogan et al., 1981; Scott and Fraser, 1984), and even less is known about the mechanisms which underlie neuronal cell death in spongiform encephalopathies. It is not clear whether apoptotic nerve cell loss is caused by progressive loss of the function of PrP^C or increasing toxicity of PrP^{Sc} in the course of the disease ("loss of function and gain of function hypotheses").

Cell culture experiments with neurotoxic prion protein fragments suggest that neuronal cell death in these diseases may be due to apoptosis. To test this hypothesis *in vivo* the *in situ* end-labeling (ISEL) technique and electron microscopy were used to study cell death in an experimental scrapie system in the mouse. ISEL, which relies on the incorporation of labeled nucleotides in fragmented DNA by terminal transferase, showed labeled nuclei in the brains and retinae of mice infected with the 79A strain of scrapie. In the brain, labeled nuclei were mainly found in the granule layer of the cerebellum of terminally ill mice. These results support the hypothesis that neuronal loss in spongiform encephalopathies is due to apoptosis.

Previous experiments have suggested that the normal cellular prion protein (PrP^C) is involved in synaptic function in the hippocampus (Collinge et al., 1994). This finding has been interpreted as an argument in favor of the "loss of function hypothesis". To investigate the synaptic function of prion protein in cerebellar Purkinje cells we utilized the controlled recording conditions of the patch clamp technique. In PrP gene-ablated mice ($Prnp^{0/0}$ mice) (Büeler et al., 1992), the kinetics of GABA- and glutamate receptor-mediated currents showed no significant deviation from those in control animals. In consequence, our findings do not confirm the hypothesis that progressive loss of synaptic function of PrP^C causes neuronal apoptosis.

A fragment of human PrP which consists of amino acids 106-126 and forms fibrils *in vitro* has been previously demonstrated to be toxic to cultured hippocampal neurones (Forloni et al., 1993). In cell culture experiments we were able to demonstrate that PrP$^{106-126}$ is toxic to cells from the cerebellum but that this toxicity requires the presence of microglia. In response to the presence of PrP$^{106-126}$, microglia increase production of oxygen radicals. However, the effect of PrP$^{106-126}$ on microglia, necessary for the toxic effect on normal neurones, is not sufficient to kill neurones from mice not expressing PrP^C. These findings are in accord with reports on the resistance of Prnp gene-ablated (Prnp$^{0/0}$) mice to scrapie.

Cell culture experiments reveal that cerebellar cells lacking PrP^C are more sensitive to oxidative stress and undergo cell death more readily than wild-type cells. This effect is reversible by treatment with vitamin E. *In vivo* studies show that the activity of Cu/Zn superoxide dismutase is reduced in Prnp$^{0/0}$ mice. Constitutively high Mn superoxide dismutase activity in these animals may compensate for this loss of responsiveness to oxidative stress. These findings suggest that PrP^C may influence the activity of Cu/Zn superoxide dismutase and may be important for cellular resistance to oxidative stress.

Taken in conjunction these experiments have begun to open new insights into the physiological and pathological functions of PrP^C and PrP^{Sc} and the pathogenesis of prion diseases.

1. Neuronal cell death in prion diseases is due to apoptosis

It is generally believed that there are two basic types and mechanisms of cell death: necrosis and apoptosis (Buja et al., 1993; Kerr et al., 1972; Searle et al., 1982). Necrosis often results from severe and sudden injury and leads to rapid cell lysis and a consecutive inflammatory response. In contrast, apoptosis proceeds in an orderly manner following a cellular suicide program involving active gene expression in response to physiological signals or types of stress. Morphologically, apoptosis is characterized by chromatin condensation and aggregation, cellular and nuclear shrinkage and formation of apoptotic bodies. It is usually not accompanied by an inflammatory response.

The use of the *in situ* end-labeling technique (ISEL) has greatly facilitated the recognition of apoptotic cells in tissue sections (Gavrieli et al., 1992; Gold et al., 1993). This technique was used to identify apoptotic cells in mice infected with the 79A strain of scrapie, a well-defined experimental scrapie model (Bruce et al., 1991; Bruce, 1993; Fraser, 1993; Bruce et al., 1994). To evaluate the pattern and time course of cell death, various brain regions were analyzed at different points of time in the course of the disease. Since cell loss in scrapie-infected rodents often is quite striking in the photoreceptor layer of the retina (Hogan et al., 1981; Kozlowski et al., 1982; Hogan et al., 1983; Buyukmihci et al.,1987a,b), the eyes were included in this study. Electron microscopy was used to independently assess apoptotic cell death (Giese et al., 1995).

Cerebellum. ISEL revealed nuclei containing fragmented DNA in the granule cell layer in scrapie-infected mice from day 120 onwards prior to onset of clinical signs of scrapie (approximately day 150). No labeling was found in control mice at any point of time and in scrapie-infected mice up to day 90 with the exception of one scrapie-infected animal showing one labeled nucleus in ten high-power visual fields 90 days after inoculation. The number of labeled cells increased from day 120 to 150 and was highest in terminally ill mice (day 166) (Fig. 1E). Corresponding to the *in situ* end-labeling assay, small, dark, round, occasionally fragmented nuclei with eosinophilic cytoplasm were found in sections from scrapie-infected mice stained with haematoxylin and eosin from day 120 onwards (Fig 1C).

Electron microscopic investigation of specimens from the cerebellum of terminally ill mice identified several cells which showed homogeneously condensed chromatin, dark cytoplasm, membrane blebbing and, occasionally, nuclear fragmentation (Fig. 1D), all of which are morphological changes characteristic of apoptosis (Kerr et al., 1972; Searle et al., 1982; Sloviter et al., 1993).

Other regions of the brain. In the rest of the brain labeling was less obvious. As described previously (Migheli et al., 1994), labeled nuclei were observed occasionally in the ependyma and in a subependymal location along the lateral ventricles both in control and scrapie-infected mice. In scrapie-infected mice at days 150 and 166 a small number of unequivocally labeled nuclei were observed in the basal ganglia, in the granule cell layer of the olfactory bulb and in the cerebral cortex.

Retina. Sections from animals at 30, 60 and 90 days after inoculation showed no pathological changes of the retina. When compared to controls, eyes removed from animals 120 days after scrapie infection showed a dissolution of rod inner and outer segment regions. This process was accompanied by the appearance of macrophages in this layer. In two out of four animals the outer nuclear layer appeared considerably thinned. No changes were detectable in the inner nuclear layer and the ganglion cell layer. At 150 and 166 days post infection the outer nuclear layer had degenerated to only a single cell layer in thickness with slight variations of two to three cells in some regions and a total loss of photoreceptor cells in other regions. The photoreceptor inner and outer segments had almost completely vanished. These changes were similar to those described previously (Hogan et al., 1981).

The ISEL assay showed that labeled nuclei were practically absent from retinae of control mice of all ages. Labeled nuclei were virtually absent from scrapie-infected mice up to day 90. At 120 days after infection several labeled nuclei were found scattered in the outer nuclear layer (Fig. 1F). Some labeled nuclei appeared shrunken, and labeling was often most intense at the periphery of the nucleus. At 150 and 166 days after infection solitary labeled nuclei were found in the remainder of the outer nuclear layer. However, due to the extreme loss of cells in this time interval, the percentage of apoptotic cells to the overall number of remaining cells increased from 0.43% (SEM=0.105, n=4) 120 days after infection to 0.91% (SEM=0.211, n=4) in terminally ill mice.

Pattern and time course of apoptotic cell death in scrapie. Judging from previous studies there are indirect indications of apoptosis occurring in prion disease *in vivo*. Using laser scanning microscopic analysis of a single cell gel assay, increased DNA damage was found in three scrapie-infected sheep brains as compared to controls, while no attempt was made to control for necrosis, autolysis or age (Fairbairn et al., 1994). Hogan et al. (1981) gave a detailed morphological description of retinal degeneration and showed an electron micrograph of a photoreceptor cell which nowadays would be interpreted as apoptosis. Additionally, the almost complete absence of an inflammatory response in the pathology of spongiform encephalopathies (Eikelenboom et al., 1991; Weissmann et al., 1994) suggests that nerve cell loss may be due to apoptosis. On the other hand, in a large study investigating apoptosis in the nervous system using the ISEL technique, apoptosis was not identified in terminally ill scrapie-infected mice (Migheli et al., 1994); however, the authors give no details concerning the strains of mice and scrapie used and the brain regions investigated. We used the ISEL technique, which has recently been employed by a great number of investigators, to identify apoptotic cells in histological sections. It appears, however, that the results obtained with this technique are not absolutely specific; ISEL is based on the incorporation of labeled nucleotides in fragmented DNA by the enzyme terminal transferase which is considered

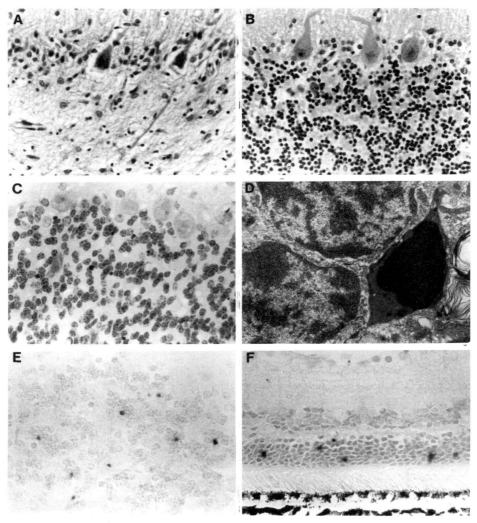

Figure 1. Neuronal cell death in prion disease. 1A. Cerebellum of a patient who died from CJD. The granule cell layer is almost completely destroyed. Hematoxylin & eosin. 1B. For comparison, this cerebellar section of an age-matched control shows normal density of granule cells. Hematoxylin & eosin. 1C. Several small dark nuclei suggestive of apoptosis are found in the granule cell layer of scrapie-infected mice. 1D. Electron micrograph of an apoptotic cell in the cerebellar granule cell layer of a mouse terminally ill with scrapie. Note dense and homogeneous aggregation of nuclear chromatin and increased density of cytoplasm. Two adjacent normal granule cell nuclei are seen to the left. 1E. *In situ* end-labeling of the cerebellum of a mouse terminally ill with scrapie shows several labeled apoptotic nuclei in the granule cell layer. 1F. *In situ* end-labeling in the retina of scrapie-infected mice identifies apoptotic cells almost exclusively in the outer nuclear layer. The highest absolute number of labeled cells is observed at 120 days after infection. Experimental procedures have been described in detail elsewhere (Giese et al., 1995)

highly characteristic of apoptosis. DNA fragmentation is also observed in necrosis or autolysis, but in these instances seems to occur at random, in contrast to the specific internucleosomal cleavage which is characteristic of apoptosis (Wyllie et al., 1984). Therefore, DNA laddering in gel electrophoresis, which shows the different size of oligonucleosomal fragments, is often used to distinguish apoptosis from mechanisms leading to random DNA cleavage. In cases where only a small percentage of cells undergoes apoptosis at any given moment, as expected in our experimental system, DNA laddering would not be applicable and hence was not attempted. Therefore classical morphological criteria were employed (Kerr et al., 1972; Searle et al., 1982) to validate data obtained by ISEL. These included chromatin condensation, nuclear fragmentation, eosinophilic and electron microscopically dark cytoplasm and membrane blebbing. Since all of these were present, our results convincingly demonstrate the occurrence of neuronal apoptosis in an experimental *in vivo* scrapie system using two different techniques.

C57Bl6 mice infected with the 79A strain of scrapie are known to show diffuse widespread accumulation of prion protein (Bruce et al., 1994) without amyloid plaques (Bruce et al., 1976; McBride et al., 1988) as well as moderate vacuolation affecting most gray matter areas including the molecular layer of the cerebellum to a fairly similar degree. Other strains of scrapie such as ME7 show comparatively little spongiform change in the cerebellum (Bruce et al., 1991). We found the most obvious labeling of apoptotic cells in the granular cell layer of the cerebellum and in the outer nuclear layer of the retina. These structures contain a high number and density of morphologically uniform neuronal cells and therefore are ideally suited for detection of a small percentage of apoptotic cells. However, other regions similar in this respect, such as the granule cell layer of the olfactory bulb or hippocampus, showed few or no labeled nuclei, indicating a different degree of cell death in these structures. ISEL offers the opportunity to detect the interesting early phase in the progression of the disease before massive cell loss and obvious clinical signs of scrapie are evident. This technique should help establish more detailed lesion profiles and elucidate the correlation between prion protein accumulation, spongiform change and neuronal cell death in the future. Our results have subsequently been confirmed by Lucassen et al. (Lucassen et al., 1995) who found different patterns of apoptotic cell death in mice infected with different strains of scrapie.

The small percentage of labeled cells that were found even in areas of massive ongoing cell death is in no way surprising since apoptosis has been shown to be completed within hours (Bursch et al., 1990) and cells can be expected to become apoptotic asynchronously. This indicates that it will be difficult to demonstrate apoptotic cells in conditions where cell loss is less rapid. The demonstration of apoptosis in an animal model of scrapie therefore is of considerable importance both for the pathophysiology of spongiform encephalopathies and the validation of cell culture experiments using neurotoxic prion protein fragments as well as in the broader context of understanding the mechanisms of neurodegenerative diseases.

2. Is PrP^C necessary for normal synaptic function?

PrP^C has been found in all mammals examined, has a high turnover rate (Caughey et al., 1989) and is widely expressed in early embryogenesis (Manson et al., 1992) as well as postnatally (Kretzschmar et al., 1986; Lieberburg, 1987; Moser et al., 1995). It is anchored to the neuronal membrane surface by glycosylphosphatidylinositol, suggesting a role in cell signalling or adhesion (Borchelt et al., 1993). Yet, PrP gene-ablated mice (Prnp$^{0/0}$) appear developmentally and behaviorally normal (Büeler et al., 1992). These mice, however, have been shown to have an altered synaptic transmission, giving rise to the hypothesis that PrP^C is necessary for normal synaptic function (Collinge et al., 1994). Intracellular single-electrode voltage clamp measurements in hippocampal CA1 neurons of Prnp$^{0/0}$ mice showed slower inhibitory postsynaptic currents (IPSCs) compared to control animals. Depolarized reversal potentials of IPSCs suggested a postsynaptic function of PrP probably due to abnormal Cl⁻ gradients or altered selectivity of GABA-gated ion channels (Collinge et al., 1994). Moreover, PrP^C may be necessary for the induction of long-term potentiation and sleep rhythm regulation (Tobler et al., 1997; Tobler et al., 1996; Manson et al., 1995; Collinge et al., 1994). The conclusion from these experiments was that the expression of PrP^C is necessary for normal synaptic function and that neuropathological changes in prion disease may be due to the lack of functionally intact prion protein ("lack of function hypothesis").

Since the cerebellum is severely affected in most cases of human prion disease as well as in many experimental models of scrapie, we investigated the synaptic role of prion protein in cerebellar Purkinje cells (Herms et al., 1995). Patch clamp experiments were performed in thin slice preparations to compare synaptic properties of Purkinje cells in wild-type and $Prnp^{0/0}$ animals (Fig. 2).

In the absence of external stimuli, IPSCs occurred in a random fashion due to spontaneous activation of GABAergic synapses (Llano et al., 1991; Vincent et al., 1992). These currents were reversibly blocked by 10 µM bicuculline, a blocker of $GABA_A$ receptors. In $Prnp^{0/0}$ animals, we observed spontaneous inhibitory currents in all cells investigated (n=23) (Fig. 2A). The relative frequency of inhibitory currents varied significantly from cell to cell but was usually found to be within 1 and 5 events per second. IPSC amplitudes in $Prnp^{0/0}$ animals ranged from 20 pA to 5 nA with an average amplitude of 300 pA . This large variation was not specific for $Prnp^{0/0}$ animals, since IPSCs in control animals displayed similar properties. Also, there was no significant difference in the relative frequency of IPSCs between $Prnp^{0/0}$ and control animals. The conclusion is that the lack of PrP^C does not affect the mean number of $GABA_A$ receptors activated by a single synaptic event and does not increase or reduce the average number of spontaneously active synapses in cerebellar Purkinje cells.

The activation time of synaptic currents was determined by fitting a linear function to the activating current as shown in Fig. 2B. The decay constant of GABA currents was determined by fitting a single exponential function to the current decay (Fig. 2A,B). An average activation time constant of 2.4 ± 0.6 ms (10 - 90% max. amplitude, n = 23 cells) and an average decay time constant of 11.3 ± 2.6 ms (n = 23) for the responses in $Prnp^{0/0}$ animals were found. The variation was notably large, but was repeatedly observed during recordings from Purkinje cells in $Prnp^{0/0}$ and wild-type animals. As shown in Fig. 2B, the average value and statistical variation of activation and decay time constants of $Prnp^{0/0}$ mice were indistinguishable from those observed in wild-type animals.

To investigate the $GABA_A$ receptor more closely, we analysed the functional parameters of isolated $GABA_A$ receptor channels in outside-out membrane patches pulled from Purkinje cells of control and $Prnp^{0/0}$ mice (Fig. 2C). Channel activation and deactivation was essentially equal in control and $Prnp^{0/0}$ mice. Moreover, GABA currents in $Prnp^{0/0}$ mice displayed a reversal potential of 2.2 ± 1.6 mV (n=5), which was close to the reversal potential of wild-type animals (3.7 ± 2.7 mV, n=4).

Earlier reports using conventional electrophysiological techniques have suggested a pivotal role for prion proteins in the function of the $GABA_A$ receptor. A positive shift in the current-voltage relation of agonist-induced $GABA_A$ receptor currents suggested that synaptic alterations might reflect modifications of the $GABA_A$ receptor level in $Prnp^{0/0}$ animals (Collinge et al., 1994; Whittington et al., 1995). However, more detailed investigations in the cerebellum (Herms et al., 1995) and in the hippocampus by Lledo et al. (1996) question the findings of Collinge and colleagues (Collinge et al., 1994; Whittington et al., 1995). The prion protein may be involved in the formation or maintenance of synapses. $GABA_A$ receptor function however, is obviously not directly dependent on PrP^C.

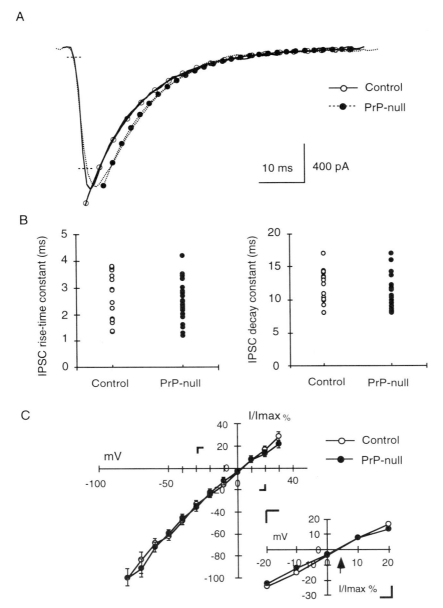

Figure 2. Synaptic properties of Prnp[0/0] cerebellar Purkinje cells. A,B Kinetic parameters of IPSCs recorded from cerebellar Purkinje cells in Prnp[0/0] and wild-type cerebellar slices. A, Two superimposed traces of spontaneous IPSCs recorded as inward currents at a holding potential of -63 mV show similar shapes; the curve fitting used single exponentials. B. The rise time constant did not differ significantly between Prnp[0/0] cerebellar slices (2.4 ± 0.6 ms, means ± S.D., n=23) and controls (2.7 ± 0.7 ms, n=18), neither did the decay time constant (11.3 ± 2.6 ms in Prnp[0/0] slices (n=23) and 12 ± 2.4 ms in control slices (n=18)). C. Kinetics of GABA-induced currents in outside-out patches from cerebellar Purkinje cells of Prnp[0/0] and control mice. Peak current was normalized to the amplitude of -73 mV (I/Imax %) as a function of membrane potential. The peak current-voltage relation was linear, reversing at 3.7 ± 2.7 mV in control (n=4) and 2.2 ± 1.6 mV in Prnp[0/0] mouse Purkinje cells (n=5) (arrow in inset).

METHODS: Patch clamp experiments were performed on Purkinje cells in thin slices of the cerebellum following standard procedures (Hamill et al., 1981; Edwards et al., 1989; Herms et al., 1995). The two inbred lines (129/Sv and C57BL/6J) from which the Prnp[0/0] mice were derived (Büeler et al., 1992) as well as the F1 cross between these were used as controls.

3. Models for the study of nerve cell loss in prion disease

PrP and resistance to oxidative stress. It has been suggested that the pathology of prion disease is related to the loss of function of PrP^C (Collinge et al., 1994). Analyses of scrapie-infected mice have indicated production of substances which may cause (Williams et al., 1994) or are biproducts of (Guan et al., 1996) oxidative stress in brain. Apoptotic cell death is also observed in scrapie-infected mice (Giese et al., 1995) and one of the known causes of apoptosis is oxidative stress. Impaired resistance to oxidative stress due to loss of activity or regulation of enzymes such as of SOD-1 may contribute to the neurodegeneration seen in prion disease.

Cultures of cells from wild-type and Prnp$^{0/0}$ cerebellum were exposed for 24 hours to varying degrees of oxidative stress as produced by xanthine oxidase. Prnp$^{0/0}$ cells show a greater susceptibility to oxidative stress induced by this enzyme. Vitamin E, an anti-oxidant, enhances survival of Prnp$^{0/0}$ cells significantly more than of wild-type cells in a dose-dependent manner (Brown and Kretzschmar, 1996). In our cell culture assays, cells that do not express PrP^C die more rapidly as a result of increased susceptibility to oxidative stress. The cause of this oxidative stress is likely to be reactive oxygen species such as superoxide or the hydroxyl radical. Experiments with pharmacological agents suggest that there is no difference in the ability of Prnp$^{0/0}$ cells to deal with increased glutamate or intracellular calcium. Furthermore, an inhibitor of catalase did not affect Prnp$^{0/0}$ cells differently, suggesting no difference in the ability of Prnp$^{0/0}$ cells to deal with hydrogen peroxide.

At present there is no other indication that the strain of Prnp$^{0/0}$ mice studied by us has a reduced survival capacity. A previous examination of Prnp$^{0/0}$ mouse phenotype demonstrated altered electrophysiologic characteristics including reduced LTP and GABA$_A$ type inhibitions (Collinge et al., 1994) which could be restored by transgenic expression of human PrP^C (Vincent et al., 1992) (see preceding chapter). The underlying biochemical mechanism of this effect has not been elucidated, and it is possible that these characteristics are due to decreased viability or increased susceptibility to oxygen radicals of *in vitro* slice preparations as we have shown for dissociated cells in our culture system.

There is increasing evidence that loss of Cu/Zn superoxide dismutase activity (SOD-1), an enzyme which breaks down superoxide, leads to apoptotic cell death (Rothstein et al., 1994). Alterations in the activity of this enzyme may be the cause of some cases of amyotrophic lateral sclerosis (Deng et al., 1993; Rosen et al., 1993). As Prnp$^{0/0}$ mouse cerebellar cell cultures have a decreased ability to deal with oxygen radicals we investigated the activity of superoxide dismutases in adult mouse brain (Brown et al., 1997). Homogenates of wild-type and Prnp$^{0/0}$ mouse brains were prepared. The SOD-1 activity was quantitated by a spectrophotometric assay based on oxygen radical formation by xanthine oxidase. The measured combined activity of superoxide dismutases is considerably lower in Prnp$^{0/0}$ mouse brain extracts. Levels of the mitochondrial enzyme Mn superoxide dismutase (MnSOD) could be determined in parallel samples by the addition of KCN. In extracts from wild-type mice the level of MnSOD is very low. In comparison, Prnp$^{0/0}$ mice have elevated MnSOD levels (Table 1). This suggests that Prnp$^{0/0}$ cells experience greater oxidative stress *in vivo* and that increased activity of MnSOD may compensate partly for loss of SOD-1 activity (Brown and Kretzschmar, 1996).The upregulation of this enzyme *in vivo* may mask the loss of activity of SOD-1 and explain why Prnp$^{0/0}$ mice show no survival problems.

Although we measured a decreased activity of Cu/Zn superoxide dismutase we could not conclude from our assay that there was a difference in the activity of extracellular superoxide dismutase (Marklund, 1982; Tibell et al., 1987), another copper binding dismutase. However, in the brain this dismutase is expressed only at low levels and its purpose is not well understood (Oury et al., 1992). Interestingly, PrP^C, also a cell surface glycoprotein, is also suggested to bind copper (Pan et al., 1992) at the N-terminal tandem repeat region (Hornshaw

Table 1. Activity of superoxide dismutase in adult mice (% inhibition of formazan production) [1]

	wild-type	Prnp$^{0/0}$
Total SOD	61 ± 5	45 ± 4
SOD-1	59 ± 5	30 ± 4
MnSOD	2 ± 2	15 ± 3

[1]The activity of superoxide dismutase in adult mice varies with PrP^C expression. The levels of total SOD activity and KCN-insensitive MnSOD activity were quantitated for the extracts of wild-type and Prnp$^{0/0}$ mouse brain. Values are expressed as percentage inhibition of formazan production. Higher inhibition of formazan production from NBT represents higher SOD activity. As a control, 1 unit of bovine SOD-1 activity was also assayed which reduced formazan production by 62 ± 5%. MnSOD activity was elevated in Prnp$^{0/0}$ mice. However, the total SOD activity was much lower. Subtracting the MnSOD activity from this indicates that Prnp$^{0/0}$ mice have greatly diminished SOD-1 activity. Shown are the mean and s.e. of assays of four mice of each strain.
METHODS: Adult mice were between 3-6 months of age. Extracts were prepared by homogenizing mouse brain in a buffer of 50 mM sodium phosphate, 100 mM NaCl and 1% NP40 (Sigma). The protein concentration was determined by a standard BCA assay. The spectrophotometric assay to quantitate both total SOD activity (both SOD-1 and MnSOD) and MnSOD activity alone was as previously described (Oberley and Spitz, 1984). 10 μg of each protein extract were assayed and compared to 1 unit of bovine SOD-1 (Sigma) activity. The assay was carried out for 2 minutes after addition of the radical producer, xanthine oxidase (Boehringer). The superoxide dismutase activity was expressed as percentage inhibition of the formazan produced in control reaction without SOD or protein extracts. For MnSOD activity, extracts were incubated at room temperature in 4 mM KCN for 20 minutes before assaying.

et al., 1995). If this is true then PrP^C would have an intrinsic antioxidant effect as sequestration of transition metals such as copper prevents them from catalysing the formation of hydroxyl radicals (Halliwell and Gutteridge, 1990).

Furthermore, although we have demonstrated that Prnp$^{0/0}$ cells have lower resistance to oxidative stress, we cannot say whether PrP^C expression directly regulates SOD-1 activity. If this were so, then PrP^C expression would have the effect of increasing SOD-1 activity and therefore increasing resistance to oxidative stress. As terminally differentiated cells such as neurones require strong defence against oxidative stress, this would explain their relatively high PrP^C levels. PC12 cells induced to differentiate by nerve growth factor also show increased PrP^C expression (Lazarini et al., 1994). Inhibition of SOD-1 in NGF treated PC12 cells leads to greater apoptosis than in untreated PC12 cells (Troy and Shelanski, 1994) although the level of SOD-1 does not change with NGF treatment.

PrP$^{106-126}$ neurotoxicity as a model of cell death in prion disease. The application of a neurotoxic synthetic prion protein fragment has now become extensively used as a model for the investigation of neurodegeneration in prion diseases. PrP$^{106-126}$, a synthetic peptide based on the human sequence of PrP^C, was first shown to be neurotoxic in rat hippocampal tissue cultures (Forloni et al., 1993). Using an MTT assay we demonstrated that this peptide at 80 μM is also toxic to mixed cultures from the cerebellum of 6 day-old wild-type mice (Brown et al., 1996). Cultures were also treated with 80 μM of a control peptide (with the amino acid sequence of PrP106-126 in a random order). This scrambled peptide was also non-toxic.

We investigated the involvement of different cell types present in mixed cerebellar cultures in the neurotoxic mechanism of PrP$^{106-126}$. Co-culturing cerebellar cells with microglia derived from wild-type mouse cortex leads to a slight increase in the percentage of cells

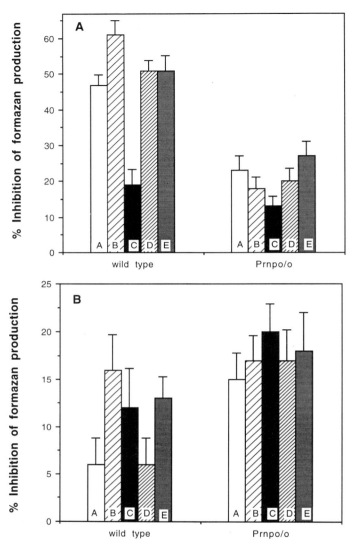

Figure 3. Activity of superoxide dismutase in cultured cerebellar cells. Extracts were prepared from cerebellar cell cultures treated with PrP[106-126] at 80 μM or with 5 μUnits of xanthine oxidase for 24h. Scrambled PrP[106-126] and βA25-35 were used as controls. The extracts were assayed for total SOD activity and MnSOD activity. SOD-1 activity (A) was calculated by the subtraction of MnSOD (B) activity from total SOD activity. The results are expressed as the percentage inhibition of the formazan reaction product formation. 100% formazan product formation is the amount of nitro blue tetrazolium (NBT) reduced by radicals formed by xanthine oxidase over a 2 minute time interval. 10 μg of extract protein were used for each assay. Increased inhibition is equivalent to increased SOD activity. Prnp[0/0] mouse extracts show high levels of MnSOD activity regardless of treatment. MnSOD activity in wild-type cerebellar cultures is induced by xanthine oxidase treatment and the PrP[106-126] and βA25-35 peptides. However, only the PrP[106-126] peptide significantly inhibits SOD-1 activity. Methods are as described in Table 1. The peptides, generated on a Milligan 9050 synthesizer, were PrP[106-126] (KTNMKHMAGAAAAGAVVGGLG), PrP[106-126] "scrambled" with the amino acids as PrP[106-126] ordered randomly (NGAKALMGGHGATKVMVGAAA) and βA25-35 based on the β-amyloid sequence (GSNKGAIIGLM). Shown are the mean and s.e. for a minimum of four experiments. Key: A = untreated, B = xanthine oxidase/xanthine, C = PrP[106-126], D = Scrambled PrP[106-126], E = βA25-35.

destroyed but when co-cultured with microglia from Prnp$^{0/0}$ mice there was no significant increase in cell loss (Brown et al., 1996). To further study the role of microglia, we used a two-hour treatment with L-leucine methyl ester (LLME) which significantly reduces the number of microglia in mixed cultures (Giulian and Baker, 1986; Giulian et al., 1988) without reducing the overall number of surviving cells. However, after this treatment, 10-day administration of 80 µM PrP$^{106-126}$ was no longer toxic to cerebellar cells, showing that the presence of microglia is a necessary requirement for the neurotoxicity of PrP$^{106-126}$.

If LLME treatment prevents the neurotoxicity of PrP$^{106-126}$ by destruction of microglia, then co-culturing LLME-treated cerebellar cultures with microglia should restore the toxic effect. Co-culturing LLME-treated cells from wild-type mice with purified microglia from cerebral cortex of either wild-type or Prnp$^{0/0}$ mice resulted in restoration of the toxic effect. This effect is not seen when LLME-treated cerebellar cells were co-cultured with astrocytes from wild-type mouse cortex. Also, co-culture with wild-type astrocytes did not enhance the toxicity of PrP$^{106-126}$ on cerebellar cultures not treated with LLME. This indicates that the neurodestructive mechanism of PrP$^{106-126}$ does not involve a direct effect on astrocytes.

When cerebellar cells from Prnp$^{0/0}$ mice were treated with PrP$^{106-126}$ using the same method as for wild-type cells the peptide showed no toxic effect. After 10 days' treatment with PrP$^{106-126}$ at 80 µM Prnp$^{0/0}$ cerebellar cells showed a slight but significant increase in number. When cerebellar cells derived from Prnp$^{0/0}$ mice were co-cultured with high numbers of microglia from wild-type mice, the response to PrP$^{106-126}$ remained unaltered (non-toxic to Prnp$^{0/0}$ cells) even when applied at very high concentrations. These findings again show that the neurotoxic effect of PrP$^{106-126}$ is based on two requirements: a specific interaction with PrP^C-expressing neurones and the presence of microglia.

If neurones are destroyed as a result of substances released by microglia stimulated by PrP$^{106-126}$, it should be possible to block the effect of these substances in culture. A strong candidate for this effect would be an oxidative substance such as the superoxide radical or a reactive by-product of it such as nitrite, both of which are known to be released by microglia under certain conditions (Giulian and Baker, 1986; Betz Corradin et al., 1993). Anti-oxidants vitamin E and N-acetyl-cysteine both blocked the toxic effect of PrP$^{106-126}$ on cultures of cerebellar neurones (Brown et al., 1996). This supports the notion that microglia place oxidative stress on neurones as a result of PrP$^{106-126}$ treatment.

Superoxide radicals and nitric oxide both can be produced by microglia (Betz Corradin et al., 1993) and induce oxidative stress. Both are short-lived molecules that readily react to produce nitrite, which is stable in solution (Beckman et al., 1990). We assayed the supernatants of microglial cultures for nitrite production and found a high increase in production of nitrite in solution after application of PrP$^{106-126}$ for 4 days. This increase was directly proportional to the log number of microglia present. However, this effect was less clearly pronounced in Prnp$^{0/0}$ microglia (Brown et al., 1996). This result provides evidence that PrP$^{106-126}$ acts directly on microglia, and gives strong support for the notion that microglia induce neuronal death by oxidative stress as a result of PrP$^{106-126}$ stimulation.

PrP$^{106-126}$ is apparently non-toxic to LLME-treated cerebellar cells, presumably because the source of oxidative stress induced by the peptide, namely microglia, have been destroyed. However, if the effect on microglia, resulting in oxidative stress, were the only effect of PrP$^{106-126}$, one would expect Prnp$^{0/0}$ mouse cerebellar cells to be destroyed in a similar manner. To test this hypothesis, xanthine oxidase was used as an alternative source to produce superoxide radicals. However, PrP$^{106-126}$ enhances cell death induced by xanthine oxidase only on LLME-treated cerebellar cells from wild-type mice. This strongly indicates that for PrP$^{106-126}$ to exert a toxic effect, cells must express PrP^C. This experiment also indicates that PrP$^{106-126}$ exerts an effect on wild-type cells reducing their ability to resist oxidative stress.

As Prnp$^{0/0}$ cells are more susceptible to oxidative stress, this in itself suggests that PrP^C expression is somehow involved in cellular resistance to the oxidative stress initiated by

$PrP^{106-126}$. As well as activating microglia to release oxidative substances, $PrP^{106-126}$ also inhibited the ability of wild-type neurones to resist oxidative stress. Cultures of cerebellar cells from 6-day old mice were examined for SOD-1 activity. Exposure of wild-type cerebellar cells to oxidative stress produced by xanthine oxidase caused an elevation in superoxide dismutase activity (Figure 3) due to increased activity of both SOD-1 and MnSOD. This reaction is as expected for cells placed under oxidative stress. However, $Prnp^{0/0}$ cerebellar cells similarly treated failed to show increased activity of SOD-1, suggesting that the phenotype of $Prnp^{0/0}$ mice is one of impaired regulation or decreased activity of SOD-1.

Cerebellar cultures were also exposed to the neurotoxic peptide $PrP^{106-126}$. Extracts of wild-type mouse cerebellar cells exposed to 80 μM $PrP^{106-126}$ for 1 day showed a decreased activity of superoxide dismutase (Figure 3). The SOD-1 activity of cerebellar cells from $Prnp^{0/0}$ mice was also decreased under these conditions (Brown and Kretzschmar, 1996). In our experiments the activity of MnSOD was increased after $PrP^{106-126}$ treatment in wild-type cells, indicating that the cells are responding to oxidative stress even though $PrP^{106-126}$ treatment inhibits SOD-1 activity. Both xanthine oxidase and βA25-35, a peptide based on the human β-amyloid sequence, induced a similar MnSOD increase in wild-type mouse cerebellar cells. MnSOD activity was always higher in $Prnp^{0/0}$ mouse cerebellar cells than in wild-type cells no matter what treatment was applied, and could not be upregulated further. In $Prnp^{0/0}$ cells MnSOD activity was also high under control conditions and after treatment with "$PrP^{106-126}$ scrambled". This suggests that $Prnp^{0/0}$ cells are experiencing high oxidative stress in culture and are exhibiting their maximal possible MnSOD response.

$PrP^{106-126}$ increases the toxicity of xanthine oxidase on wild-type cells (Brown et al., 1996). Oxidative stress caused by xanthine oxidase in the present experiments caused increased SOD-1 activity while $PrP^{106-126}$ inhibited SOD-1 activity. $Prnp^{0/0}$ cells lacking PrP^C showed a slight decrease in SOD-1 activity when treated with $PrP^{106-126}$ but no significant response to xanthine oxidase. This indicates that $Prnp^{0/0}$ cells have lost the ability to respond to xanthine oxidase-produced oxidative stress by upregulating SOD-1 and explains why this oxidative stress is more toxic to $Prnp^{0/0}$ cells.

$PrP^{106-126}$ reduced the activity of SOD-1 in cerebellar cell cultures. However, this does not imply that a direct interaction between these molecules occured in our cell culture system. Since it has been demonstrated that a similar PrP fragment can bind to PrP^C (Kaneko et al., 1995) cellular inactivation of SOD-1 may come about by binding of the peptide to PrP^C.

In our experiments the activity of MnSOD was increased after $PrP^{106-126}$ treatment in wild-type cells, indicating that the cells are responding to oxidative stress even though $PrP^{106-126}$ treatment inhibits SOD-1 activity. Both xanthine oxidase and βA25-35 induced a similar MnSOD increase in wildtype mouse cerebellar cells. MnSOD activity was always higher in $Prnp^{0/0}$ mouse cerebellar cells than in wild-type cells no matter what treatment was applied. In $Prnp^{0/0}$ cells MnSOD activity was also high under control conditions and after treatment with "$PrP^{106-126}$ scrambled". This suggests that $Prnp^{0/0}$ cells are experiencing high oxidative stress in culture.

Taken together these results show that $Prnp^{0/0}$ cells in culture are unable to respond to oxidative stress by upregulating SOD-1 but that other enzymes such as MnSOD are upregulated to cope with the basal level of oxidative stress present in the culture system. These results also suggest a further explanation of why $Prnp^{0/0}$ cells are resistant to the toxicity of $PrP^{106-126}$. The action of $PrP^{106-126}$ on wild-type neurones may be to inhibit SOD-1 activity. As $PrP^{106-126}$ is unable to greatly inhibit SOD-1 in $Prnp^{0/0}$ cells then these cells are not killed by the peptide's action.

$PrP^{106-126}$ appears to have the following mechanism to bring about neuronal cell death *in vitro*: $PrP^{106-126}$ interacts with neurones and microglia. To interact with neurones, $PrP^{106-126}$ requires PrP^C expression by those neurones. $PrP^{106-126}$ causes production of neurotoxic substances by microglia such as superoxide. The effect of $PrP^{106-126}$ on neurones reduces their ability to resist oxidative stress. One consequence of this is reduced activity of SOD-1. As a result of this effect on neurones, microglia-derived toxic substances are able to kill neurones.

CONCLUSIONS

Prion research in the past was focused on the biochemistry of the infectious agent. Questions concerning cellular pathogenesis and the processes underlying neuronal cell death have only recently been addressed. It appears that nerve cell death in prion disease follows apoptotic mechanisms *in vitro* and *in vivo*. While some groups hold that neuronal apoptosis is caused by progressive loss of function of PrP^C in disease, others argue that PrP^{Sc} has toxic effects on host neurones. We have demonstrated that PrP$^{106-126}$, a synthetic PrP peptide, has toxic effects on neurones expressing PrP^C while neurones devoid of PrP^C are resistant. The toxic effect of PrP$^{106-126}$ requires the presence of microglia in culture, which respond by producing oxygen radicals. It will be crucial to test these cell culture findings *in vivo*. PrP^C expression may be related to or necessary for resistance to oxidative stress as SOD-1 expression is altered in PrP^C deficient mice. Loss of this PrP^C "function" may be the reason for cell death in prion disease.

Acknowledgements

The authors thank C. Weissmann, University of Zürich, for providing Prnp$^{0/0}$ mice. We thank Roswitha Fischer, Wolfgang Dröse and Brigitte Maruschak for their technical assistance as well as Cynthia Bunker for editing of the manuscript. This study was supported by research grants from the Bundesministerium für Bildung, Wissenschaft, Forschung und Technologie, the Deutsche Forschungsgemeinschaft (DFG) and the Wilhelm-Sander-Stiftung.

References

Beckman, J.S., Beckman, T.W., Chen, J., Marshall, P.A., Freeman, B.A., 1990, Apparent hydroxyl radical production by peroxynitrite: Implications for endothelial injury from nitric oxide and superoxide, *Proc. Natl. Acad. Sci. USA* 87:1620.

Betz Corradin, S., Mauel, J., Donini, S.D., Quattrocchi, E., Ricciardi-Castagnoli, P., 1993, Inducible nitric oxide synthase activity of cloned murine micro glial cells, *Glia* 7:255.

Borchelt, D.R., Rogers, M., Stahl, N., Telling, G., Prusiner, S.B., 1993, Release of the cellular prion protein from cultured cells after loss of its glycoinositol phospholipid anchor, *Glycobiology* 3:319.

Brown, D.R., Schmidt, B., Kretzschmar, H.A., 1996, Role of microglia and host prion protein in neurotoxicity of prion protein fragment, *Nature* 380:345.

Brown, D.R., Schulz-Schaeffer, W.J., Schmidt, B., Kretzschmar, H.A., 1997, Prion protein-deficient cells show altered response to oxidative stress due to decreased SOD-1 activity, *Exp. Neurol.* in press:

Brown, D.R., Kretzschmar, H.A., 1996, Role of prion protein in resistance to oxidative stress and regulation of superoxide dismustase, *Society for Neuroscience* 22:708.(Abstract)

Bruce, M.E., Dickinson, A.G., Fraser, H., 1976, Cerebral amyloidosis in scrapie in the mouse: effect of agent strain and mouse genotype. *Neuropathol. Appl. Neurobiol.* 2:471.

Bruce, M.E., McConnell, I., Fraser, H., Dickinson, A.G., 1991, The disease characteristics of different strains of scrapie in *Sinc* congenic mouse lines: implications for the nature of the agent and host control of pathogenesis. *J. Gen. Virol.* 72:595.

Bruce, M.E., 1993, Scrapie strain variation and mutation, *Br. Med. Bull.* 49:822.

Bruce, M.E., McBride, P.A., Jeffrey, M., Scott, J.R., 1994, PrP in pathology and pathogenesis in scrapie-infected mice, *Mol. Neurobiol.* 8:105.

Buja, L.M., Eigenbrodt, M.L., Eigenbrodt, E.H., 1993, Apoptosis and necrosis. Basic types and mechanisms of cell death, *Arch. Pathol. Lab. Med.* 117:1208.

Bursch, W., Paffe, S., Putz, B., Barthel, G., Schulte-Hermann, R., 1990, Determination of the length of the histological stages of apoptosis in normal liver and in altered hepatic foci of rats, *Carcinogenesis* 11:847.

Buyukmihci, N.C., Goehring-Harmon, F., Marsh, R.F., 1987a, Photoreceptor degeneration during infection with various strains of the scrapie agent in hamsters, *Exp. Neurol.* 97:201.

Buyukmihci, N.C., Goehring-Harmon, F., Marsh, R.F., 1987b, Photoreceptor degeneration in experimental transmissible mink encephalopathy of hamsters, *Exp. Neurol.* 96:727.

Büeler, H., Fischer, M., Lang, Y., Bluethmann, H., Lipp, H.-P., DeArmond, S.J., Prusiner, S.B., Aguet, M., Weissmann, C., 1992, Normal development and behaviour of mice lacking the neuronal cell-surface PrP protein. *Nature* 356:577.

Caughey, B., Race, R.E., Ernst, D., Buchmeier, M.J., Chesebro, B., 1989, Prion protein biosynthesis in scrapie-infected and uninfected neuroblastoma cells. *J. Virol.* 63:175.

Collinge, J., Whittington, M.A., Sidle, K.C.L., Smith, C.J., Palmer, M.S., Clarke, A.R., Jefferys, J.G.R., 1994, Prion protein is necessary for normal synaptic function. *Nature* 370:295.

Deng, H.X., Hentati, A., Tainer, J.A., Iqbal, Z., Cayabyab, A., Hung, W.Y., Getzoff, E.D., Hu, P., Herzfeldt, B., Roos, R.P., Warner, C., Deng, G., Soriano, E., Smyth, C., Parge, H.E., Ahmed, A., Roses, A.D., Hallewell, R.A., Pericak-Vance, M.A., Siddique, T., 1993, Amyotrophic lateral sclerosis and structural defects in Cu,Zn superoxide dismutase, *Science* 261:1047.

Edwards, F.A., Konnerth, A., Sakmann, B., Takahashi, T., 1989, A thin slice preparation for patch-clamp recordings from neurones of the mammalian central nervous system. *Pflugers Arch.* 414:600.

Eikelenboom, P., Rozemuller, J.M., Kraal, G., Stam, F.C., McBride, P.A., Bruce, M.E., Fraser, H., 1991, Cerebral amyloid plaques in Alzheimer's disease but not in scrapie-affected mice are closely associated with a local inflammatory process. *Virchows Arch. [B]* 60:329.

Fairbairn, D.W., Carnahan, K.G., Thwaits, R.N., Grigsby, R.V., Holyoak, G.R., ONeill, K.L., 1994, Detection of apoptosis induced DNA cleavage in scrapie-infected sheep brain, *FEMS Microbiol. Lett.* 115:341.

Forloni, G., Angeretti, N., Chiesa, R., Monzani, E., Salmona, M., Bugiani, O., Tagliavini, F., 1993, Neurotoxicity of a prion protein fragment, *Nature* 362:543.

Fraser, H., 1993, Diversity in the neuropathology of scrapie-like diseases in animals, *Br. Med. Bull.* 49:792.

Gavrieli, Y., Sherman, Y., Ben-Sasson, S.A., 1992, Identification of programmed cell death in situ via specific labeling of nuclear DNA fragmentation, *J. Cell Biol.* 119:493.

Giese, A., Groschup, M.H., Hess, B., Kretzschmar, H.A., 1995, Neuronal cell death in scrapie-infected mice is due to apoptosis, *Brain Pathol.* 5:213.

Giulian, D., Young, D.G., Woodward, J., Brown, D.C., Lachman, L.B., 1988, Interleukin-1 is an astroglial growth factor in the developing brain, *J. Neurosci.* 8:709.

Giulian, D., Baker, T.J., 1986, Characterization of ameboid microglia isolated from developing mammalian brain, *J. Neurosci.* 6:2163.

Gold, R., Schmied, M., Rothe, G., Zischler, H., Breitschopf, H., Wekerle, H., Lassmann, H., 1993, Detection of DNA fragmentation in apoptosis: application of in situ nick translation to cell culture systems and tissue sections, *J. Histochem. Cytochem.* 41:1023.

Guan, Z., Söderberg, M., Sindelar, P., Prusiner, S.B., Kristensson, K., Dallner, G., 1996, Lipid Composition in Scrapie-Infected Mouse Brain: Prion Infection Increases the Levels of Dolichyl Phosphate and Ubiquinone, *J. Neurochem.* 66:277.

Halliwell, B., Gutteridge, M.C., 1990, The antioxidants of human extracellular fluids, *Arch. Biochem. Biophys.* 280:1.

Hamill, O., Marty, A., Neher, E., Sakmann, B., Sigworth, F.J., 1981, Improved patch-clamp techniques for high resolution current recording from cells and cell-free membrane patches. *Pflugers Arch.* 391:85.

Herms, J.W., Kretzschmar, H.A., Titz, S., Keller, B.U., 1995, Patch-clamp analysis of synaptic transmission to cerebellar Purkinje cells of prion protein knockout mice, *Eur. J. Neurosci.* 7:2508.

Hogan, R.N., Baringer, J.R., Prusiner, S.B., 1981, Progressive retinal degeneration in scrapie-infected hamsters: a light and electron microscopical analysis, *Lab. Invest.* 44:34.

Hogan, R.N., Kingsbury, D.T., Baringer, J.R., Prusiner, S.B., 1983, Retinal degeneration in experimental Creutzfeldt-Jakob disease. *Lab. Invest.* 49:708.

Hornshaw, M.P., McDermott, J.R., Candy, J.M., Lakey, J.H., 1995, Copper binding to the N-terminal tandem repeat region of mammalian and avian prion protein: Structural studies using synthetic peptides, *Biochem. Biophys. Res. Commun.* 214:993.

Jeffrey, M., Halliday, W.G., Goodsir, C.M., 1992, A morphometric and immunohistochemical study of the vestibular nuclear complex in bovine spongiform encephalopathy, *Acta Neuropathol. (Berl.)* 84:651.

Jeffrey, M., Fraser, J.R., Halliday, W.G., Fowler, N., Goodsir, C.M., Brown, D.A., 1995, Early unsuspected neuron and axon terminal loss in scrapie-infected mice revealed by morphometry and immunocytochemistry, *Neuropathol. Appl. Neurobiol.* 21:41.

Kaneko, K., Peretz, D., Pan, K.M., Blochberger, T.C., Wille, H., Gabizon, R., Griffith, O.H., Cohen, F.E., Baldwin, M.A., Prusiner, S.B., 1995, Prion protein (PrP) synthetic peptides induce cellular PrP to acquire properties of the scrapie isoform, *Proc. Natl. Acad. Sci. USA* 92:11160.

266

Kerr, J.F.R., Wyllie, A.H., Currie, A.R., 1972, Apoptosis: A basic biological phenomenon with wide-ranging implications in tissue kinetics, *Br. J. Cancer* 26:239.

Kozlowski, P.B., Moretz, R.C., Carp, R.I., Wisniewski, H.M., 1982, Retinal damage in scrapie mice. *Acta Neuropathol. (Berl.)* 56:9.

Kretzschmar, H.A., Prusiner, S.B., Stowring, L.E., DeArmond, S.J., 1986, Scrapie prion proteins are synthesized in neurons. *Am. J. Pathol.* 122:1.

Lazarini, F., Castelnau, P., Chermann, J.-F., Deslys, J.-P., Dormont, D., 1994, Modulation of prion protein gene expression by growth factors in cultured mouse astrocytes and PC-12 cells, *Mol. Brain Res.* 22:268.

Lieberburg, I., 1987, Developmental expression and regional distribution of the scrapie-associated protein mRNA in the rat central nervous system. *Brain Res.* 417:363.

Llano, I., Marty, A., Armstrong, C.M., Konnerth, A., 1991, Synaptic and agonist-induced currents of Purkinje cells in rat cerebellar slices. *J. Physiol.* 434:183.

Lledo, P.M., Tremblay, P., De Armond, S.J., Prusiner, S.B., Nicoll, R.A., 1996, Mice deficient for prion protein exhibit normal neuronal excitability and synaptic transmission in the hippocampus, *Proc. Natl. Acad. Sci. USA* 93:2403.

Lucassen, P.J., Williams, A., Chung, W.C.J., Fraser, H., 1995, Detection of apoptosis in murine scrapie, *Neurosci. Lett.*, 198:185.

Manson, J., West, J.D., Thomson, V., McBride, P., Kaufman, M.H., Hope, J., 1992, The prion protein gene: a role in mouse embryogenesis? *Development* 115:117.

Manson, J.C., Hope, J., Clarke, A.R., Johnston, A., Black, C., MacLeod, N., 1995, PrP gene dosage and long term potentiation, *Neurodegeneration* 4:113.

Marklund, S.L., 1982, Human copper-containing superoxide dismutase of high molecular weight, *Proc. Natl. Acad. Sci. USA* 79:7634.

Masters, C.L., Richardson, E.P., Jr., 1978, Subacute spongiform encephalopathy (Creutzfeldt-Jakob disease). The nature and progression of spongiform change. *Brain* 101:333.

McBride, P.A., Bruce, M.E., Fraser, H., 1988, Immunostaining of scrapie cerebral amyloid plaques with antisera raised to scrapie-associated fibrils (SAF). *Neuropathol. Appl. Neurobiol.* 14:325.

Migheli, A., Cavalla, P., Marino, S., Schiffer, D., 1994, A study of apoptosis in normal and pathologic nervous tissue after in situ end-labeling of DNA strand breaks, *J. Neuropathol. Exp. Neurol.* 53:606.

Moser, M., Colello, R.J., Pott, U., Oesch, B., 1995, Developmental expression of the prion protein gene in glial cells, *Neuron* 14:509.

Oury, T.D., Ho, Y.S., Piantadosi, C.A., Crapo, J.D., 1992, Extracellular superoxide dismutase, nitric oxide, and central nervous system O_2 toxicity, *Proc. Natl. Acad. Sci. USA* 89:9715.

Pan, K.M., Stahl, N., Prusiner, S.B., 1992, Purification and properties of the cellular prion protein from Syrian hamster brain. *Protein Sci.* 1:1343.

Rosen, D.R., Siddique, T., Patterson, D., Figlewicz, D.A., Sapp, P., Hentati, A., Donaldson, D., Goto, J., O'Regan, J.P., Deng, H.X., Rahmani, Z., Krizus, A., McKenna-Yasek, D., Cayabyab, A., Gaston, S.M., Berger, R., Tanzi, R.E., Halperin, J.J., Herzfeldt, B., van den Bergh, R., Hung, W.Y., Bird, T., Deng, G., Mulder, D.W., Smyth, C., Laing, N.G., Soriano, E., Pericak-Vance, M.A., Haines, J., Rouleau, G.A., Gusella, J.S., Horvitz, H.R., Brown, R.H., 1993, Mutations in Cu/Zn superoxide dismutase gene are associated with familial amyotrophic lateral sclerosis, *Nature* 362:59.

Rothstein, J.D., Bristol, L.A., Hosler, B., Brown, R.H., Kuncl, R.W., 1994, Chronic inhibition of superoxide dismutase produces apoptotic death of spinal neurons, *Proc. Natl. Acad. Sci. USA* 91:4155.

Scott, J.R., Fraser, H., 1984, Degenerative hippocampal pathology in mice infected with scrapie, *Acta Neuropathol. (Berl.)* 65:62.

Searle, J., Kerr, J.F.R., Bishop, C.J., 1982, Necrosis and apoptosis: distinct modes of cell death with fundamentally different significance, *Path. Ann.* 17:229.

Sloviter, R.S., Dean, E., Neubort, S., 1993, Electron microscopic analysis of adrenalectomy-induced hippocampal granule cell degeneration in the rat: apoptosis in the adult central nervous system, *J. Comp. Neurol.* 330:337.

Tibell, L., Hjalmarsson, K., Edlund, T., Skogman, G., Engström, A., Marklund, S.L., 1987, Expression of human extracellular superoxide dismutase in Chinese hamster ovary cells and characterization of the product, *Proc. Natl. Acad. Sci. USA* 84:6634.

Tobler, I., Gaus, S.E., Deboer, T., Achermann, P., Fischer, M., Rülicke, T., Moser, M., Oesch, B., McBride, P.A., Manson, J.C., 1996, Altered circadian activity rhythms and sleep in mice devoid of prion protein, *Nature* 380:639.

Tobler, I., Deboer, T., Fischer, M., 1997, Sleep and sleep regulation in normal and prion protein-deficient mice, *J. Neurosci.* 17:1869.

Troy, C.M., Shelanski, M.L., 1994, Down-regulation of copper/zince superoxide dismutase causes apoptotic death in PC 12 neuronal cells, *Proc. Natl. Acad. Sci. USA* 91:6384.

Vincent, P., Armstrong, C.M., Marty, A., 1992, Inhibitory synaptic currents in rat cerebellar Purkinje cells: modulation by postsynaptic depolarization, *J. Physiol.* 456:453.

Weissmann, C., Bueler, H., Fischer, M., Sailer, A., Aguzzi, A., Aguet, M., 1994, PrP-deficient mice are resistant to scrapie, *Annals New York Acad. Sci.* 724:235.

Whittington, M.A., Sidle, K.C.L., Gowland, I., Meads, J., Hill, A.F., Palmer, M.S., Jefferys, J.G.R., Collinge, J., 1995, Rescue of neurophysiological phenotype seen in PrP null mice by transgene encoding human prion protein, *Nature Genet.* 9:197.

Williams, A.E., van Dam, A.-M., Man-A-Hing, W.K.H., Berkenbosch, F., Eikelenboom, P., Fraser, H., 1994, Cytokines, prostaglandins and lipocortin-1 are present in the brains of scrapie-infected mice, *Brain Res.* 654:200.

Wyllie, A.H., Morris, R.G., Smith, A.L., Dunlop, D., 1984, Chromatin cleavage in apoptosis: association with condensed chromatin morphology and dependence on macromolecular synthesis, *J. Pathol.* 142:67.

CLINICAL, PATHOLOGICAL, AND MOLECULAR CHARACTERIZATION OF GERSTMANN-STRÄUSSLER-SCHEINKER DISEASE IN THE INDIANA KINDRED (*PRNP* F198S)

B. Ghetti, F. Tagliavini, S.R. Dlouhy, P. Piccardo, K. Young, R.D. Yee, G.D. Hutchins, G. Giaccone, F. Prelli, B. Frangione, F. Unverzagt, O. Bugiani, and M.R. Farlow

Indiana University School of Medicine, Indianapolis, IN
Istituto Neurologico "Carlo Besta", Milano, Italy
New York University Medical Center, New York, NY

INTRODUCTION

An adult-onset, autosomal dominant neurodegenerative disorder characterized clinically by cerebellar ataxia, parkinsonism and dementia, has been studied in a large Indiana kindred (IK)[1-3]. The clinical signs of affected individuals have similarities with those observed in patients from an Austrian family ("H" family) described by Gerstmann Sträussler and Scheinker[4] and further characterized by Braunmühl[5], Seitelberger[6,7], Budka et al.[8], and Kretzschmar et al[9]. In the "H" family, the salient clinical features were adult-onset ataxia, pyramidal signs and dementia. The disease in the "H" family was inherited as an autosomal dominant trait. Neuropathologically, amyloid deposits in cerebellum and cerebrum as well as atrophy in spinal cord, brain stem and cerebellum, were consistently observed.

Familial disorders, that have clinical and pathologic features similar to those observed in the patients from the "H" family, have been given the eponym "Gerstmann-Sträussler-Scheinker (GSS) disease"; for review see Farlow et al[10]. The pathogenesis of GSS is incompletely understood. In some cases, spongiform changes similar to those seen in Creutzfeldt-Jakob disease (CJD) are present and it has been shown that laboratory animals inoculated with brain tissue obtained from such patients may develop a spongiform encephalopathy[11]. Similarly to other transmissible spongiform encephalopathies, such as CJD and kuru in humans as well as scrapie and bovine spongiform encephalopathy in animals, the presence of fragments of proteinase-K (PK) resistant prion protein (PrP) are detected in GSS disease. Genetic studies in numerous families have revealed that at least five mutations of the prion protein gene (*PRNP*) are associated with GSS disease. For review see Hsiao and Prusiner[12], Prusiner[13], and Ghetti et al.[14]

The pathologic phenotype of GSS in the Indiana Kindred (GSS-IK) is characterized by amyloid deposits immunoreactive with antibodies to PrP[3,14-20]; however, one of the most striking features is the consistent presence of neurofibrillary tangles (NFT) in many areas of the central nervous system [1,3,14-18,21-23]. These lesions are comparable to those seen in Alzheimer disease (AD) and make the phenotype of the disease in the IK novel in the spectrum of GSS disease and among those disorders belonging to the group of prion diseases.

RESULTS AND DISCUSSION

We have recently re-examined and updated the information related to the pedigree. The pedigree of the IK spans eight generations and includes over 3,000 members; records indicate that at least 57 individuals are known to have been affected. Inheritance occurs with an autosomal dominant pattern and with an apparently high degree of penetrance; there is no gender predominance. There are currently living: three affected, 89 fully at-risk (children of affected individuals) and 228 half at-risk (children of at-risk subjects) family members. From 89 at-risk and half-at-risk family members examined over the last nine years, 15 were determined to be affected.

Genetic Studies

Prion Protein gene (*PRNP*). The first missense mutation of the *PRNP* gene associated with GSS disease was found at codon 102 (P102L)[24]. GSS disease in the IK is caused by a missense mutation at codon 198 of the *PRNP* gene; this mutation consists in a thymine to cytosine transition in the second position of codon 198 and results in the substitution of serine (S) for phenylalanine (F) at residue 198 of PrP[25,26]. The F198S mutation was reported for the first time in the IK. This mutation can be detected by direct se quencing[26], allele-specific oligonucleotide hybridization[25] or by EcoRI digestion of DNA fragments amplified with primers designed for that purpose (unpublished). The mutation F198S is in coupling with a codon for valine at codon 129. Even though the kindred is large, this coupling relationship has been true for all family members who carry the mutation, (i.e., there has not been any recombination detected between codons 129 and 198). Thus, in theory, for all GSS-IK patients studied so far, the mutant PrP protein produced in the brain should contain serine at position 198 and valine at position 129.

To date, all affected members of GSS-IK have been heterozygous for the F198S mutation, i.e., no homozygous F198S patients have been found. Thus, each patient has one *PRNP* allele that produces mutant PrP and one allele that produces normal PrP. Although, as discussed above, the mutant allele is the same (129 valine - 198 serine) throughout the kindred, the normal *PRNP* allele is variable. Some family members have a normal allele with methionine (M) at position 129 whereas others have a normal allele with valine (V) at position 129. Thus, there are two genetically distinct types of affected individuals, those that are heterozygous M/V at codon 129 and those that are homozygous V/V at codon 129. We have previously reported (based upon nine affected family members) that V/V homozygotes had an earlier age of onset of the disease than M/V heterozygotes[25]. Since then, three additional mutation-carrying members of the kindred have become clinically affected. Of these, two are M/V heterozygotes (ages of onset 49 and 60 years) and one is a V/V homozygote (age of onset 47 years). The patient with onset at 49 is the youngest M/V heterozygote to be diagnosed as symptomatic in this kindred. Despite the relatively early age of onset for that M/V patient, overall, there continues to be a difference in age of onset for V/V homozygotes (44.5 years) compared with M/V heterozygotes (58.9 years).

Clinical studies

Neurology. The age of onset of symptoms varies from the late-30s to the mid-60s. From serial neurological examinations in at-risk and affected patients, a composite clinical picture emerges: initial symptoms are a gradual loss of short-term memory and progressive gait ataxia, exaggerated or first noticed when the individual is under stress. These symptoms may progress slowly over as long as five or more years or rapidly over as little as one year. Rigidity and bradykinesia generally occur late in the disease and dementia worsens with their onset. Tremor is mild or absent. Psychotic depression has been seen in several patients. Without carbidopa/levodopa treatment, rapid weight loss and death usually occur within one year of onset of parkinsonism. Bradykinesia and rigidity improve with treatment; however death still occurs from pneumonia or other illness within two years after the onset of parkinsonian signs. Early in the course, testing reveals deficits in short term memory and cognition, as well as abnormalities in eye movement and mild cerebellar incoordination. Later, there is dysarthria and progressively more severe bradykinesia, rigidity, and gait abnormality. In the end-stage, patients have severe extrapyramidal abnormalities, dysphagia with choking spells and global dementia[2]. Following the onset of the initial symptoms, death occurs within five to seven years, on the average.

Neuropsychology. Three affected and 41 at-risk individuals have been studied[27-30]. Two affected subjects, one symptomatic for three years and the other for seven, showed generalized cerebral dysfunction consistent with global dementia. These subjects had clinically significant impairments in intelligence, memory, attention, cognitive processing speed, executive ability, and manual motor skills. A third subject who had been symptomatic for less than one year had a more selective pattern of neuropsychological impairment. In this individual, impairments in memory, verbal fluency, and motor skills suggested a pattern consistent with subcortical dysfunction. Compared to eight AD patients at similar levels of overall cognitive dysfunction, these three affected subjects from the IK showed greater impairment in constructional ability, verbal fluency, and motor skills. The neuropsychological differences between the affected IK and AD patients may reflect: a more extensive subcortical neuropathologic involvement, differences in the laminar distribution of amyloid plaques, and/or differences in the biologic activity of PrP versus that of amyloid precursor protein.

We have also compared clinically unaffected members of the family who carry the mutation with family members free of the mutation[27]. Subjects with the mutation (n = 14) scored significantly below non-mutation carrying subjects (n = 27) on a word list learning test and a measure of verbal fluency. Borderline significant differences were found on measures of attention (digit repetition) and psychomotor speed (digit symbol substitution), with mutation carrying subjects scoring lower. Despite the fact that ataxia is a prominent sign in clinically affected subjects from this kindred, finger tapping speed and manual dexterity were no different between carriers and non-carriers.

Brain Imaging. We have studied seven patients by magnetic resonance imaging (MRI) of the brain and preliminary correlations between MRI abnormalities and clinical signs and symptoms have been made[31]. All patients show mild to moderate cerebellar atrophy that correlates with ataxia. Marked decrease in T2 signal intensity is seen in the basal ganglia, particularly the putamen and caudate nucleus, in four of the seven imaged individuals. The presence of those findings seems to be associated with moderate to severe parkinsonian signs. MRI and neuropathologic findings suggest that iron accumulation takes place in the caudate nucleus and substantia nigra.

Positron Emission Tomography (PET) measurements of glucose metabolism based upon the uptake of 2-[F-18]Fluoro-2-Deoxy-D-Glucose (FDG) in the brain of eight at-risk and five affected (three early stage, two late stage) members of GSS-IK were performed. Standardized uptake values (SUV) of FDG at 45 minutes post injection, normalized for plasma glucose levels, were used as a quantitative index of metabolism in these subjects. Both severely affected members of GSS-IK showed widespread metabolic abnormalities throughout the cerebral and cerebellar hemispheres. One of the early stage patients had several areas of reduced metabolism relative to the at-risk group with the most prominent reductions in the basal ganglia and cerebellum, regions known from neuropathology and neuroimaging to be affected early in the disease. In one subject, affected for 12 years, who died seven weeks following the PET study, neuropathologic studies using antibodies to PrP and to the tau protein (τ) were correlated to FDG SUVs. The topography of functionally altered regions in this patient matched the distribution of the two neuropathologic markers.

Neuro-ophthalmology. Five affected and 11 unaffected at-risk family members have been studied[32]. The affected individuals show abnormal eye movements that are characteristic of cerebellar damage. They include 1. gaze-evoked and rebound nystagmus, 2. upbeat nystagmus, 3. saccade hypometria (undershooting), 4. impaired smooth pursuit and optokinetic nystagmus, and 5. abnormal visual-vestibular interactions, including inability to suppress vestibulo-ocular responses by fixation.

Three of the five affected persons could not produce voluntary upward saccades of more than 30 degrees in amplitude. However, oculocephalic maneuvers increase upward eye movements. Supranuclear palsy of upward gaze is similar to that seen in parkinsonian patients.

Most of the at-risk subjects have normal eye movements. However, two have pathologic nystagmus, although they did not have definite abnormalities in the remainder of their neurologic examination. One of them showed progression of eye-movement abnormalities over a six-month period. Three other at-risk members have slight abnormalities of pursuit, optokinetic nystagmus, or visual-vestibuloocular responses. Progression of abnormalities has been found in one out of three at-risk individuals that were examined at a six to 10 months interval.

Neuropathologic Studies

Affected individuals

The central nervous system of 12 patients studied so far has consistently shown moderate cerebral atrophy and moderate to severe cerebellar atrophy. Other main macroscopic findings are marked loss of pigment of both the substantia nigra and the locus coeruleus. Microscopically, the following lesions are observed: amyloid deposits with or without a neuritic component, NFT, nerve cell loss, gliosis, and iron deposition[1,3,14-18].

PrP-Amyloid and PrP deposits. Amyloid deposits are immunolabelled by antibodies against PrP and are distributed throughout the gray structures of the cerebrum, cerebellum, and midbrain. They are intertwined at the periphery with astrocytic processes. Microglial elements and macrophages are present. The size of the amyloid deposits ranges between 10 and 100 μm in the cerebellum, and between 15 and 160 μm in the cerebral cortex. One or more amyloid cores are observed in each plaque. In thioflavin-S treated sections, amyloid deposits are strongly fluorescent when observed under ultraviolet light. Fluorescent deposits are not seen in or around the walls of blood vessels; however, they may be present in the parenchyma adjacent to vessels. In the cerebral cortex, the highest concentration of amyloid deposits of any type is found in the deeper layers; however deposits in clusters are often found in subpial regions of the cerebral and cerebellar cortex[1,3,14-18]. In the cerebellar and cerebral cortex, poorly circumscribed fluorescent deposits are also seen. Their fluorescence is less intense than that of amyloid cores. These deposits may correspond to an early stage of amyloid formation.

In the cerebral cortex, amylod deposits are in most instances associated with a crown of degenerating neurites, so that when the lesions are studied with classical stains they are morphologically similar to the neuritic plaques of AD. They are invariably found in most areas of the cerebral cortex and hippocampus of symptomatic individuals.

The mean number of amyloid deposits in vertical strips of cerebral cortex comprising the entire cortical thickness is 16/sq mm in the parahippocampal gyrus, 12/sq mm in the insular gyrus, and 11/sq mm in the superior frontal gyrus. Fewer amyloid deposits are found in the hippocampus. Within the subcortical gray matter, PrP amyloid cores are abundant in the thalamus but not in the corpus striatum. Within the brainstem, the mesencephalic tegmentum and the periaqueductal gray matter are severely involved with PrP deposition.

By electron microscopy, the amyloid deposits appear to be composed of bundles of fibrillar structures, measuring 9-10 nm, and radiating out from a central core. Surrounding the mass of fibrils, neuritic processes are found in the cerebral cortex, but are generally absent in the cerebellum.

The neurites associated with the amyloid cores immunoreact with antibodies to τ, ubiquitin, and to N- and C-terminal domains of the ß-amyloid precursor protein. The accumulation of the ß-amyloid precursor protein in nerve cell processes is not associated with extracellular deposition of ß-amyloid, except in older patients where ß-amyloid immunoreactivity also may be observed around PrP-amyloid deposits[17,33].

Neurofibrillary tangles and τ Pathology. NFT in cell bodies and in neurites are a major feature of the neuropathologic phenotype in GSS-IK. Regions particularly rich in NFT are the cingulate, the frontal, the orbital, the insular, the temporal, and the parietal cortex. In the occipital cortex NFT are also present but less numerous. In the hippocampus, NFT are not found in the CA1 through CA3 areas, but are numerous in CA4. Among the subcortical nuclei, the corpus striatum, the thalamus, the amygdala, and the substantia innominata show a variable. but also severe degree of involvement with NFT and neuropil threads. In midbrain and pons, the substantia nigra, the griseum centrale, and the locus coeruleus have high numbers of NFT[1,3,14-18].

NFT are birefringent after Congo red and fluorescence under ultraviolet light after thioflavin S staining. By electron microscopy, paired helical filaments appear as the main constituents of the NFT found in cell bodies and neurites. Each member of the pair is a filament about 10 nm in diameter. The pair of filaments measures about 22-24 nm at its maximum width, and the helical twist has a period of about 70-80 nm. NFT are also found in large numbers in GSS patients with the Q217R mutation[16,17] and in PrP cerebral amyloid angiopathy with a stop codon mutation at codon 145 (Y145Stop)[14,35]. NFT are less consistently found in patients with GSS disease caused by the mutation *PRNP* A117V [36,37].

Immunocytochemical studies in GSS-IK using antibodies Alz50, Tau-46 and Tau-1, recognizing epitopes at the N- and C-termini and in the mid-region of the τ protein, respectively, show immunopositivity of neurons with NFT[21-23]. Tau-1 requires dephosphorylation before immunostaining, since the epitope recognized by this antibody is phosphorylation dependent. Furthermore, phosphorylation dependent anti-τ monoclonal antibodies AT8 and PHF1, which recognize phosphorylated Ser-202/Thr-205 and phosphorylated Ser-396/Ser-404 respectively, label strongly the NFT and neuropil threads. The NFT also may be stained with antibodies against ubiquitin. In addition to the NFT, such antibodies also reveal the neuritic component surrounding the PrP amyloid deposits of the neocortex as well as the neuropil threads. Thus, the cytoskeletal alterations observed in nerve cell bodies and neurites in GSS-IK are indistinguishable from those observed in AD. By immunogold electron microscopy, Alz50, Tau-46 and PHF1 recognize paired helical filaments[23].

Spongiform changes. Rarely, areas of the hippocampal pyramidal layer show vacuolation of the neuropil.

Nerve cell loss. This is observed in almost all cortical regions, with varying degree of severity in the various patients studied. In the cerebellar cortex, rarefaction of Purkinje cells is severe; swollen axons of Purkinje cells are found in the granule cell layer. Neuronal loss is evident in the red nucleus, substantia nigra, dentate nucleus, inferior olivary nucleus, and various other nuclear groups of the medulla. The oculomotor nuclei and the nucleus hypoglossus are apparently normal. In the substantia nigra, abundant pigment is found in the neuropil.

Perls stain shows that deposition of iron is prominent in globus pallidus, caudate nucleus, putamen, red nucleus and substantia nigra.

At-risk individuals

A neurologically asymptomatic sister of an affected individual died suddenly at age 42. The neuropathologic examination showed that amyloid deposits in the cerebellar molecular and granule cell layers were already prominent. The amyloid deposits were immunolabelled by antibody to PrP. NFT were not found[37].

Biochemical Studies

Prion protein isoforms. Western blots of brain extracts probed with monoclonal antibody 3F4, which recognizes an epitope corresponding to residues 109-112 of human PrP showed that non treated samples (i.e., no PK digestion or deglycosylation) contained four major PrP species of approximately 33-35, 28-30, 19-20, and 9 kDa, and multiple immunoreactive bands of high molecular weight. The latter may correspond to PrP polymers.

Proteinase K-resistant prion protein isoforms. PK digestion of the brain extracts generated three prominent broad bands of ca. 27-29, 18-19, and 8 kDa, and a weaker but sharp band at 33 kDa, as detected with antibody 3F4. The latter band (33 kDa) may be attributed to PrP and/or to cross reactivity of antibody 3F4 with residual PK. The stoichiometry among the PrP species differed from that of the undigested peptides for a notable increase in the signal of the low molecular weight band. No PrP signal was observed in PK-treated and non-treated brain extracts when antibody 3F4 was absorbed against a PrP peptide spanning residues 102-114.

N-deglycosylation of non PK-treated extracts with PNGase F resulted in disappearance of the 33-35 kDa band accompanied by an increased signal of the 28-30 kDa band[20]. The 28-30 kDa band, seen with antibody 3F4, is consistent with the molecular weight of deglycosylated full-length PrP, as shown by a similar species detected with antibodies raised against synthetic peptides homologous to residues 23-40 and 220-231 of human PrP (PrP23-40 and PrP220-231)[20].

In non PK-treated brain extracts, the electrophoretic mobility of the 19-20 and 9 kDa bands was not modified by deglycosylation. The combination of PK and enzymatic deglycosylation with PNGase F treatment generated a pattern similar to that of PK treatment alone with prominent fragments at ca. 27-29, 18-19 and 8 kDa . These PrP fragments were immunoreactive with antibody 3F4 and with antisera AS 6800 (raised against a synthetic peptide homologous to residues 89-104 of human PrP), but not with PrP23-40 and PrP220-231.

To investigate the sensitivity of PrP to PK, samples from the cerebellum were exposed to PK under non-denaturing and denaturing conditions. Under non-denaturing conditions, the major PrP isoforms present in GSS-IK retained partial PK resistance, the intensity of the signal after 4 hours being similar to that observed after one hour of enzyme treatment. Conversely, denatured PrP was completely degraded after 30 min digestion with PK at 37°C in the presence of sodium dodecyl sulphate[20].

PK-resistant PrP in areas with predominantly fibrillar or non-fibrillar deposits. PK-resistant PrP fragments of similar electrophoretic mobility were seen in all brain regions examined (frontal cortex, caudate nucleus and cerebellum) of two IK patients analyzed. In semi-quantitative experiments (similar amounts of total protein loaded), comparable signals were observed in samples from the cerebellum and caudate nucleus, two areas that have high and low amount of amyloid respectively[20].

In immunoblot studies, the strongest signal was obtained from tissue corresponding to a patient who had the longest clinical course of the disease (12 years). To follow up on this observation, we also studied PK-resistant PrP obtained from the cerebella of five additional patients of the IK, who had a duration of clinical signs varying from two to seven years. Similar electrophoretic patterns were observed and high amounts of PrP were present in all cases regardless of the duration of disease. Nevertheless, in repeated experiments the patient with 12 years duration of clinical signs always showed the most intense signal.

Subcellular distribution of PrP. PrP was localized in the microsomal fraction. Sarkosyl-soluble and PK-sensitive PrP isoforms from this fraction were seen as prominent bands of approximately 33-35 kDa in both control (familial AD) and GSS-IK[20]. In addition, Sarkosyl-insoluble PrP was present as four major PrP species of approximately 33-35, 28-30, 19-20 and 9 kDa[20]. PK digestion of these samples generated three prominent bands of 27-29, 18-19, and 8 kDa in GSS-IK, comparable to the PK-resistant species present in brain homogenates. As expected, no Sarkosyl-insoluble PrP was present in control extracts[20].

Amyloid protein isolation and characterization. The biochemical composition of PrP amyloid was first determined in brain tissue samples obtained from patients of the Indiana kindred[39,40] carrying the F198S mutation in coupling with 129V[25,26]. Amyloid cores were isolated by a procedure combining buffer extraction, sieving, collagenase digestion and sucrose gradient centrifugation. Proteins were extracted from amyloid fibrils with formic acid, purified by gel filtration chromatography and reverse-phase HPLC, analyzed by SDS-PAGE and immunoblot, and sequenced. The amyloid preparations contained two major peptides of ≈ 11 and ≈ 7 kDa spanning residues 58-150 and 81-150 of PrP, respectively. The amyloid peptides had ragged N- and C-termini.

The finding that the amyloid protein was an N- and C-terminal truncated fragment of PrP was verified by immunostaining brain sections with antisera raised against synthetic peptides homologous to residues 23-40, 90-102, 127-147 and 220-231 of human PrP. The amyloid cores were strongly immunoreactive with antibodies that recognized epitopes located in the mid-region of the molecule while only the periphery of the cores was immunostained by antibodies to N- or C-terminal domains. In addition, antisera to the N- and C-termini of PrP labeled large areas of the neuropil that did not possess the tinctorial and optical properties of amyloid. Immunogold electron microscopy showed that antibodies to the mid-region of PrP decorated fibrils of amyloid cores while antisera to N- and C-terminal epitopes labeled amorphous material at the periphery of the cores or dispersed in the neuropil. These data suggest that amyloid deposition in GSS is accompanied by accumulation of PrP peptides without amyloid characteristics[19].

In GSS-IK, the amyloid protein does not include the region containing the amino acid substitution. To establish whether amyloid peptides originate from mutant protein alone or both mutant and wild-type PrP, we analyzed patients heterozygous M/V at codon 129 and used 129V as a marker for protein from the mutant allele. Amino acid sequencing and electrospray mass spectrometry of peptides generated by digestion of the amyloid protein with endoproteinase Lys-C showed that the samples contained only peptides with 129V, suggesting that only mutant PrP was involved in amyloid formation[40].

<u>Characterization of paired helical filaments.</u> Paired helical filament-enriched fractions obtained from the neocortex of IK patients contained SDS-soluble τ isoforms with electrophoretic mobility and immunochemical profile corresponding to the τ isoforms extracted from the brain of patients with AD[23]. These proteins migrate between 60 and 68 kDa, immunoreact with antibodies to the N- and C-termini of τ, and require dephosphorylation to be accessible to Tau-1. Thus, the immunocytochemical findings are consistent with those of the western blot analysis showing that significant similarity exists between GSS-IK and AD as to the Alz50, T46 and Tau-1 immunostaining of NFT.

Animal Transmission

Transmission to animals has been demonstrated for some cases of GSS with *PRNP* P102L mutation. Masters et al. originally showed that the inoculation of brain homogenates from three patients affected by GSS into non-human primates in some instances induced a spongiform encephalopathy in the recipient animals[11]. Subsequent experiments carried out by inoculating intracerebrally brain tissue into marmosets and mice resulted in the development of spongiform degeneration, confirming the original observation by Masters et al[11].

In the case of the Indiana kindred, brain tissue and buffy coat from one affected individual were inoculated into hamsters in two experiments in the laboratory of Drs. E. and L. Manuelidis. No pathologic changes were observed in the primary transmission attempt, nor in the second and third serial passage (Dr. L. Manuelidis, personal communication). Tissue homogenates from another F198S patient have been inoculated into hamsters and mice and no transmission has occurred[3,26]. Amyloid enriched fractions and tissue homogenates from IK patients have been inoculated into marmosets and into transgenic mice containing a normal human *PRNP* gene; no transmission has occurred 24 and 12 months, respectively, after innoculation (Baker, personal communication; Collinge, personal communication).

CONCLUSIONS

The following are the characteristics of the Indiana variant of Gerstmann-Sträussler-Scheinker disease.

1. GSS-IK is inherited with an autosomal dominant pattern and is caused by a mutation in the *PRNP* gene at codon 198, resulting in a serine for phenylalanine substitution.

2. Ataxia, parkinsonism, and presenile dementia are the main neurological symptoms.

3. The duration of the disease from the onset of ataxia is two to twelve years.

4. Homozygosity V/V at codon 129 is associated with an earlier onset of symptoms.

5. PrP-amyloid deposition occurs in cerebrum and cerebellum; deposition of PrP amyloid occurs in the cerebellum before the onset of clinical signs.

6. PrP-immunopositive amyloid deposits of the cerebral cortex are surrounded by abnormal neurites.

7. Alzheimer NFT are present in cerebral cortex, hippocampus and subcortical nuclei.

8. PrP amyloid deposition may be associated with amyloid ß-protein in aged patients.

9. Distinct PK-resistant PrP isoforms are present in the cerebrum and cerebellum.

10. The major components of the amyloid are 11 and 7 kDa fragments of prion protein spanning residues 58-150 and 81-150 respectively.

ACKNOWLEDGMENTS

These studies were supported by grants PHS R01 NS 29822, P30 AG 10133 and the Italian Ministry of Health.

REFERENCES

1. B. Azzarelli, J. Muller, B. Ghetti, M. Dyken, and P.M. Conneally, Cerebellar plaques in familial Alzheimer's disease (Gerstmann-Sträussler-Scheinker variant?). *Acta Neuropathol.* 65:235-246 (1985).
2. M. Farlow, R.D. Yee, S.R. Dlouhy, P.M. Conneally, B. Azzarelli, and B. Ghetti, Gerstmann-Sträussler-Scheinker disease. I. Extending the clinical spectrum. *Neurology* 39:1446-1452 (1989).
3. B. Ghetti, F. Tagliavini, K. Hsiao, S.R. Dlouhy, R.D. Yee, G. Giaccone, P.M. Conneally, M.E. Hodes, O. Bugiani, S.B. Prusiner, B. Frangione, and M.R. Farlow, Indiana variants of Gerstmann-Sträussler-Scheinker disease, in *Prion diseases of humans and animals*, S. Prusiner, J. Collinge, J. Powell, and B. Anderton, eds, Ellis Horwood, New York (1992).
4. J. Gerstmann, E. Sträussler, and I. Scheinker, Über eine eigenartige hereditär-familiäre Erkrankung des Zentralnervensystems. Zugleigh ein Beitrag zur Frage des vorzeitigen lokalen Alterns. *Zeitschr Neurol Psych* 154:736-762 (1936).
5. A. von Braunmühl, Über eine eigenartige hereditär-familiäre erkrankung des Zentral nervensystems. *Archiv Für Psychiatrie und Zeitschrift Neurologie* 191:419-449 (1954).
6. F. Seitelberger, Eigenartige familiär-hereditäre Krankheit des Zentralennerven-systems in einer niederösterreichischen Sippe. *Wien Klin Wochenschr* 74:687-691 (1962).
7. F. Seitelberger F, Sträussler's disease. *Acta Neuropathol (Berl)* 7:341-343 (1981).
8. H. Budka, F. Seitelberger, M. Feucht, P. Wessely, and H.A. Kretzschmar, Gerst mann-Sträussler-Scheinker syndrome (GSS): rediscovery of the original Austrian family. *Clin Neuropathol* 10:99 (1991).
9. H. Kretzschmar, G. Honold, F. Seitelberger, M. Feucht, P. Wessely, P. Mehraein, and H. Budka, Prion protein mutation in family first reported by Gerstmann, Sträussler, and Scheinker. *Lancet* 337:1160 (1991).
10. M. Farlow, F. Tagliavini, O. Bugiani, and B. Ghetti, Gerstmann-Sträussler-Scheinker disease, in *Handbook of Clinical Neurology*, J.M.B.V. de Jong, ed., Elsevier, Amsterdam (1992).
11. C. Masters, D. Gajdusek, and C. Gibbs, Creutzfeldt-Jakob disease virus isolations from the Gerstmann-Sträussler syndrome with an analysis of the various forms of amyloid plaque deposition in the virus-induced spongiform encephalo-pathies. *Brain* 104:559-588 (1981).
12. K. Hsiao and S.B. Prusiner, Inherited human prion diseases. *Neurology* 40:1820-1827 (1990).
13. S.B. Prusiner, Molecular Biology of Prion Diseases. *Science* 252:1515-1522 (1991).
14. B. Ghetti, P. Piccardo, B. Frangione, O. Bugiani, G. Giaccone, K. Young, F. Prelli, M.R. Farlow, S.R. Dlouhy, F. Tagliavini, Prion Protein Amyloidosis. *Brain Pathol* 6:172-145 (1996)
15. B. Ghetti, F. Tagliavini, C.L. Masters, K. Beyreuther, G. Giaccone, L. Verga, M.R. Farlow, P.M. Conneally, S.R. Dlouhy, B. Azzarelli, and O. Bugiani, Gerstmann-Sträussler-Scheinker disease. II. Neurofibrillary tangles and plaques with PrP-amyloid coexist in an affected family. *Neurology* 39:1453-1461 (1989).
16. B. Ghetti, F. Tagliavini, G. Giaccone, O. Bugiani, B. Frangione, M.R. Farlow, and S.R. Dlouhy, Familial Gerstmann-Sträussler-Scheinker disease with neurofibril lary tangles. *Mol Neurobiol* 8:41-48 (1994).
17. B. Ghetti, S.R. Dlouhy, G. Giaccone, O. Bugiani, B. Frangione, M.R. Farlow, and F. Tagliavini, Gerstmann-Sträussler-Scheinker Disease and the Indiana Kindred. *Brain Pathol* 5:61-75 (1995).

18. B. Ghetti, P. Piccardo, B. Frangione, O. Bugiani, G. Giaccone, K. Young, F. Prelli, M.R. Farlow, S.R. Dlouhy, and F. Tagliavini, Prion protein hereditary amyloidosis: parenchymal and vascular. *Sem Virol* 7:189-200 (1996).

19. G. Giaccone, L. Verga, O. Bugiani, B. Frangione, D. Serban, S.B. Prusiner, M.R. Farlow, B. Ghetti, and F. Tagliavini, Prion protein preamyloid and amyloid deposits in Gerstmann-Sträussler-Scheinker disease, Indiana kindred. *Proc Natl Acad Sci (USA)* 89:9349-9353 (1992).

20. P. Piccardo, C. Seiler, S.R. Dlouhy, K. Young, M.R. Farlow, F. Prelli, B. Frangione, O. Bugiani, F. Tagliavini, and B. Ghetti, Proteinase-K-resistant prion protein isoforms in Gerstmann-Sträussler-Scheinker disease (Indiana kindred). *J Neuropathol Exp Neurol* 55:1157-1163 (1996).

21. G. Giaccone, F. Tagliavini, L. Verga, B. Frangione, M.R. Farlow, O. Bugiani, B. Ghetti, Neurofibrillary tangles of the Indiana kindred of Gerstmann-Sträussler-Scheinker disease share antigenic determinants with those of Alzheimer disease. *Brain Res* 530:325-329 (1990).

22. G. Giaccone, F. Tagliavini, L. Verga, B. Frangione, M.R. Farlow, O. Bugiani, and B. Ghetti, Indiana Kindred of Gerstmann-Sträussler-Scheinker syndrome: Neurofibrillary tangles and neurites of plaques with PrP amyloid share antigenic determinants with those of Alzheimer's disease. In *Alzheimer's Disease: Basic Mechanisms, Diagnosis and Therapeutic Strategies*, K. Iqbal, D.R.C. McLachlan, B. Winblad and H.M. Wisniewski, eds, John Wiley & Sons Ltd (1991).

23. F. Tagliavini, G. Giaccone, F. Prelli, L. Verga, M. Porro, J. Trojanowski, M. Farlow, B. Frangione, B. Ghetti, and O. Bugiani, A68 is a component of paired helical filaments of Gerstmann-Sträussler-Scheinker disease, Indiana kindred. *Brain Res* 616:325-328 (1993).

24. K. Hsiao, H.F. Baker, T.J. Crow, M. Poulter, F. Owen, J.D. Terwillinger, D. Westaway, J. Ott, and S.B. Prusiner, Linkage of a prion protein missense variant to Gerstmann-Sträussler syndrome. *Nature* 338:342-345 (1989).

25. S.R. Dlouhy, K. Hsiao, M.R. Farlow, T. Foroud, P.M. Conneally, P. Johnson, S.B. Prusiner, H.E. Hodes, and B. Ghetti, Linkage of the Indiana kindred of Gerstmann-Sträussler-Scheinker disease to the prion protein gene. *Nature Genet* 1:64-67 (1992).

26. K. Hsiao, S.R. Dlouhy, M.R. Farlow, C. Cass, M. DaCosta, P.M. Conneally, M.E. Hodes, B. Ghetti, and S.B. Prusiner, Mutant prion proteins in Gerstmann-Sträussler-Scheinker disease with neurofibrillary tangles. *Nature Genet* 1:68-71 (1992).

27. F.W. Unverzagt, M.R. Farlow, S.R. Dlouhy, J. Norton, N. Mercado, A.M. Torke, and B. Ghetti, Presymptomatic neuropsychological deficits in Gerstmann-Sträussler-Scheinker disease. *Ann Neurol* 36:272-273 (1994).

28. F.W. Unverzagt, M.R. Farlow, S.R. Dlouhy, A.M. Torke, and B. Ghetti, Neuropsychological functioning in affected and at-risk Gerstmann-Sträussler-Scheinker disease patients. *J Intl Neuropsychol Soc* 1:385 (1995).

29. F.W. Unverzagt, M.R. Farlow, A. Torke, N. Mercado, and B. Ghetti, Neuropsychological functioning in Gerstmann-Sträussler-Scheinker disease and Alzheimer's disease: A Preliminary report. *J Intl Neuropsychol Soc* 1:385 (1995).

30. F.W. Unverzagt, M.R. Farlow, J. Norton, S.R. Dlouhy, K. Young, and B. Ghetti, Neuropsychological function in patients with Gerstmann-Sträussler-Scheinker disease from the Indiana Kindred (F198S). *J Intl Neuropsychol Soc* (In Press) (1996).

31. M.R. Farlow, M.K. Edwards, M. Kuharik, G. Giaccone, F. Tagliavini, O. Bugiani and B. Ghetti, Magnetic resonance imaging in the Indiana kindred of Gerstmann-Sträussler-Scheinker disease. *Neurobiol Aging* 11:265 (1990).

32. R.D. Yee, M.R. Farlow, D.A. Suzuki, K.F. Betelak, and B. Ghetti, Abnormal eye movements in Gerstmann-Sträussler-Scheinker disease. *Arch Opthalmol* 110:68-74 (1992).

33. O. Bugiani, G. Giaccone, L. Verga, B. Pollo, B. Frangione, M.R. Farlow, F. Tagliavini, and B. Ghetti, ßPP participates in PrP-amyloid plaques of Gerstmann-Sträussler-Scheinker disease, Indiana kindred. *J Neuropathol Exp Neurol* 52:64-70 (1993).

34. S. Ikeda, N. Yanagisawa, G.G. Glenner and D. Allsop, Gerstmann-Sträussler-

Scheinker disease showing ß-protein amyloid deposits in the peripheral regions of PrP-immunoreactive amyloid plaques. *Neurodegeneration* 1:281-288 (1992).

35. B. Ghetti, P. Piccardo, M.G. Spillantini, Y. Ichimiya, M. Porro, F. Perini, T. Kitamoto, J. Tateishi, C. Seiler, B. Frangione, O. Bugiani, G. Giaccone, F. Prelli, M. Goedert, S.R. Dlouhy, and F. Tagliavini, Vascular variant of prion protein cerebral amyloidosis with τ-positive neurofibrillary tangles: The phenotype of the stop codon 145 mutation in *PRNP*. *Proc Natl Acad Sci USA* 93:744-748 (1996).

36. K. Hsiao, C. Cass, G.D. Schellenberg, T. Bird, E. Devine-Gage, H. Wisniewski, and S.B. Prusiner, A prion protein variant in a family with the telencephalic form of Gerstmann-Sträussler-Scheinker syndrome. *Neurology* 41:681-184 (1991).

37. D. Nochlin, S.M. Sumi, T.D. Bird, A.D. Snow, C.M. Leventhal, K. Beyreuther, and C.L. Masters, Familial dementia with PrP positive amyloid plaques: a variant of Gerstmann-Sträussler syndrome. *Neurology* 39:910-918 (1991).

38. M.R. Farlow, O. Bugiani, G. Giaccone, F. Tagliavini, and B. Ghetti, Neuropathology of presymptomatic Gerstmann-Sträussler-Scheinker disease of the Indiana kindred. *Neurology* 41:(Suppl. 1) 119 (1991).

39. F. Tagliavini, F. Prelli, J. Ghiso, O. Bugiani, D. Serban, S.B. Prusiner, M.R. Farlow, B. Ghetti, and B. Frangione, Amyloid protein of Gerstmann-Sträussler-Scheinker disease (Indiana kindred) is an 11 kd fragment of prion protein with an N-terminal glycine at codon 58. *EMBO J* 10:513-519 (1991).

40. F. Tagliavini, F. Prelli, M. Porro, G. Rossi, G. Giaccone, M.R. Farlow, S.R. Dlouhy, B. Ghetti, O. Bugiani, and B. Frangione, Amyloid fibrils in Gerstmann-Sträussler-Scheinker disease (Indiana and Swedish kindreds) express only PrP peptides, encoded by the mutant allele. *Cell* 79:695-703 (1994).

STUDIES ON THE PATHOGENESIS OF SCRAPIE AND THE PURIFICATION OF SCRAPIE AGENT IN THE HAMSTER MODEL

Heino Diringer,[1] Elizabeth Baldauf,[1] Michael Beekes,[1] and Muhsin Özel[2]

Robert Koch-Institut
Bundesinstitut für Infektionskrankheiten und nicht übertragbare Krankheiten
[1] FG 123
[2] FG 122
Nordufer 20
13353 Berlin
Germany

TERMINOLOGY

Scrapie-associated and related fibrils[1,2,3] (sometimes also referred to as prion rods[4]) are a pathological hallmark of all known transmissible spongiform encephalopathies (TSE). In some cases the fibrils form classical amyloid plaques, but very often they do not show pronounced super-aggregation in situ, thus eluding histological detection (for a review see: Diringer[5]). Independently of their presentation in the light microscope, i.e. their degree of super-aggregation (microdisperse, amorphous or plaque-like) the physico-chemical and structural properties of the fibrils and of the protein of which they consist entirely fit the molecular criteria[6] for an amyloid[4,7,8].

TSE-specific fibrils can be extracted with or without protease treatment from infected tissue of the central nervous system of all mammalian species which are susceptible to TSE. Once concentrated, they display the classical characteristic of all amyloids, i. e. green birefringence after staining with Congo red.

The TSE-specific amyloid fibrils are composed of a host-derived pathological protein, formerly referred to as SAF-protein or protease-resistant protein, now very often designated as prion protein (PrP27-30, PrP^{SC}) (for a review see: Gajdusek[9] and Collinge & Prusiner[10]). However, the PrP-nomenclature is based on a hypothetical etiological concept, the prion hypothesis[11,12], thus anticipating the still unknown nature of the infectious agent. Therefore we prefer a more descriptive terminology and refer to the pathological protein which constitutes the TSE-specific amyloid fibrils as TSE-specific amyloid protein (TSE-AP).

PATHOGENESIS OF SCRAPIE IN HAMSTERS AFTER ORAL AND INTRAPERITONEAL INFECTION

As reported previously, TSE-specific amyloid protein and infectivity show a parallel accumulation in the central nervous system of hamsters orally infected with scrapie[13]. The ratio between the two disease markers was found to be relatively constant at a mean value of about 10^6 TSE-AP molecules per infectious unit during the course of infection. Therefore we traced the spread of infection by monitoring the appearance of the pathological protein in our animal model after oral administration of scrapie agent. Our initial study revealed an onset of the pathogenetic process in the spinal cord between vertebrae T4-T9. Then the process of TSE-AP formation and TSE-AP accumulation showed an anterograde and retrograde spread to the brain and to the lumbar spinal cord, respectively. In the brain TSE-AP could be found only after all spinal cord segments between vertebrae C1 and T9 had been positive for the pathological protein. However, as we analyzed only small samples of the homogenate of entire brains, we possibly missed the onset of TSE-AP formation in the brain.

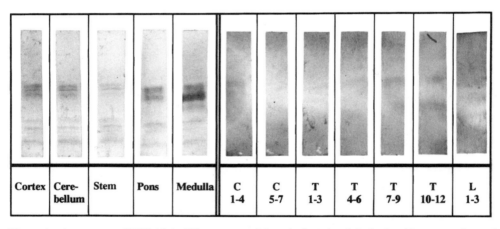

Cortex	Cerebellum	Stem	Pons	Medulla	C 1-4	C 5-7	T 1-3	T 4-6	T 7-9	T 10-12	L 1-3

Figure 1a. Appearance of TSE-AP in different parts of the spinal cord and the brain of hamsters after oral infection with scrapie: Western blot detection with MAb 3F4 at 89 dpi; 20-50 mg of tissue for the different spinal cord, medulla and pons samples; up to 100 mg for samples of the cerebellum, cortex and remaining stem (together with the diencephalon).

In a subsequent study, different spinal cord segments and the medulla oblongata, pons, cerebellum, cortex and remaining brain stem (together with the diencephalon) were analyzed separately after oral infection. This revealed an onset of TSE-AP formation as shown in Figure 1a and 1b (Baldauf et al., submitted for publication).

The pathogenetic process occurred simultaneously in the spinal cord and in the brain, sometimes even earlier in the latter. This demonstrates the existence of an alternative route of access for the infection from the periphery to the brain other than via the spinal cord. Similar findings were observed in another study on the pathogenesis of scrapie after intraperitoneal administration of agent[14] (Baldauf et al., submitted for publication).

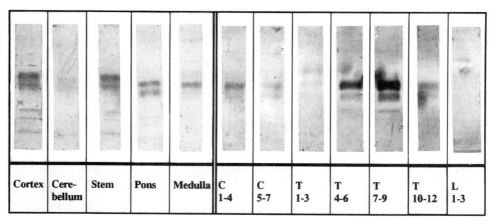

Cortex	Cere-bellum	Stem	Pons	Medulla	C 1-4	C 5-7	T 1-3	T 4-6	T 7-9	T 10-12	L 1-3

Figure 1b. As for Figure 1a, but samples were taken from a different animal at 96 dpi.

PURIFICATION OF THE SCRAPIE AGENT

Because some of the presentations in this symposium have dealt with studies involving fractions enriched for infectivity, let us continue with some comments on the purification of the scrapie agent.

Scrapie in hamsters has not only become an important model to study the pathogenesis of TSE but has also been used to develop techniques for the isolation of the scrapie agent, presently the most difficult problem to solve in the research on these diseases. At present, three basic purification protocols are available for the concentration and enrichment of infectivity. One large-scale protocol[15] has been said to result in samples of essentially the

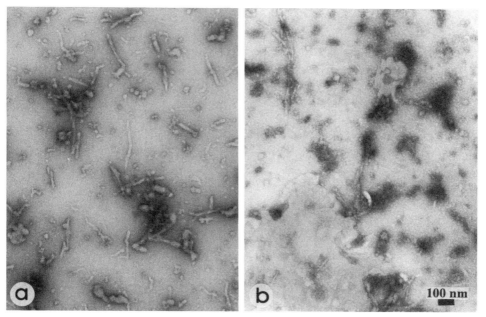

Figure 2. (a) Preparation of SAF from a brain homogenate without detergent[17] and (b) from a brain homogenate containing 10% sarcosyl[18]. Magnification: x 50 000.

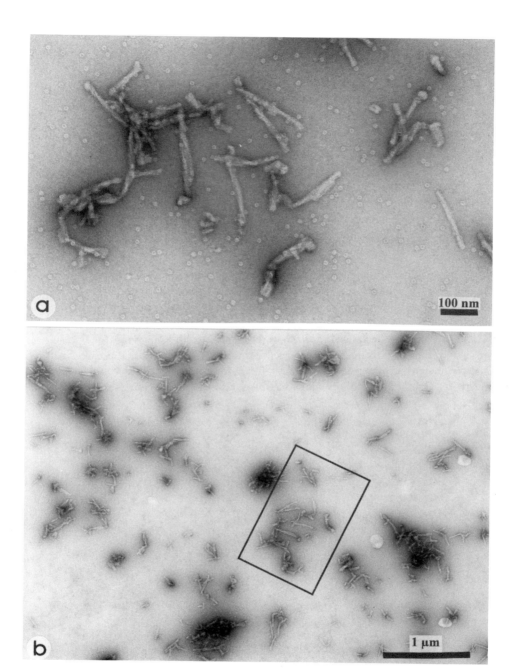

Figure 3. Preparation of SAF by an unpublished technique (Diringer et al., in preparation); (a) represents the area marked in (b) at a higher magnification. Magnification: (a) x 150 000; (b) x 28 000.

same grade of purity[16] as has been reported for two small-scale protocols[17,18], which have gained world wide-application. It has also be claimed that fractions obtained by the large scale protocol consist almost entirely of the TSE-specific amyloid protein (prion protein)[12].

Electron micrographs of fractions obtained with the two small scale techniques are presented in Figure 2a and 2b. It is obvious that in such fractions one cannot exclude the presence of a virus. Therefore we have continued our purification efforts.

Figure 3 shows electron micrographs of a preparation obtained with a hitherto unpublished protocol. But even these infectious fractions of TSE-AP are not pure enough to allow the exclusion of an undetected virus.

Firstly, certain impurities like for example ferritin (Figure 3a) still cannot be removed completely, thus constituting a background, in which a small virus could be hidden. Furthermore one can conclude from previous experiments that in our highly purified fraction one infectious unit (LD_{50}) is associated with at least 10^6 molecules of the TSE-specific amyloid protein[13]. About 150 - 300 of the TSE-AP molecules represent 100 nm of a scrapie-associated fibril (this has been estimated by counting subunits in fibril preparations at high magnifications).

Figure 3a contains about 50 x 100 nm of SAF or 50 x 150 - 50 x 300 molecules of the TSE-specific amyloid protein, i.e. a total of 7 500 - 15 000 TSE-AP molecules. Figure 3b represents an area covered by roughly 10 times more amyloid. As 1 LD_{50} is equivalent to 10^6 molecules of TSE-AP, an electron microscopist would have to scan an area 10-fold larger than that presented in Figure 3b in order to find a single virus particle. Then he would have to exclude the presence of such a particle in uninfected control samples. This demonstrates the difficulties encountered in the search for the scrapie agent by electron microscopy.

At present, similar problems still hinder the detection (or definitive exclusion) of a viral nucleic acid or a viral protein. Therefore it is important to continue the efforts for a rigorous purification of the scrapie agent. This will most easily provide a positive proof for the puzzling nature of the TSE agent, whether it is a virus, a virino, or a prion.

REFERENCES

1. P.A. Merz, R.A. Somerville, H.M Wisniewski, and K Iqbal, Abnormal fibrils from scrapie-infected brain, *Acta Neuropathol.* 54:63-74 (1981).
2. P.A. Merz, R.A. Somerville, H.M. Wisniewski, L. Manuelidis, and E.E. Manuelidis, Scrapie-associated fibrils in Creutzfeldt-Jakob disease, *Nature.* 306:474-476 (1983).
3. P.A. Merz, R.G. Rohwer, R. Kascsak, H.M. Wisniewski, R.A. Somerville, C.J. Gibbs, and D.C. Gajdusek, Infection-specific particle from the unconventional slow virus diseases, *Science.* 225:437-440 (1984).
4. S.B. Prusiner, M.P. McKinley, K.A. Bowman, D.C. Bolton, P.E. Bendheim, D.F. Groth, and G.G. Glenner, Scrapie prions aggregate to form amyloid-like birefringent rods, *Cell.* 35:349-358 (1983).
5. H. Diringer, Hidden amyloidoses, *Exp Clin Immunogenet.* 9:212-229 (1992).
6. G.G. Glenner, Amyloid deposits and amyloidosis: the β-fibrilloses, *New Engl J Med.* 302:1283-1292 and 1333-1343 (1980).
7. B.W. Caughey, A. Dong, K.S. Bhat, D. Ernst, S.F. Hayes, and W.S. Caughey, Secondary structure analysis of the scrapie-associated protein PrP 27-30 in water by infrared spectroscopy, *Biochemistry.* 30:7672-7680 (1991).

8. M. Gasset, M.A. Baldwin, R.J. Fletterick, and S.B. Prusiner, Perturbation on the secondary structure of the scrapie prion protein under conditions that alter infectivity, *Proc Natl Acad Sci USA*. 90:1-5 (1993).

9. D.C. Gajdusek, Subacute spongiform encephalopathies: transmissible cerebral amyloidoses caused by unconventional viruses, in: *Virology*, B.N. Fields and D.M. Knipe, ed., Raven Press, New York (1990).

10. J. Collinge and S.B. Prusiner, Terminology of prion diseases, in: *Prion Diseases of Humans and Animals*, S.B. Prusiner, J. Collinge, J. Powell, and B. Anderton, ed., Ellis Horwood, New York (1992).

11. S.B. Prusiner, Novel proteinaceous infectious particles cause scrapie, *Science*. 216:136-144 (1982)

12. S.B. Prusiner, The prion diseases, *Scientific American*. January:30-37 (1995).

13. M. Beekes, E. Baldauf, and H. Diringer, Sequential appearance and accumulation of pathognomonic markers in the central nervous system of hamsters orally infected with scrapie, *J Gen Virol*. 77:1925-1934 (1996).

14. M. Beekes, E. Baldauf, and H. Diringer, Pathogenesis of scrapie in hamsters after oral and intraperitoneal infection, in: *Prion: Jenner, Pasteur and their Successors*, S. Plotkin and B. Fantini, ed., Elsevier, Paris (in print).

15. S.B. Prusiner, D.C. Bolton, D.F. Groth, K.A. Bowman, S.P. Cochran, and M.P. McKinley, Further purification and characterization of scrapie prions, *Biochemistry*. 21:6942-6950 (1982).

16. S.B. Prusiner, K.A. Bowman, and D.F. Groth, Purification of scrapie prions, in: *Prions - Novel Infectious Pathogens causing Scrapie and Creutzfeldt-Jakob Disease*, S.B. Prusiner and M.P. McKinley, ed., Academic Press, San Diego (1987).

17. H. Diringer, H. Hilmert, D. Simon, E. Werner, and B. Ehlers, Towards purification of the scrapie agent, *Eur J Biochem*. 134:555-560 (1983).

18. H. Hilmert and H. Diringer, A rapid and efficient method to enrich SAF-protein from scrapie brains of hamsters, *Biosci Rep*. 4:165-170 (1984)

PRP PEPTIDES AS A TOOL TO INVESTIGATE THE PATHOGENESIS
OF PRION PROTEIN AMYLOIDOSES

Fabrizio Tagliavini,[1] Giorgio Giaccone,[1] Frances Prelli,[2] Blas Frangione,[2] Mario Salmona,[3] Gianluigi Forloni,[3] Bernardino Ghetti,[4] and Orso Bugiani[1]

[1]Istituto Nazionale Neurologico Carlo Besta, via Celoria 11, 20133 Milano, Italy
[2]Department of Pathology, New York University Medical Center, 530 First Avenue, New York, NY 10016, USA
[3]Istituto di Ricerche Farmacologiche Mario Negri, via Eritrea 62, 20157 Milano, Italy
[4]Department of Pathology and Laboratory Medicine, Indiana University School of Medicine, Indianapolis, IN 46202-5251, USA

INTRODUCTION

In prion diseases, abnormal isoforms of the prion protein (PrP) and PrP amyloid accumulate in the brain.[1,2] At difference with normal PrP, the disease-specific PrP is resistant to protease digestion.[1,2] The abnormal protein likely issues from protease-sensitive precursors following a post-translational modification involving a shift from α-helix to ß-sheet secondary structure.[3-5]

Amyloid formation and deposits mostly occur in Gerstmann-Sträussler-Scheinker disease (GSS) and in PrP cerebral amyloid angiopathy (PrP-CAA). GSS is transmitted with an autosomal dominant trait and is clinically heterogeneous, as it presents with either ataxia or spastic paraparesis, parkinsonism and dementia. GSS is associated with variant *PRNP* genotypes resulting from mutations at different codons [102 (P>L), 105 (P>L), 117 (A>V), 198 (F>S) and 217 (Q>R)] combined with polymorphisms at codon 129 (M/V) and 219 (E/K).[6] PrP-CAA is a cerebrovascular amyloidosis associated with a mutation at *PRNP* codon 145 (Y>stop).[7]

The clinical variability of GSS is related to the site and amount of amyloid deposition and to neurodegeneration as well.[6] Unicentric and multicentric amyloid deposits are widely distributed throughout the brain, and are abundant in the cerebellar cortex. In some patients with the codon 117 mutation, cerebellum is spared, however. In addition to amyloid deposits, spongiform changes in the cerebral cortex and neurofibrillary tangles with paired helical filaments have been found in patients with P>L substitution at codon 102 and with mutations at codon 145, 198 and 217, respectively.[6-10]

Amyloid deposition is consistently accompanied by hypertrophy and proliferation of astrocytes and microglial cells, neuritic changes and nerve cell loss.[6] The close relationship between amyloid deposits and tissue changes suggests that PrP amyloid is somehow involved in the pathogenesis of nerve cell degeneration and glial cell hypertrophy and proliferation. Accordingly, we investigated (i) the composition of amyloid fibrils in GSS families with different *PRNP* mutations, (ii) the PrP sequences significant to amyloid formation, and (iii) synthetic PrP peptides for their effect on nerve and glial cells *in vitro*.

THE AMYLOID PROTEIN IN GSS

PrP amyloid was assessed biochemically in the brain tissue of patients belonging to the Indiana kindred of GSS,[11,12] carrying a F>S substitution at PrP residue 198 (GSS 198).[13,14] The amyloid analysis issued two major peptides having a molecular weight of ~11 and ~7 kDa and spanning residues 58-150 and 81-150 of PrP, respectively.[11,12]

The finding that the amyloid protein was an N- and C-terminal truncated fragment of PrP was verified on brain sections immunostained with antisera to synthetic peptides homologous to PrP residues 23-40, 90-102, 127-147 and 220-231. Antisera to the mid-region of the molecule labeled the core of amyloid deposits, while those to N- and C-terminal domains labeled the peripheral ring.[15]

In GSS 198, the region containing the amino acid substitution was not included in the amyloidogenic fragments. To establish whether the latter originate from the mutant PrP or from both mutant and wild-type PrP, we analyzed patients heterozygous M/V at codon 129 using V129 as a marker for the mutant allele. This was made possible by the fact that V129 in this family is in phase with mutant S198. Following digestion of the amyloid protein with endoproteinase Lys-C, amino acid sequencing and mass spectroscopy showed that only peptides with V129 were contained in the amyloid molecule, suggesting that amyloid formation was due to the mutant PrP.[12]

Subsequently, the analysis of the amyloid protein in GSS kindreds carrying such different *PRNP* mutations, as A>V at codon 117 and Q>R at codon 217, disclosed the smallest amyloid subunit corresponding to a ~7 kDa N- and C-terminal truncated fragment of PrP, similar in size and sequence to that found in GSS 198. In these cases the amyloid protein derived from the mutant allele,[12] and it contained the variant V117 in patients with GSS 117.[16]

ASSEMBLY AND CONFORMATION OF PRP PEPTIDES IN VITRO

To identify the PrP residues relevant to conformational change and aggregation, and the conditions favoring this process, we investigated fibrillogesis using synthetic peptides homologous to serial segments of the GSS amyloid protein, i.e., the octapeptide repeat region, and residues 89-106, 106-126 and 127-147. Peptides within the PrP sequence 106-147 were able promptly to assemble into fibrils, while peptides corresponding to the N-terminal region were not.[17] Peptide PrP 106-126 was the most fibrillogenic, raising meshworks of straight filaments ultrastructurally similar to GSS *in situ* amyloid. PrP 106-126 fibrils not only were partly resistant to proteinase K digestion, but also exhibited optical and X-ray diffraction properties of *in situ* amyloid (i.e., birefringence under polarized light after Congo red staining, fluorescence after thioflavine S treatment, and reflections corresponding to H-bonds between neighboring polypeptide chains in a cross ß-configuration).[17,18] Furthermore, circular dichroism spectroscopy showed peptide PrP 106-126 able to adopt such different conformations following microenvironment variations as ß-sheet secondary structure in phosphate buffer pH 5.0, ß-sheet combined to random coil in phosphate buffer pH 7.0, random coil in deionized water, and α-helical structure in the presence of micelles formed by a 5% SDS solution.[19] Adding α-helix stabilizing solvents (e.g., trifluoroethanol or hexafluoropropanol) to PrP 106-126 in deionized water induced random coil conformation of the peptide to shift to α-helix, but did not modify the ß-sheet structure of the peptide, if the latter was previously suspended in phosphate buffer, pH 5.0.[19] These data suggest that the PrP region including residues 106-126 may feature in the conformational transition from normal to abnormal PrP.

BIOLOGICAL EFFECTS OF PRP PEPTIDES IN VITRO

To test the view that neuronal loss and astrocytic hypertrophy and proliferation in GSS and PrP-CAA follow PrP amyloid deposition in the brain, we studied several synthetic PrP peptides for the effect on primary cultures of neurons and astrocytes.

A prolonged exposure of rat hippocampal neurons to micromolar concentrations of PrP 106-126 resulted in a severe cell loss.[20] On the contrary, exposure to PrP peptides different from PrP 106-126 or to a scrambled sequence of PrP 106-126 was ineffective. The neurotoxicity of PrP 106-126 was dose-dependent, as it appeared at a concentration of 20 μM, was statistically significant at 40 μM, and resulted in almost complete cell loss at 60 or 80 μM. Fluorescence microscopy following DNA-binding fluorochromes (e.g., Hoechst 33258) and electron microscopy as well showed that neurons exposed to PrP 106-126 underwent condensation of the chromatin and fragmentation of the nucleus. Such apoptotic figures were confirmed by agarose gel electrophoresis of DNA extracted from cells at 7-day exposure demonstrating DNA fragmentation following the cleavage of nuclear DNA in internucleosomal regions.[20]

In rat astroglial cultures chronically treated with PrP 106-126, both size and density of the cell processes were remarkably increased.[21] On the contrary, cultures treated with control peptides were unrevealing. Northern blot analysis showed that hypertrophy of astrocytes was associated with a prominent increase in glial fibrillary acidic protein (GFAP) transcripts. Densitometric assessment of GFAP mRNA normalized for the level of ß-actin message demonstrated that mRNA increase was dependent on peptide concentration, reaching significant levels at 10 μM and resulting in three to fivefold increase at 25 μM to 50 μM. GFAP transcripts run parallel with GFAP, as determined by Western blot analysis.[21] Furthermore, astroglial hypertrophy was associated to a slight increase in cell number at the highest peptide concentration (50 μM). The proliferation rate was found substantially increased by using serum-free culture medium for the whole extent of experiment.[22] The proliferative effect of PrP 106-126 was abolished by nicardipine, a blocker of L-type voltage-sensitive calcium channels. Microfluorimetric analysis of intracellular calcium levels showed that exposure of astrocytes to PrP 106-126 was followed by rapidly increasing cytosolic calcium concentrations, whereas exposure to the scrambled peptide was not. This phenomenon was abolished removing calcium from the medium, and was prevented by preincubation with nicardipine. These findings suggest that astroglial proliferation due to PrP 106-126 follows the activation of L-type voltage-sensitive calcium channels that causes the intracellular calcium level to raise.[22] Furthermore, abnormally high cytosol calcium levels have been detected in leukocytes exposed to the peptide. This phenomenon was prevented by the *H. pertussis* toxin acting on receptor-related G proteins involved in the cytosol calcium balance,[23] illustrating the complexity of the peptide effect on the plasma membrane. The latter was investigated through the assessment of the membrane microviscosity in several cell lines and liposomes, which was found increased up to 18-53%.[24]

In summary, our studies on GSS amyloid protein showed that this protein is an N- and C-terminal truncated fragment of PrP originating from mutant molecules. This fragment contains a sequence (i.e., residues 106-126) that is able to adopt different conformations according to microenvironment conditions, but has a high propensity to form stable ß-sheet structures. When synthesized as a peptide, the sequence PrP 106-126 is fibrillogenic and partially resistant to protease digestion; it is able to interact with the plasma membrane and to be toxic to neurons and trophic to astroglial cells *in vitro*. As the sequence 106-126 is integrated in PrP peptides that accumulate in the central nervous system of patients with prion diseases, it might influence both molecular conformation and biologic properties of disease-specific PrP isoforms and PrP amyloid.

ACKNOWLEDGEMENTS

This work was supported by the Italian Ministry of Health, Department of Social Services, by the U.S. National Institutes of Health (Grant NS29822), by Telethon-Italy (Grant E.250) and by European Community (Biomed 2, PL 960601).

REFERENCES

1. S.B. Prusiner, Molecular biology of prion diseases, *Science* 252:1515 (1991).
2. S.B. Prusiner, Genetic and infectious prion diseases, *Arch. Neurol.* 50:1129 (1993).
3. B.W. Caughey, A. Dong, K.S. Bhat, D. Ernst, S.F. Hayes, and W.S. Caughey, Secondary structure analysis of the scrapie-associated protein PrP 27-30 in water by infrared spectroscopy, *Biochemistry* 30:7672 (1991).
4. K.-M. Pan, M. Baldwin, J. Nguyen, M. Gasset, A. Serban, D. Groth, I. Mehlhorn, Z. Huang, R.J. Fletterick, F.E. Cohen, and S.B. Prusiner, Conversion of α-helices into ß-sheets features in the formation of the scrapie prion protein, *Proc. Natl. Acad. Sci. USA* 90:10962 (1993).
5. J. Safar, P.P. Roller, D.C. Gajdusek, and C.J. Gibbs, Conformational transitions, dissociation, and unfolding of scrapie amyloid (prion) protein, *J. Biol. Chem.* 268:20276 (1993).
6. B. Ghetti, P. Piccardo, B. Frangione, O. Bugiani, G. Giaccone, K. Young, F. Prelli, M. Farlow, S.R. Dlouhy, and F. Tagliavini, Prion protein amyloidosis, *Brain. Pathol.* 6:127 (1996).
7. B. Ghetti, P. Piccardo, M.G. Spillantini, Y. Ichimiya, M. Porro, F. Perini, T. Kitamoto, J. Tateishi, C. Seiler, B. Frangione, O. Bugiani, G. Giaccone, F. Prelli, M. Goedert, S.R. Dlouhy, and F. Tagliavini, Vascular variant of prion protein cerebral amyloidosis with τ-positive neurofibrillary tangles: The phenotype of the stop codon 145 mutation in *PRNP, Proc. Natl. Acad. Sci. USA* 93:744 (1996).
8. B. Ghetti, F. Tagliavini, G. Giaccone, O. Bugiani, B. Frangione, M.R. Farlow, S.R. Dlouhy, Familial Gerstmann-Sträussler-Scheinker disease with neurofibrillary tangles, *Mol. Neurobiol.* 8:41 (1994).
9. G. Giaccone, F. Tagliavini, L. Verga, B. Frangione, M.R. Farlow, O. Bugiani, and B. Ghetti, Neurofibrillary tangles of the Indiana kindred of Gerstmann-Sträussler-Scheinker disease share antigenic determinants with those of Alzheimer disease, *Brain Res.* 530:325 (1990).
10. F. Tagliavini, G. Giaccone, F. Prelli, L. Verga, M. Porro, J. Trojanowski, M.R. Farlow, B. Frangione, B. Ghetti, and O. Bugiani, A68 is a component of paired helical filaments of Gerstmann-Sträussler-Scheinker disease, Indiana kindred, *Brain Res.* 616:325 (1993).

11. F. Tagliavini, F. Prelli, J. Ghiso, O. Bugiani, D. Serban, S.B. Prusiner, M.R. Farlow, B. Ghetti, and B. Frangione, Amyloid protein of Gerstmann-Sträussler-Scheinker disease (Indiana kindred) is an 11 kd fragment of prion protein with an N-terminal glycine at codon 58, *EMBO J.* 10:513 (1991).

12. F. Tagliavini, F. Prelli, M. Porro, G. Rossi, G. Giaccone, M.R. Farlow, S.R. Dlouhy, B. Ghetti, O. Bugiani, and B. Frangione, Amyloid fibrils in Gerstmann-Sträussler-Scheinker disease (Indiana and Swedish kindreds) express only PrP peptides encoded by the mutant allele, *Cell* 79:695 (1991).

13. S.R. Dlouhy, K. Hsiao, M.R. Farlow, T. Foroud, P.M. Conneally, P. Johnson, S.B. Prusiner, M.E. Hodes, and B. Ghetti, Linkage of the Indiana kindred of Gerstmann-Sträussler-Scheinker disease to the prion protein gene, *Nature Genet.* 1:64 (1992).

14. K. Hsiao, S.R. Dlouhy, M.R. Farlow, C. Cass, M. DaCosta, P.M. Conneally, M.E. Hodes, B. Ghetti, and S.B. Prusiner, Mutant prion proteins in Gerstmann-Sträussler-Scheinker disease with neurofibrillary tangles, *Nature Genet.* 1:68 (1992).

15. G. Giaccone, L. Verga, O. Bugiani, B. Frangione, D. Serban, S.B. Prusiner, M.R. Farlow, B. Ghetti, and F. Tagliavini, Prion protein preamyloid and amyloid deposits in Gerstmann-Sträussler-Scheinker disease, Indiana kindred, *Proc. Natl. Acad. Sci. USA* 89:9349 (1992).

16. F. Tagliavini, F. Prelli, M. Porro, G. Rossi, G. Giaccone, T.D. Bird, S.R. Dlouhy, K. Young, P. Piccardo, B. Ghetti, O. Bugiani, and B. Frangione, Only mutant PrP participates in amyloid formation in Gerstmann-Sträussler-Scheinker disease with Ala>Val substitution at codon 117, *J. Neuropathol. Exp. Neurol.* 54:416 (1991).

17. F. Tagliavini, F. Prelli, L. Verga, G. Giaccone, R. Sarma, P. Gorevic, B. Ghetti, F. Passerini, E. Ghibaudi, G. Forloni, M. Salmona, O. Bugiani, and B. Frangione, Synthetic peptides homologous to prion protein residues 106-147 form amyloid-like fibrils *in vitro*, *Proc. Natl. Acad. Sci. USA* 90:9678 (1993).

18. C. Selvaggini, L. De Gioia, L. Cantù, E. Ghibaudi, L. Diomede, F. Passerini, G. Forloni, O. Bugiani, F. Tagliavini, and M. Salmona, Molecular characteristics of a protease-resistant, amyloidogenic and neurotoxic peptide homologous to residues 106-126 of the prion protein, *Biochem. Biophys. Res. Commun.* 194:1380 (1993).

19. L. De Gioia, C. Selvaggini, E. Ghibaudi, L. Diomede, O. Bugiani, G. Forloni, F. Tagliavini, and M. Salmona, Conformational polymorphism of the amyloidogenic and neurotoxic peptide homologous to residues 106-126 of the prion protein, *J. Biol. Chem.* 269:7859 (1994).

20. G. Forloni, N. Angeretti, R. Chiesa, E. Monzani, M. Salmona, O. Bugiani, and F. Tagliavini, Neurotoxicity of a prion protein fragment, *Nature* 362:543 (1993).

21. G. Forloni, R. Del Bo, N. Angeretti, R. Chiesa, S. Smiroldo, R. Doni, E. Ghibaudi, M. Salmona, M. Porro, L. Verga, G. Giaccone, O. Bugiani, and F. Tagliavini, A neurotoxic prion protein fragment induces rat astroglial proliferation and hypertrophy, *Eur. J. Neurosci.* 6:1415 (1994).

22. T. Florio, M. Grimaldi, A. Scorziello, M. Salmona, O. Bugiani, F. Tagliavini, G. Forloni, and G. Schettini, Intacellular calcium rise through L-type calcium channels, as molecular mechanism for prion protein fragment 106-126-induced astroglial proliferation, *Biochem. Biophys. Res. Commun.* 228:397 (1996).

23. L. Diomede, S. Sozzani, W. Luini, M. Algeri, L. De Gioia, R. Chiesa, P. M.-J. Lievens, O. Bugiani, G. Forloni, F. Tagliavini, and M. Salmona, Activation effect of a prion protein fragment (PrP 106-126) on human leukocytes, *Biochem. J.* 320:563 (1996).

24. M. Salmona, G. Forloni, L. Diomede, M. Algeri, L. De Gioia, N. Angeretti, G. Giaccone, F. Tagliavini, and O. Bugiani, A neurotoxic and gliotrophic fragment of the prion protein increases plasma membrane microviscosity. *Neurobiol. Dis.* in press (1997).

PROPERTIES OF THE PRION PROTEINS IN CREUTZFELDT JAKOB DISEASE PATIENTS HETEROZYGOUS FOR THE E200K MUTATION

Ruth Gabizon[1], Glenn Telling[2], Zeev Meiner[1],

Michele Halimi[1], Irit Kahana[1], and Stanley B. Prusiner[2,3]

1) Department of Neurology, Hadassah University Hospital, Jerusalem

Israel 2) Department of Neurology and 3) Biochemistry and Biophysics,

University of California, San Francisco, CA 94143-0518.

INTRODUCTION

Three groups of inherited prion diseases are recognized based on the clinical manifestations of these diseases: familial CJD Gerstmann-Straussler-Scheinker (GSS) disease and fatal familial insomnia (FFI) [1]. In Libyan Jews suffering from CJD, a point mutation which results in the substitution of K for E at residue 200 of PrP was identified [2,3]. This mutation was found in all definitely affected individuals and yields a maximum lod score of 4.85[4] The same mutation is also present in other clusters of the disease [5].

Many lines of evidence argue that the abnormal isoform of the prion protein (PrP^{Sc} or PrP^{CJD}) is necessary for both the transmission and pathogenesis of these diseases[6]. The fundamental event underlying prion propagation features a conformational change in PrP[7]. Some of the helical structures present in PrP^C is lost and replaced in PrP^{Sc} by a large amount of β-sheet structures[8]. The pathological PrP mutations may help destabilize some of the alpha-helix present in PrP^C resulting in spontaneous conversion. Studies with Tg mice led to the proposal that prion propagation involves the formation of a complex between PrP^{Sc} and PrP^C, probably in the presence of a still unidentified protein X, to form a heterodimeric intermediate which is transformed into two molecules of PrP^{Sc}[9,10].

Most patients with inherited prion diseases are heterozygous and produce, we presume, both wt and mutant PrP^C. Whether mutant PrP can be produced in some cultured cells with properties resembling those of PrP^{Sc} is today under scientific debate. In brains of the E200K mutation carriers , however, mutant PrP, although different in its glycosilation and metabolic properties, seems to be produced as PrP^C.[11,12] Whether E200K PrP^C or any other mutant PrP can function of PrP^C is unknown. This issue will no be resolved untill the function of PrP^C is elucidated and a reliable assay for such a function is produced. At some point late in life, E200K PrP^C molecules may spontaneously convert into mutant PrP^{Sc}. The spontaneously formed mutant PrP^{Sc} molecules may theoretically recruit both wt and mutant PrP to form new PrP^{Sc} molecules for the propagation of the disease in the brain. However, in most cases, only mutant PrP^C seems to transform into PrP^{Sc}. This was shown by us for the E200K mutation as well as by others for a number of pathogenic PrP [11, 13, 14] mutations.

RESULTS

PrPE200K is produced as sensitive to Proteinase K

In the brains of CJD patients carrying the the E200K mutation, the digestion pattern of the PrP protein in the presence of PK (until a concentration of 200 µg/ml) is the typical pattern of PrP^{Sc} digestion, resulting in a truncated PrP protein denominated PrP 27-30.

In order to investigate whether E200K PrP protein in other tissues of CJD patients and even in brains of carriers presents with PrP^{Sc} properties, we digested extracts of fibroblasts and EBV transformed B-cells from an homozygous E200K CJD patient, as well as a brain sample from an homozygous codon 200 individual suffering from an undefined chronic neurological disorder with 10 µg/ml PK for 30 min at 37°C. No protease resistant PrP was present in any of those samples, suggesting that even in brain, mutant PrP is not necessarily produced as PrP^{Sc} (fig1). We have previously shown that PrP in fibroblasts and monocytes of E200K CJD patients was released by PIPLC, another hallmark of PrP^{C}[12]. These results suggest that even in genetic patients, which like sporadic CJD patients and unlike scrapie infected animals develop CJD spontaneously, the mutation by itself is not enough for disease pathogenesis. An additional trigger, probably aged related is necessary to produce the conditions necessary for conversion of mutant PrP^C into PrP^{Sc}. Although the penetrance of CJD among E200K mutation carriers seems to be 100%, it is age related and probably less than 100% at a normal life span.

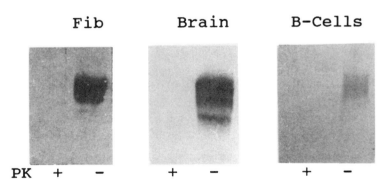

Fig 1. Immunoblots of brain, fibroblasts and B-cells of an homozygous E200K individual.

Cultured fibroblasts, brain extract and EBV transformed B-cells of an homozygous E200K individual with undefined neurological symptoms were digested in the presence and absence of 20 µg/ml PK for 40 min at 37°C. the samples were then immunoblotted with the 3F4 αPrP mAb.

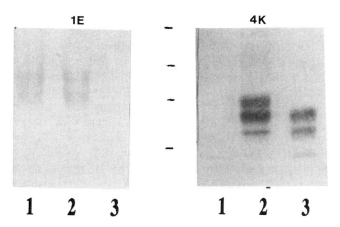

FIG 2. Immunoblot of 3F4mAb immunopurified PrP from control and CJD brain samples with 1E and 4K antisera.

200 μl of brain microsomes from a control brain(lanes 1), CJD (lanes 2) and CJD digested with proteinase K (lanes 3) brains were denatured for 10min by the addition of a total concentration of 3MGuSCN. The samples were precipitated with methanol and the pellets dissolved in TNE with 1% Sarkosyl. 10μl of ascitic fluid of monoclonal 3F4 were added for overnight incubation at 4°C. 100μl of sepharose antimouse IgG was added for 1h incubation followed by precipitation and 5 washes in TNE Sarkosyl. The immunoprecipitates were then analized by SDS page and immunoblotted with either 1E and 4K antisera. To avoid cross reaction of anti rabbit with the mouse antibody presence on the nitrocellulose paper due to the immunoprecipitation, 100μl of normal mouse serum was added during the incubation with the secondary antibody. Lane 1 normal brain, lane 2, CJD brain, lane 3 CJD brain digested with proteinase K. M.W. markers: 34.9,28.7,20.5 kDa

Only E200K PrP transforms into PrPSc in E200K patients

In order to investigate which of the PrP proteins is comprised in the PrP^{Sc} present in the brains of heterozygous E200K CJD patients, we prepared rabbit antisera to sets of peptides with sequences around the codon 200 of the PrP gene, each set with either K or E at residue 200. Two of these antisera, 1E (raised against KLH coupled to peptide GENFTETDVKMMERVVEQM) and 4K (raised against KLH coupled to peptide KQHTVTTTTKGENFTKTDVKMMER) reacted specifically with wtPrP and PrPE200K, respectively, both by immunoprecipitation and by western blotting [11]. In the experiments described bellow, the CJD brain samples used were from individuals heterozygous for the E200K mutation, while control brains were from homozygous wild type individuals.

To establish whether the proteins reacting with the new antipeptide antisera are indeed PrP proteins. PrP from a control brain sample as well as from a sample of CJD E200K brain was immunoprecipitated with α PrP mAb3F4[15], before being analyzed by SDS page and immunoblotting with either 1E or 4K antisera (Figure 2). mAb 3F4 is directed against an epitope composed of amino acids 108-111 of the hamster and human PrP gene.

In this experiment, membrane preparations of a control wt brain (lines 1) and from an heterozygous CJD (E200K) brain were solubilized in 1% Sarkosyl followed by the incubation in the presence or absence of proteinase K (10 μg/ml). After denaturation with 3M GuSCN, the proteins were precipitated with methanol and resuspended in TNE with 1% Sarcosyl prior to immunoprecipitation of PrP with the 3F4mAb. The immunopurified proteins were immunoblotted with antisera 1E (fig 1, left panel) and 4K (right panel). As can be seen in the figure, antisera 4K reacts with PrP present in the CJD brain in the presence and absence of PK but not with PrP from control brain, showing that it does not react with wt PrP. On the other hand, antisera 1E reacted with PrP from control brain as well as from brain derived from an heterozygous CJD E200K patient (which presumably contains wild type and mutant PrP) but not with PrP present in the protease digested sample of CJD. The fact that the 1E antisera did not react with PK digested PrP although it reacts with wt PrP in undigested brain shows that no wt PrP is present in PK treated brain samples of E200K heterozygous CJD patients.

DISCUSSION

We have shown that PrPE200K is not produced with PrP^{Sc} like properties in the cells of E200K mutation carriers. Although the degradation rate of the mutant PrP protein is slower, no truncated PrP; PrP27-30, could be detected even after digesting brain homogenates of E200K patients as well as homogenates of other cells with very low PK concentrations. Keeping in mind that PrPE200K in CJD patients is resistant to high concentrations of PK (> 200 µg/ml), this suggest mutant PrP undergoes a configurational change late in life of the carriers. The mechanism of this spontaneous conversion is probably age dependent since the penetrance of CJD in E200K carriers is age dependent[16]. It has been reported that mutant PrP in CHO cells is produced with properties which resemble PrP^{Sc}[17]. It will be interesting to investigate whether CHO cells metabolize PrP proteins by pathways different then those of cells which do not spontaneously convert mutant PrP^C into mutant PrP^{Sc}. PrP proteins have been shown to be present in novel lipid microdomeins denominated DIGS or RAFTS [18].

As opposed to mutant PrP^C, which converts into PrP^{Sc} and accumulates as such during the course of the disease, wt PrP in the brains of heterozygous E200K CJD remains sensitive to protease digestion[11]. Similar results were observed in other heterozygous genetic CJD patients[13, 14]. This results suggests there is some species barrier which inhibits the conversion of wt PrP to PrP^{Sc} by interaction with mutant PrP^{Sc}. It is possible that the explanation for such a barrier lies in the fact that each PrP mutation may result in a different strain of prion disease.

Development of treatments for CJD and specifically for genetic CJD may depend on a better understanding of the pathogenic mechanism that causes the disease. In particular, it is crucial to investigate and elucidate the differences in biochemical properties and metabolic pathways by which wt and mutant PrP evolve in heterozygous patients, as well as the interaction between these two proteins. The use of our new antibodies, which can distinguish wt from E200K mutant PrP, will be instrumental in this research.

REFERENCES

1.   Prusiner, S.B. Inherited prion diseases. *Proc Natl Acad Sci U S A* **91**, 4611-4 (1994).
2.   Goldfarb, L.G., Korczyn, A.D., Brown, P., Chapman, J. & Gajdusek, C.G. Mutation in codon of scrapie amyloid precursor gene linked to Creutzfeldt-Jakob disease in Sepharadic Jews of Lybian and Non Lybian origin. *Lancet* **336**, 514-515 (1990).

3.   Hsiao, K. *et al.* Mutation of the prion protein in Libyan Jews with Creutzfeldt-Jakob disease. *N Engl J Med* **324**, 1091-7 (1991).

4.   Gabizon, R. *et al.* Mutation and polymorphism of the prion protein gene in Libyan Jews with Creutzfeldt-Jakob disease (CJD). *Am J Hum Genet* **53**, 828-35 (1993).

5.   Goldfarb, L.G. *et al.* Creutzfeldt-Jacob disease associated with the PRNP codon 200Lys mutation: an analysis of 45 families. *Eur J Epidemiol* **7**, 477-86 (1991).

6.   Prusiner, S.B. Molecular biology of prion diseases. *Science* **252**, 1515-22 (1991).

7.   Safar, J. The folding intermediate concept of prion protein formation and conformational links to infectivity. *Curr Top Microbiol Immunol* **207**, 69-76 (1996).

8.   Pan, K.M. *et al.* Conversion of alpha-helices into beta-sheets features in the formation of the scrapie prion proteins. *Proc Natl Acad Sci U S A* **90**, 10962-6 (1993).

9.   Prusiner, S.B. *et al.* Transgenetic studies implicate interactions between homologous PrP isoforms in scrapie prion replication. *Cell* **63**, 673-86 (1990).

10.   Scott, M. *et al.* Transgenic mice expressing hamster prion protein produce species-specific scrapie infectivity and amyloid plaques. *Cell* **59**, 847-57 (1989).

11.   Gabizon, R. *et al.* Insoluble wild-type and protease-resistant mutant prion protein in brains of patients with inherited prion disease. *Nat Med* **2**, 59-64 (1996).

12.   Meiner, Z., Halimi, M., Polakiewicz, R.D., Prusiner, S.B. & Gabizon, R. Presence of prion protein in peripheral tissues of Libyan Jews with Creutzfeldt-Jakob disease. *Neurology* **42**, 1355-60 (1992).

13.   Tagliavini, F. *et al.* Amyloid fibrils in Gerstmann-Straussler-Scheinker disease (Indiana and Swedish kindreds) express only PrP peptides encoded by the mutant allele. *Cell* **79**, 695-703 (1994).

14.   Kitamoto, T., Iizuka, R. & Tateishi, J. An amber mutation of prion protein in Gerstmann-Straussler syndrome with mutant PrP plaques. *Biochem Biophys Res Commun* **192**, 525-31 (1993).

15.   Kascsak, R.J. *et al.* Mouse polyclonal and monoclonal antibody to scrapie-associated fibril proteins. *J. Virol* **61**, 3688-3693 (1987).

16.   Spudich, S. *et al.* Complete penetrance of Creutzfeldt-Jakob disease in Libyan Jews carrying the E200K mutation in the prion protein gene. *Mol Med* **1**, 607-13 (1995).

17.   Lehmann, S. & Harris, D.A. Two mutant prion proteins expressed in cultured cells acquire biochemical properties reminiscent of the scrapie isoform. *Proc Natl Acad Sci U S A* **93**, 5610-4 (1996).

18.   Taraboulos, A. *et al.* Cholesterol depletion and modification of COOH-terminal targeting sequence of the prion protein inhibit formation of the scrapie isoform [published erratum appears in J Cell Biol 1995 Jul;130(2):501]. *J Cell Biol* **129**, 121-32 (1995).

STRAIN VARIATION IN SCRAPIE AND BSE

Moira E. Bruce

Institute for Animal Health
BBSRC & MRC Neuropathogenesis Unit
Ogston Building, West Mains Road
Edinburgh EH9 3JF, UK

Previous studies have demonstrated that there are numerous laboratory strains of scrapie, that can be distinguished on the basis of their disease characteristics in panels of inbred mouse strains. The main criteria used in these strain typing studies are the incubation periods of the disease and the distribution of pathological changes seen in the brains of these mice, expressed in the form of a "lesion profile". Studies of scrapie strain characteristics on passage in hosts of different species or genotypes have shown that the agent contains an informational component that is independent of the host. In recent years strain typing methods have been used to explore epidemiological links between spongiform encephalopathies occurring in different species.

BSE has been transmitted to mice, directly from cattle and indirectly from experimentally-infected sheep, goats and pigs. Closely similar patterns of incubation periods and neuropathology were seen in panels of mouse strains infected from each of these species. Transmissions to mice have also confirmed that domestic cats and two species of antelope have been accidentally infected with the BSE strain of agent. However, transmissions from sheep with natural scrapie, collected either before or after the start of the BSE epidemic, have given variable results, with no individual sources resembling BSE. There has also been no overlap so far in the mouse-passaged strains isolated from BSE and scrapie. These studies have now being extended to include transmissions from recent cases of Creutzfeldt-Jakob disease where there is a possible occupational or dietary link with BSE.

BSE has been transmitted to sheep by either intracerebral or oral challenge, the incidence and incubation period of the disease depending on the PrP genotype of the sheep. The most susceptible genotype (Alanine/Alanine$_{136}$:Glutamine/Glutamine$_{171}$) was from the "negative line" of the NPU Cheviot flock, in which we see no natural scrapie. Infectivity was detected by mouse bioassay in the brains and spleens of these sheep, and was confirmed to be BSE by strain typing in a panel of mouse strains.

In contrast, infectivity in naturally BSE-affected cows has been detected only in CNS tissue; mouse bioassays of a large number of tissues, including spleen and lymph nodes have been negative.

The above results show that the BSE phenotype has been retained on passage through seven different species and, in the case of sheep, this phenotype has also been seen in transmissions from spleen. If strain determination depends on specific post-translational modification of host PrP, then these modifications must be "replicated" faithfully in PrP protein with different amino acid sequences. However, it remains possible that a separate informational molecule is involved.

TRANSMISSION STUDIES OF
FATAL FAMILIAL INSOMNIA

Katie Sidle[1], Bob Will[2], Peter Lantos[3], and John Collinge[1]

[1] Prion Disease Group,
Imperial College School of Medicine at St Mary's,
London, UK.
[2] CJD Surveillance Unit, Edinburgh, UK.
[3] Institute of Psychiatry, London, UK.

The prion diseases, which include Creutzfeld-Jacob disease (CJD) and Gerstmann-Sträussler-Scheinker disease in humans, and scrapie and bovine spongiform encephalopathy in animals, are a group of neurodegenerative diseases that are characterised histologically by spongiform neuronal degeneration, neuronal astrocytosis and the accumulation of an abnormal form of the cellular prion protein (PrP^C), PrP^{Sc}, which may form amyloid deposits (Beck and Daniel, 1987). Around 15% of human prion disease is inherited and these cases can be diagnosed by detection of pathogenic mutations in the prion protein gene *(PRNP)* (Palmer and Collinge, 1993). These diagnostic molecular markers have allowed recognition of a wider spectrum of human prion disease, encompassing cases of atypical pre-senile dementia without the characteristic histological features of CJD and GSS (Collinge et al. 1990).

Fatal Familial Insomnia (FFI) is a familial neurological disorder, characterised by a progressive untreatable insomnia with endocrine and autonomic dysfunction and motor signs (Lugaresi et al. 1986; Medori et al. 1992b; Medori et al. 1992a). The sleep disorder consists of a marked reduction or loss of both slow wave and rapid eye movement phases with patients developing complex hallucinations, followed by stupor and coma in the terminal stages. Histologically, these patients show selective atrophy of the anterior ventral and medio-dorsal thalamic nuclei and only occasionally spongiform change in the cerebral cortex. However, FFI has been re-classified as an inherited prion disease following the demonstration of PrP^{Sc} on Western blotting and a *PRNP* coding mutation at codon 178 resulting in an aspartic acid to asparagine PrP substitution (Medori et al. 1992b). Interestingly, this same missense mutation is also seen in some families, presenting with a CJD like illness (Goldfarb et al. 1991). However, although these two diseases share the same point mutation, the mutant alleles differ with respect to a common coding polymorphism at codon 129: familial CJD kindreds with the PrP asparagine 178 mutation encode valine at codon 129 on the same allele, whereas FFI kindreds encode a methionine at codon 129 of the same allele (Goldfarb et al. 1992). It is suggested that this polymorphism, known to determine genetic susceptibility to both iatrogenic and sporadic CJD (Palmer et al. 1991; Collinge et al. 1991), accounts for these phenotypic differences (Goldfarb et al. 1992).

Despite several transmission studies demonstrating transmission of CJD to experimental laboratory animals, attempts to transmit FFI similarly, including to 18 sub-human primates (Brown et al. 1994), have been negative, leading to speculation that FFI may represent a non-transmissible form of prion disease or "prion protein disease". Certainly, whilst PrP is detectable in brain homogenates from human FFI cases on Western blotting, the signals are typically much weaker than in classical CJD (Medori et al. 1992b).

We attempted to transmit two cases of confirmed FFI to two lines of transgenic mice expressing human PrP, designated Tg110 and Tg152 (Telling et al. 1994). The Tg110 transgenic line has 2-4 copies of the human PrP transgene and expresses at a level of 50% of that seen in human brain, whilst the Tg152 line has 30-50 copies and expresses at a level of 200% (Whittington et al. 1995). Brain homogenates from both FFI cases were analyzed by Western blot using anti-PrP monoclonal antibody 3F4 and PrP^{Sc} was only weakly detectable in one case.

Both FFI cases transmitted to both transgenic lines with incubation periods varying from 431-633 days (Collinge et al. 1995). The mice developed a rapidly progressive neurodegenerative syndrome, but the clinical features differed to some extent from those seen in CJD transmissions into the same transgenic lines. In particular, motor signs were milder and usually occurred only in the terminal stages. In early disease the mice appeared non-specifically unwell and became thin, dehydrated and showed loss of condition. Progression from the first onset of neurological signs to terminal stages occurred in 1-9 days (mean 3.8 ± 1.11). Age matched control transgenic mice, of both Tg110 and Tg152 lines, inoculated at the same time with PBS alone remained well >700 days post-inoculation. Neuropathological examination revealed spongiform change and severe astrocytosis but immunohistochemistry for PrP was negative. In addition, PrP^{Sc} was not detectable on immunoblots despite using a tenfold lower concentration of proteinase K than routinely used. Under these conditions, PrP^{Sc} was detectable in CJD-inoculated mouse brains a 1/300 dilution.

While an extensive body of experimental data suggests that PrP^{Sc} is the principal, if not the sole, component of the transmissible agent, recent observations have challenged this association and indicated that PrP^{Sc} and infectivity could be uncoupled in particular experimental paradigms (Xi et al. 1992). It is known that the ratio of infectious units to PrP^{Sc} molecules is around $1:10^5$ (Weissmann, 1991). For this reason the terminology "PrP*" has been suggested to designate the infectious fraction, while PrP^{Sc} refers to protease resistant PrP, which may or may not be infectious (Weissmann, 1991). It is possible that protease resistance is only a marker of infectivity; non-protease resistant forms of PrP*, protected from proteolytic cleavage by association with PrP^{Sc}, may exist. For this reason, it will be important to establish the titre of infectivity in the transgenic mice infected following inoculation with FFI, as they do not appear to have significant amounts of PrP^{Sc}. Transgenic mice expressing a mutant PrP analogous to one form of human inherited prion disease develop spontaneous and lethal spongiform neurodegeneration but also produce little or no detectable PrP^{Sc} (Hsiao et al. 1990; Prusiner, 1994). These observations suggest that PrP^{Sc} may not itself be directly neurotoxic and may be more consistent with models for prion neurodegeneration based on loss of function of PrP (Collinge et al. 1994).

References

Beck, E. and Daniel, P.M. Neuropathology of transmissible Spongiform Encephalopathies. In: *Prions: Novel infectious pathogens causing scrapie and Creutzfeldt-Jakob disease.* edited by Prusiner, S.B. and McKinley, M.P. San Diego: Academic Press, 1987, p. 331-385.

Brown, P., Gibbs, C.J.J., Rodgers Johnson, P., et al. Human spongiform encephalopathy: the National Institutes of Health series of 300 cases of experimentally transmitted disease. *Ann Neurol* 35:513-529, 1994.

Collinge, J., Owen, F., Poulter, M., et al. Prion dementia without characteristic pathology. *Lancet* 336:7-9, 1990.

Collinge, J., Palmer, M.S. and Dryden, A.J. Genetic Predisposition to Iatrogenic Creutzfeldt-Jakob disease. *Lancet* 337:1441-1442, 1991.

Collinge, J., Whittington, M.A., Sidle, K.C.L., et al. Prion protein is necessary for normal synaptic function. *Nature* 370:295-297, 1994.

Collinge, J., Palmer, M.S., Sidle, K.C.L., et al. Transmission of fatal familial insomnia to laboratory animals. *Lancet* 346:569-570, 1995.

Goldfarb, L.G., Haltia, M., Brown, P., et al. New mutation in scrapie amyloid precursor gene (at codon 178) in Finnish Creutzfeldt-Jakob kindred. *Lancet* 337:4251991.

Goldfarb, L.G., Petersen, R.B., Tabaton, M., et al. Fatal familial insomnia and familial Creutzfeldt-Jakob disease: disease phenotype determined by a DNA polymorphism. *Science* 258:806-808, 1992.

Hsiao, K.K., Scott, M., Foster, D., Groth, D.F., DeArmond, S.J. and Prusiner, S.B. Spontaneous neurodegeneration in transgenic mice with mutant prion protein. *Science* 250:1587-1590, 1990.

Lugaresi, E., Medori, R., Baruzzi, P.M., et al. Fatal familial insomnia and dysautonomia, with selective degeneration of thalamic nuclei. *N.Engl.J Med.* 315:997-1003, 1986.

Medori, R., Montagna, P., Tritschler, H.J., et al. Fatal familial insomnia: A second kindred with mutation of prion protein gene at codon 178. *Neurology* 42:669-670, 1992a.

Medori, R., Tritschler, H.J., LeBlanc, A., et al. Fatal familial insomnia, a prion disease with a mutation at codon 178 of the prion protein gene [see comments]. *N.Engl.J Med* 326:444-449, 1992b.

Palmer, M.S., Dryden, A.J., Hughes, J.T. and Collinge, J. Homozygous prion protein genotype predisposes to sporadic Creutzfeldt-Jakob disease. *Nature* 352:340-342, 1991.

Palmer, M.S. and Collinge, J. Mutations and Polymorphisms in the Prion Protein Gene. *Human Mutation* 2:168-173, 1993.

Prusiner, S.B. Molecular biology and genetics of prion diseases. *Philos.Trans.R.Soc.Lond.[Biol.]* 343:447-463, 1994.

Telling, G.C., Scott, M., Hsiao, K.K., et al. Transmission of Creutzfeldt-Jakob disease from humans to transgenic mice expressing chimeric human-mouse prion protein. *Proc Natl Acad Sci USA* 91:9936-9940, 1994.

Weissmann, C. Spongiform encephalopathies. The prion's progress. *Nature* 349:569-571, 1991.

Whittington, M.A., Sidle, K.C.L., Gowland, I., et al. Rescue of neurophysiological phenotype seen in PrP null mice by transgene encoding human prion protein. *Nature Genetics* 9:197-201, 1995.

Xi, Y.G., Ingrosso, L., Ladogana, A., Masullo, C. and Pocchiari, M. Amphotericin B treatment dissociates *in vivo* replication of the scrapie agent from PrP accumulation. *Nature* 356:598-601, 1992.

MOLECULAR,GENETIC AND TRANSGENETIC STUDIES OF HUMAN PRION DISEASE

J Collinge, J Beck, TA Campbell, M Desbuslais, I Gowland,
A Hill, S Mahal, J Meads, KCL Sidle and JGR Jefferys[1]

Prion Disease Group, Imperial College School of Medicine at St. Mary's
and Department of Neurology, St. Mary's Hospital, London
and [1]Department of Physiology, University of Birmingham

Human prion diseases occur in inherited, sporadic and acquired forms. The inherited forms are caused by coding mutations in the PrP gene; genetic susceptibility to acquired and sporadic disease is determined by a common protein polymorphism (at PrP residue 129), homozygotes being at higher risk. Recently a "new variant" of human prion disease (vCJD) has been reported in the UK and putatively linked to BSE transmission to humans. Molecular genetic analysis of *PRNP* coding and promotor sequence in eight vCJD cases has not to date provided evidence of genetic susceptibility factors with the exception that all are homozygous for methionine at residue 129 of PrP.

Transmission of prion diseases between mammalian species occurs inefficiently, limited by a "species barrier" determined in part by PrP sequence homology between inoculum and recipient and also by prion strain type. Transgenic mice expressing human PrP but not murine PrP appear to lack a species barrier to human prions. Sporadic and iatrogenic CJD cases of all three codon 129 genotypes transmit efficiently to these mice with short incubation periods usually in the range 180-220 days, irrespective of codon 129 genotype of inoculum. Inherited prion diseases that failed to transmit to primates have also been transmitted using this model. As these mice are competent to produce human PrP^{Sc} and human prions, they can be utilised for experimental assessment of the ability of animal prions to induce production of human PrP^{Sc}. Studies of the transmission characteristics of a range of human cases with widely varying clinicopathological phenotype, including vCJD, are now in progress to determine whether human prion diseases are caused by a single or a number of prion strains.

PrP null mice demonstrate abnormalities of inhibitory neurotransmission relevant to the epileptic-like discharges seen in the EEG in CJD and in scrapie-infected mice, raising the possibility that prion neurodegeneration is due, at least in part, to PrP loss of function. This phenotype can be rescued with PrP transgenes, confirming its specificity for PrP and indicating that human PrP is functional in mouse brain. Further neurophysiological abnormalities include altered seizure thresholds and an absence of late afterhyperpolarisation in CA1 pyramidal cells in hippocampal slices. Loss of PrP^C in an adult nervous system may have more severe effects than in a developing one, where compensatory adaptation can be expected at many levels during neurodevelopment. Such a mechanism for prion neurodegeneration is more consistent with the lack of correlation between neuropathological features (including the amount of PrP amyloid) and clinical severity in these diseases.

MOLECULAR BIOLOGY OF PRION PROPAGATION

Stanley B Prusiner, Kiyotoshi Kaneko, Martin Vey,
Glenn Telling, Michael Scott, Ruth Gabizon,
Albert Taraboulos, Stephen J De Armond and Fred E Cohen

University of California, San Francisco, CA 94143
and Hadassah Medical Centre, Hebrew University, Jerusalem, Israel

Prions are novel pathogens and are distinct from both viroids and viruses. They are composed largely, if not entirely, of the scrapie isoform of the prion protein (PrP) designated PrP^{Sc}. Prions cause neurodegenerative diseases including scrapie of sheep, mad cow disease, and Creutzfeldt-Jakob disease (CJD) of humans. That prion diseases are manifest as genetic, infectious and sporadic illnesses is unprecedented. The conversion of PrP^C into PrP^{Sc} is a post-translational process involving a profound conformational change which is the fundamental event underlying the propagation of prions. PrP^C contains -40% α-helix and virtually no ß-sheet; in contrast, PrP^{Sc} has -30% α-helix and -40% ß-sheet. These data argue that the conversion of α-helices into ß-sheets underlies the formation of PrP^{Sc}. Efficient formation of PrP^{Sc} requires targeting PrP^C by glycosylphosphatidyl inositol (GPI) anchor to a caveolae-like membrane domain (CLD) which is detergent insoluble and enriched for cholesterol and glycosphingolipids. Redirecting PrP^C to clathrin-coated pits by creating chimeric PrP molecules with four different C-terminal targeting domains prevented the formation of PrP^{Sc}. To determine if these C-terminal transmembrane segments prevented PrP^C from refolding into PrP^{Sc} by altering the structure of the polypeptide, we fused the 28 amino acid C-termini from the Qa protein. Two C-terminal Qa segments differing by a single residue direct the trans-membrane protein to clathrin coated pits or the BPI form to CLDs. The CLD targeted PrP^C was converted into PrP^{Sc} while the transmembrane PrP^C was not. Transgenic (Tg) mice expressing human (Hu) prion protein (PrP) and chimeric Hu/mouse (Mo) PrP genes were inoculated with brain extracts from humans with inherited or sporadic prion disease. Although Tg(HuPrP) mice were resistant to Hu prions, they became susceptible upon ablation of the MoPrP gene ($Prnp^{0/0}$). Mice harbouring the chimeric transgene were susceptible to Hu prions and registered a modest increase in susceptibility with the $Prnp^{0/0}$ background. Amino acid mismatches between PrP^{Sc} in the inoculum and PrP^C encoded by the transgene at residues 102 and 129 resulted in delayed onset of CNS dysfunction; whereas, a mismatch at either position 178 or 200 did not. Our studies suggest that the ß-sheet of PrP^{Sc} binds to the first and second putative α-helices of PrP^C while the PrP^C/PrP^{Sc} complex binds to a macromolecule provisionally designated protein X through residues of PrP^C at the C-terminal end of the fourth putative α-helix. The formation of PrP^{Sc} seems to be a specific, tightly regulated process mediated by protein X which might function as a molecular chaperone in the refolding of PrP^{C} into PrP^{Sc}. Studies of two prion strains with similar incubation times in mice exhibited distinct incubation periods and different neuropathological profiles upon serial transmission to transgenic (Tg) mice expressing chimeric (MH2M) PrP genes, and subsequent transmission to Syrian hamsters (SHa). After transmission to SHa, the Me7 strain was indistinguishable from a previously established SHa strain Sc237, despite being derived from an independent ancestral source. This apparent convergence suggests that prion diversity may be limited. The Me7 mouse strain could also be transmitted directly to SHa, but when derived in this way, its properties were distinct from Me7 passaged through Tg(MH2M) mice. Prion strains seem to be generated and maintained by the sequence of the PrP substrates that are converted into PrP^{Sc} which is the only known component of the infectious prion; thus, strain specific characteristics appear to be propagated through PrP interactions. Our subcellular targeting studies argue that the formation of PrP^{Sc} occurs within a specific subcellular compartment and, as such, it is likely to involve auxiliary macromolecules such as protein C that are found within DIG micro-domains. The unprecedented mechanism of prion propagation promises to provide important insights into the aetiologies and pathogenesis of the more common neurodegenerative diseases.

ULTRASTRUCTURAL AND IMMUNOCYTOCHEMICAL STUDIES ON PRION PATHOGENESIS

Lajos László

Department of General Zoology,
Eötvös University, Budapest, Hungary

The main pathological hallmarks of transmissible spongiform encephalopathies or prion diseases are: accumulation of membrane bound vacuoles within neuronal processes, mainly dendrites; dystrophic neurites containing aggregations of neurofilaments, mitochondria and electron-dense, multi-lamellar bodies; and hypertrophic astrocytes. An enhanced endosomal-lysosomal activity and extensive neuronal autophagy can be seen in experimental models of scrapie and CJD *in vivo*.

Currently available scrapie prion protein (PrP^{Sc}) expressing cell lines originate from the peripheral nervous system, and do not show significant cytopathological changes related to spongiform degeneration. We produce a prion-infected, well differentiated hypothalamic cell line which secretes gonadotropin hormone releasing hormone and was transfected with the trk-A gene encoding the NGF-receptor (ScGT1-1-trk9). These cells persistently infected with mouse RML prions show intense vacuolation and characteristic features of programmed cell death. Prion-infection caused a prominent expansion (over 10-fold) of autophagic-lysosomal compartments in ScGT1-1 cells. The appearance of giant autophagic vacuoles (never can be seen in control cells) indicate focal cytoplasmic degeneration and isolation of injured cytoplasmic areas and organelles. Fine ultrastructural analysis shows that the mitochondrial system might be the main target for this selective segregation and elimination by macroautophagy in scrapie-infected neurones. The significant loss of mitochondria may cause energy depletion and ultimately cell death. The extremely elevated autophago-lysosomal activity may be related to a disturbance of intracellular protein catabolism during the production and/or processing of PrP^{Sc}. Long-term NGF treatment increased the viability of ScGT1-1-trk cells and reduced vacuolation and morphological signs of apoptosis. However, on the basis of our ultrastructural findings alone, it is unclear whether the expansion and dysfunction of the endosomal-lysosomal system, and the possible abnormalities in energy metabolism are playing a primary 'causative' role in neurodegeneration and cell death, or are merely secondary correlates to damage caused other pathological mechanisms, eg oxidative stress.

Immunoelectron microscopic analysis on the distribution of PrP in scrapie-infected mouse brain and cultured cells demonstrate clusters of gold particles corresponding to

PrP not only on the surface of the cells, but in membrane invaginations, small intracellular vesicles, early and late endosomes and secondary lysosomes. The presence of PrP in the subcompartments of endosomal-lysosomal system of scrapie-infected cells may indicate that these structures play a substantial role in the processing, accumulation, 'enzymatic purification' and moreover 'secretion' of scrapie agent.

Our ultrastructural and immunocytochemical results on prion diseases support the notion that these studies may provide further insight into the cell biology and pathogenesis of neurodegenerative diseases including Alzheimer's disease, Parkinson's disease and motor neurone disease as well.

Note: Detailed description of the new results on the scrapie infected hypothalamic cell line established in Stan Prusiner's laboratory can be found in the November 1997 issue of *J Virology*: Schatzl H.M., Laszlo, L., Holtzman, D.M., Tatzelt, J., DeArmond, S.J., Weiner, R.I., Mobley, W.C., and Prusiner, S.B. (1997) A Hypothalamic Neuronal Cell Line Persistently Infected with Scrapie Prions Exhibits Apoptosis. *J. Virology* **71**: 8821-8831.

OVERVIEW OF THE BSE EPIDEMIC

John Wilesmith

Epidemiology Unit
Central Veterinary Laboratory
New Haw, Weybridge
Surrey KT15 3NB, UK

Bovine spongiform encephalopathy (BSE) was first recognised in Great Britain in November 1986. Epidemiological studies suggested that the first cases occurred around April 1995 and that the most likely vehicle of a scrapie-like agent was meat and bone meal, used as a protein supplement in cattle feedstuffs. Subsequent case-control studies and the effects of the legislative action, enacted in July 1988, prohibiting the feeding of ruminant derived protein to ruminants have provided strong supporting evidence that the feed borne source has been responsible for the majority of cases of BSE. The occurrence of cases of BSE in animals born after July 1988 has stimulated a number of epidemiological studies. These have revealed that incomplete compliance with the specified bovine offal ban, introduced for animals in September 1990, and various means of cross contamination of cattle feedstuffs with meat and bone meal or products containing this material were the main reason for the continued exposure of the cattle population, albeit at a very reduced rate. In addition, interim analyses of a cohort study, initiated in July 1989 specifically to examine the risk of maternal transmission, have recently been possible. The results of these which indicate a low risk of maternal transmission will be discussed. However, the main finding of the epidemiological studies is that the epidemic of BSE is rapidly declining and the incidence will be insignificant at the turn of the millennium even if no further control measures are instigated.

ASSESSING RISKS OF BSE TRANSMISSION TO HUMANS

John Collinge

Prion Disease Group
Imperial College School of Medicine at St. Mary's
St. Mary's Hospital, London, UK

The prion diseases or transmissible spongiform encephalopathies are a group of neurodegenerative conditions affecting humans and animals which are transmissible between mammalian species by inoculation with, or in some cases by dietary exposure to, infected tissues. The transmissible agent or prion consists principally of a modified form of a host encoded glycoprotein, the prion protein (PrP). This disease related isoform, PrP^{Sc}, differs from the cellular isoform, PrP^{C}, by a post-translational modification which appears to be conformational rather than covalent. The epidemic of a novel animal prion disease, BSE, in the UK has led to fears of transmission to humans through dietary exposure to beef or beef products. Recently, the occurrence of cases of sporadic Creutzfeldt-Jakob disease (CJD) in young people has caused renewed alarm and these unusual cases appear to represent a novel form of human prion disease. A causal link with BSE is unproven but remains a likely explanation. Transmission of prion diseases between species is usually inefficient and associated with extremely prolonged incubation periods at primary passage, the so-called "species barrier". The effectiveness of the species barrier between cattle and humans is unknown but the ability of bovine PrP^{Sc} to induce conversion of human PrP^{C} can be studied in transgenic mice expressing human PrP, which lack a species barrier to human prions,. These mice can also be utilised in strain typing studies to try and identify whether the "new variant" CJD cases arose from exposure to bovine prions.

Prions and Brain Diseases in Animals and Humans
Edited by Morrison, Plenum Press, New York, 1998

HUMAN ACTIVITIES ARE CAUSING SELECTION OF PATHOGENIC AGENTS

Luc Montagnier

Fondation Mondiale Recherche
 et Prevention de Sida
Paris, France

Human activities are causing selection of pathogenic agents in various ways:

- transmission by sex or blood, leading to long lasting infections
  (for example: HIV)
- transmission of air-borne agents (crowding in megapoles, chemical pollution)
- heat and detergent treatments selecting resistant microorganisms (spores)

The BSE inducing agent may fall into the latter category.

Although there is strong evidence that brain pathology is caused by self-transformation of the prion protein, there remains a possibility that an exogenous agent can trigger the whole process, particularly in oral transmission of the disease. Therefore there is a need to develop a strong research effort along this line, to isolate and characterize the agent and use it as a diagnostic tool.

The action of such an agent on neurones may be indirect, so that it does not need to be present in these cells. An example of such a situation is given by the encephalopathy and encephalitis caused by HIV.

Another example will be described: it is the protein oxidation induced in blood platelets by an as yet unidentified, tetracyclin sensitive factor.

PATHOGENIC SIMILARITY OF SLOW INFECTIONS, INDUCED BY PRIONS AND VIRIONS

V A Zouev

Gamaleya Research Institute
  for Epidemiology and Microbiology
Moscow

It is known that one of the main signs of pathogenesis of subacute spongiform encephalopathies is the development of status spongiosis in the grey and/or white CNS matter. Such a picture is the result of the action of prion protein.

We observed the spongiform encephalopathy among a progeny born of mice intranasally infected with influenza virus A during pregnancy. A small part of this progeny showed three weeks after the birth signs of disease marked by progressively retarded growth and weight motor uncoordination, a wadding gait and increasing emaciation. Disease was invariably fatal.

Pathohistological changes in such mice were characterised by primary degenerative changes in different organs, particularly expressed in the CNS, including the signs of spongiosis in the white matter of the truncal part of the cerebrum and the cerebellum. In some instances there was a motoneuronal loss in the spinal cord accompanied by pronounced signs of spongiosis in the white matter of the spinal cord.

Later a similar pathological picture was demonstrated in a progeny born of a female mice that had received intravenously the inactivated influenza virus or neuraminidase preparation in high concentration.

On the base similarity of the damage actions of free (PrP) and structural (Viral neuraminidase) proteins expressed supposition about the origin of the PrP.

PROTEIN FOLDING AND MISFOLDING

A. R. Clarke

Dept of Biochemistry
University of Bristol
School of Medical Sciences
University Walk, Bristol BS8 1TD, UK

At one level the prion diseases will be better understood when we have an adequate description of the structural rearrangements which underlie conversion of the cellular form of PrP to the infectious agent. The recent elucidation of the structure of soluble, recombinant PrP by NMR marks an important first stage in this aim and will focus attention on the dynamics of folding of this molecule, in vitro.

The process of misfolding of PrP to generate the scrapie agent, or particle, is likely to share mechanistic properties with the misfolding and aggregation events in other proteins and has prompted an interest in cellular factors which may assist in the folding of PrP and also in its delivery to the target site in the cell. In this respect it is constructive to assess how our understanding of the spontaneous folding of some experimentally accessible model proteins, in vitro, has developed with regard to the structure and stability of intermediate states and the nature of rate-limiting steps in the pathway. Alongside these studies, attempts to define the mechanism of action of chaperonins, which assist in locating the native structure amongst the colossal number of alternatives, has stimulated interest in the molecular mechanism of unproductive folding and in the physical structure of misfolded states.

As a general background to these issues and their relation to prion diseases, the competition between folding and misfolding of a chaperone-dependent, imported mitochondrial protein, malate dehydrogenase, will be discussed with emphasis on the following: (1) the molecular events in productive and unproductive pathways, where they diverge and how they compete; (2) the influence of chaperonins on the relative rates of these processes and therefore on the yield of the native state and (3) the nature of the committed step in productive folding and on the properties of well populated intermediate states.

Prions and Brain Diseases in Animals and Humans
Edited by Morrison, Plenum Press, New York, 1998

CJD RISK FACTORS:
ANALYSIS OF 104 PATIENTS

Eva Mitrova

Ambassador Extraordinary and Plenipotentiary
 of the Slovak Republic
Stefania ut 22-24, H-1143 Budapest, Hungary

The first, hypothetical explanation of the codon 200 mutation of the PrP gene in Slovak clustering CJD patients (1991) presumed a genetically increased susceptibility to exogenous (possibly zoonotic) infection. Later, worldwide-found patients with codon 200 mutation of PrP gene became generally accepted as "familial CJD", regardless of certain non-explained observations, as an incomplete penetrance of the disease or phenotype, identical with sporadic CJD. To challenge this view a re-evaluation of data on CJD in Slovakia (annual occurrence of CJD/mill./year since 1975 varied from 1.2 - 0.8 - 2.0 - 1.0) has been done. A total of 104 patients, 65 with and 39 without kodon 200 mutation, have been analysed and compared from the histopathological, epidemiological and molecular biological point of view. Special attention has been paid to patients with long duration of clinical stage and age at death below 40, to a professional risk, as well as to codon 129 polymorphism. Obtained results are discussed in relation to the possible role of exogenous risk factors.

CONCLUDING TALK:
PRIONS FROM A PHYSICIST'S VIEWPOINT –
IS THE 'PROTEIN ONLY' HYPOTHESIS CORRECT?

Douglas R.O. Morrison

CH-1296 Coppet
Switzerland

ABSTRACT
The talk considers prion science from a physicist's standpoint. Arguments for and against the "Protein Only" hypothesis are reviewed, the result of a vote is given, and a judgement is made in favour. The origin of the BSE epidemic is discussed and it is concluded it did not originated with sheep but with cows - previously called "staggers". The unusually long incubation period is used to favour a replication mechanism based on the necessity of forming a seed of critical size, and on the importance of concentrations. It is suggested that sporadic diseases, e.g. CJD, are the result of a series of random transformations and then a slow accumulation. The age-dependency of the new variant of human Creutzfeldt-Jakob Disease is shown to be significantly different from sporadic CJD and probably reflects eating habits. It is pointed out that the minimum incubation period is about 10 years which is twice the 5 years for kuru. Comments are made on the future epidemic of the nvCJD epidemic. The roles of mutations and polymorphisms which are considered to be natural, are discussed. This talk has been updated to October 1997.

SUBJECTS
1. Introduction
2. List of Questions
3. Is the "Protein Only" Hypothesis Correct?
4. Protein Only Hypothesis is Several Hypotheses
5. Mutations and Polymorphisms - List for Humans
6. Origin of BSE - Sheep or Cows?
7. Replication Mechanism - Why is the Incubation Period So Long?
8. Mechanism of Sporadic Prion Disease - Why is the Average Age so High?
9. Age Distribution of New Variant of CJD Cases
10. New Variant of CJD - Possible Future
11. Is nvCJD an Appropriate Name for the new Human Bovine Disease?
12. Drugs and the Need for Simple Early Detection
13. Prion Proteins Purpose? The Brain
14. Some Highlights
15. Claimed Link of Farmers and CJD
16. Final Discussions and Conclusions
17. Overall Conclusions to Workshop

Prions and Brain Diseases in Animals and Humans
Edited by Morrison, Plenum Press, New York, 1998

1. INTRODUCTION

Although this is the concluding talk, it is not a full summary of the meeting - rather it is one physicist's perception. A more objective summary of the talks is given in the Introduction at the beginning of the book.

Here a series of questions are posed and a number of points are made from a personal standpoint. A few of the highlights of the meeting are described which reflect a personal choice which is probably different from those that would be chosen by an expert on prions. A major subject is whether the "Protein Only" hypothesis is correct. It was a surprise to find that several participants at this prion meeting, did not believe in the prion hypothesis and thought a nucleic acid was the infectious agent. There was an anonymous vote on this. This is discussed and a comparison of the similar difficulties that the Quark hypothesis had to face, are described.

2. LIST OF QUESTIONS

1. Is the "Protein Only" Hypothesis Correct?

2. How does the abnormal prion protein, PrP^{sc}, convert the normal cellular prion protein, PrP^c, to an infectious, PrP^{sc}, form?

3. Why does it take 15 years to go from the stomach to the brain? Why is the Incubation period so long?

4. Is the infectious form PrP^{sc}, or is it the residue PrP-res?

5. How many species have prion proteins?

6. Are prions only one example of a new general class of diseases which do not require nucleic acids?

7. What does the prion protein do?

8. What is the role of aggregates? Is there a threshold?

9. Is Sporadic CJD the result of a mutation?

10. Is a polymorphism just a partly established mutation?

11. Strains?

12. Did BSE come from sheep (scrapie) or from cows?

13. Is the age distribution of the new variant of CJD, nvCJD, different from that of sporadic CJD? Why?

14. Is there any connection between Alzheimer's Disease and prion diseases?

3. IS THE "PROTEIN ONLY" HYPOTHESIS CORRECT?

Before this Erice meeting, my reading of the recent literature had shown many experiments which broadly supported the "Prion Only" hypothesis, or more exactly, the "Protein Only" hypothesis, though I was made aware of contrary opinions. So it was a surprise at Erice to find a group of people at this prion conference who did not believe in prions!

3.1 Vote on Hypothesis

At the suggestion of Charles Weissmann, an anonymous vote was held just before this talk where people were asked to express their belief in the "Prions Only" hypothesis as a percentage. The original graph of the result is shown in fig. 1. Three groups can be seen;

a) 0 to 10% confidence - 5 votes

b) Mildly in favour, 50 to 80% - 6 votes

c) Strongly in favour, 90 to 110% - 15 votes.

(One voter gave 110% saying the hypothesis had been proved over and over again! Stan Prusiner said he wished he had thought of that).

Previously the Conventional Wisdom was that all diseases replicated using nucleic acids - DNA, RNA. However very early it was found that the

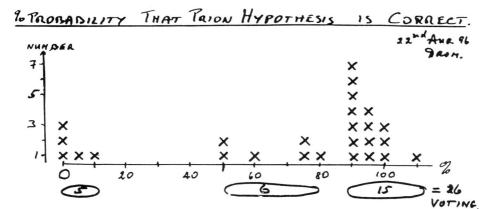

Fig.1. Original graph of the results of a vote as to the percentage belief of participants in the "Protein Only" Hypothesis.

infectious agent of Transmissible Spongiform Encephalopathies, TSE, was extremely small and was remarkably resistant to destruction by chemicals, heat or radiation. In particular, it was still infectious after exposure to UV light of about 260 nm which is a frequency especially effective in destroying nucleic acids. This led Alper et al.[1] in 1967 to propose "Does the agent of scrapie replicate without nucleic acid?"

This mechanism was further described theoretically, also in 1967, by Griffiths[2] who was a mathematician, but who then made no further contribution.

But after 1967, there was a noticeable absence of papers suggesting an absence of nucleic acids, though there was private discussion, Fifteen years later, in 1982, Stan Prusiner wrote[3] that the agent was a protein and no nucleic acid was needed. Further he named the agent as a prion protein. He proposed the "Prion Only" hypothesis where an abnormal prion, PrP^{sc}, could transform a chemically identical naturally occurring cellular prion protein, PrP^c, into the infectious agent PrP^{sc} which could then transform further normal PrP^c into agents. Thus this was a new method of replication.

There was considerable disbelief, even ridicule, of Stan's ideas but as he explains in his talk, more and more experimental results appeared to support the hypothesis, and a large majority now believes in prions.

3.2 Arguments against and for the Hypothesis

However some believe that the agent contains nucleic acids and is a slow virus or a very small virus called a virino in analogy with a small neutron being called a neutrino (in fact a neutrino is not a small neutron - they are very different particles as a neutron is a hadron which is a strongly interacting object which is composite containing quarks etc., while a neutrino is considered to be a fundamental particle which is leptonic, that is, weakly interacting).

The arguments in favour of a slow virus or virino seem to be of two kinds. Firstly, by showing that the evidence in favour of prions can also be interpreted as evidence for a nucleic acid. Thus for the small size of the agents, some viruses have been found which are equally small. And when experiments have shown evidence for prions, frequently Heino Diringer has gently managed to get the experimenter to admit that the agent could not have been purified enough to exclude the presence of nucleic acids - it is very difficult to purify completely. However Detlev Riesner has stated that in his experiment, the amount of residual nucleic acids are too small to give the results observed. In addition in this book, several accounts are given of synthetic peptides equivalent to part of the prion protein, e.g. from codon 106 to 121, behaving like as a Transmissible Spongiform Encephalopy, TSE, agent - and in this case, there is no nucleic acid present.

Secondly, the disbelievers claim that there is positive evidence which the "Protein Only" hypothesis cannot reasonably explain. The main example of this is the question of strains of agents. In sheep about 20 distinct strains of scrapie have been identified. If several different strains of the abnormal agent are introduced into one type of sheep, then it is found that the infectious agent produced in the sheep has taken the characteristic strain of the invader rather than the host even though there are few invaders and many host prions in each sheep. This is claimed to be unreasonable and contrary to the "Protein Only" hypothesis. However Charles Weissmann et al.[4] point out that there are at least two hypotheses that would explain this result; (a) the Conformational hypothesis which suggests that PrP-res converts PrP-sen into a form that takes the characteristics of PrP-res so that each strain reflects the features of PrP-res. The experimental justification of this hypothesis is given from experiments with drowsy and hyper hamsters; (b) the Targeting hypothesis proposes that the cells have a variety of forms and each invading PrP-res carries a special modification which targets a subset of the cells. These cells would then produce PrP-res with the same modification. Again some experimental evidence is quoted in favour of this hypothesis.

In conclusion, this strain question does not appear such a strong argument.

The evidence which first convinced many, came from Charles Weissmann and his colleagues[5] in Zurich who created mice which had no prions at all! These are transgenic mice, called knock-out mice. When the infectious agent was introduced into the k-o mice, no infection develops because the abnormal PrPsc cannot find any normal PrPc to transform. At this meeting Adriano Aguzzi reported experiments with k-o mice where grafts containing PrPsc were placed in different regions - the extensive results were consistent with the expectations based on the "Protein Only" hypothesis.

3.3 Comparison with the Quark Hypothesis

In 1964 Zweig at CERN and Gell-Mann at Caltech made the revolutionary proposal of the "Quark Hypothesis" that protons and neutrons were not fundamental particles but were composed of smaller particles called quarks which had fractional electrical charge, 1/3 and 2/3 of the electron charge.

This new theory allowed many calculations and predictions to be made fairly easily and of the predictions verifiable, all turned out to be correct. However there were a strong group of theorists who did not believe that these new quark particles were necessary and they were able to repeat the calculations and finally obtain the same predictions - but some time later. This situation continued for many years, the quark believers making new and often surprising predictions and these were confirmed experimentally, and then the disbelievers calculating the same result - very frustrating for the quark believers.

Finally in 1973 decisive evidence was found for a new quark called charm and then all disbelieve ceased. The situation with the Prion Only hypothesis has some resemblance.

3.4 Conclusion on the Protein Only Hypothesis

The sheer weight of evidence in favour is very convincing. A complex structure has been built up depending on the hypothesis and it works and makes predictions that have been verified. The contrary evidence seems unconvincing and generally has an alternative explanation in terms of prions - and no virus or virino has been identified despite extensive searching. The fact that as Charles Weissmann showed, mice with no natural prions do not get ill when infected, is very strong evidence as is the evidence from Yeast where there probably is no unknown virus.

The overall conclusion is that the "Protein Only" hypothesis is almost certainly correct and should be preferentially used unless spectacular new evidence is produced.

4. "PROTEIN ONLY" IS SEVERAL HYPOTHESIS

It is probably more accurate and more general to talk of "Protein Only" rather than "Prion Only".
It is helpful to consider the "Protein Only" hypothesis as three hypotheses
4.1 Infectious Agent consists of Proteins only
This is the original basic hypothesis
4.2 Replication Mechanism

It is assumed that replication occurs by conformal transformation of the normal cellular PrP^c protein by an abnormal PrP^{sc} protein. But then it is hard to understand the extremely long incubation period. A third component has been suggested which could be a chaperone which helps in protein folding. It has been called PrP* or PrPX. But as discussed below it is probably more complicated.

4.3 PrP-res as Infectious Agent

It is assumed that all mammals have prion proteins naturally - Stan called them PrP^c. He named the infectious agent PrP^{sc} where the "sc" comes from scrapie in sheep which at that time was one of the best known TSE agents. Now there is a strong tendency to replace PrP^c and PrP^{sc} with the more general names PrP-sen and PrP-res respectively.

Normally natural detergents break down the prion protein. Thus PrP^c is dissolved by (is sensitive to) proteinase K, PK, and is named PrP-sen. On the other hand when PrP^{sc} is treated with PK, residues remain, i.e. it is resistant to proteinase K, and these residues are named PrP-res. At first they were named PrP 27-30 because first measurements gave the mass of PrP-res as 27 to 30 kDaltons (where one Dalton is an atomic mass unit) while the original PrP^c has 35 kDa. Many measurements since have shown that there is a rich spectrum of masses of the residues going down to 7 kDa. In the work of Collinge et al.[6], use of the Western Blot analysis showed for nvCJD, three bands of very roughly 22, 26 and 30 kDa (the 3 numbers fluctuate considerably and are not given with errors) corresponding to the addition (glycosylation) of zero, one or two sugars respectively.
It has been found that PrP-res is an infective agent. This is reasonable as it allows replication of the disease.

While the exact mechanism by which PrP-res transforms PrP-sen into an infectious form is unclear, some parts of the prion protein molecule must be more important than others since PrP-res has fewer kDaltons than PrP-sen.

5. MUTATIONS AND POLYMORPHISMS - LIST FOR HUMANS

5.1 Mutations are a Fact of Life - Age of Disease

Mutations are occurring all the time. They are the cause of new species. This is used to estimate the age of species by counting the number of mutations and then, since the rate of mutations is known, the age can be derived.

In sheep significant mutations in prions are empirically called strains when they yield different 'lesion profiles'(see section 6.6.7). Since there are about 20 strains of scrapie, it may be deduced that scrapie is an ancient disease. Kuru appears to have only one strain and hence it may be considered as a recent disease in agreement with Fore tribal memories indicating a year of origin of about 1900 to 1920.

Since BSE has only one strain even when passaged through other animals (sheep, pig, cat, etc.), it must also be considered a recent disease.

5.2 Polymorphism

If a mutation ceases to be rare and becomes important, there will exist a period of time in which the old and the new amino acids can co-exist as alternatives. This is the probable explanation of polymorphisms, for example at codon 129 either Methione or Valine can exist.

5.3 Mutations Correlate with Disease

In the papers presented to this meeting, many examples have been given where a specific mutation in the prion correlates with a specific set of symptoms, changing for example, the incubation period, clinical or pathological symptoms.

This indicates that the specific mutation may be involved in the mechanism of replication of the disease. This supports the Protein Only hypothesis

5.4 Double correlation

The polymorphism at codon 129 with human prions which gives the possibility of a Methione, M, or a Valine, V, is very important. The maternal and paternal alleles provide 37% MM, 52% MV, and 11% VV. It was reported by Bernardino Ghetti that with GSS in the Indiana kindred, the dominant mutation was a Serine for Phenylalanine at codon 198 but all the patients had a Valine at codon 129.

This suggests that the mechanism of replication should involve both codons 129 and 198 simultaneously, and also both the paternal and the maternal alleles. Then the mechanism of replication chosen should be capable of explaining these multiple requirements.

5.5 How Many Strains in Humans?

The best studied prion protein is the human one. New strains are continually being found (one was reported at this meeting by Andrea LeBlanc) and it is possible at least another one will come from the disease giving BSE.

At this meeting there were mentioned ten mutations (Pro102Leu, Pro105Leu, Ala117Val, Tyr145Stop, Asp178Asn, Thr183Ala, Phe198Ser, Glu200Lys, Val210Ile, Gln217Arg), two Polymorphisms (129 M/V and 219 E/K), and deletions and insertions in the N-terminal region. It has been found that different combinations can result in different diseases - the 178 mutation gives FFI when on the same allele, codon 129 is Methione, and gives Famial CJD when 129 is valine. Thus potentially there are many possible combinations - e.g. 10 mutations and 2 polymorphisms could give 2^{12} = 4096 double combinations. But the fact that fewer are found suggests that not all combinations are possible.

Note added; In Science of 10 October 1997, Stan Prusiner says there are 20 different mutations in the human prion gene (this does not include Tyr145Stop), so that even more combinations are possible.

Another possibility is that the same chemical prion protein could exist in two different geometrical shapes (or forms). One example of this has been found so far - Hyper, HY, and Drowsy, DY, hamsters. Although the amino acid sequences are identical, the illnesses are different.

It is concluded that there is potentially many strains and possibly more will be discovered in time.

6. ORIGIN OF BSE - SHEEP OR COWS?

6.1 Proof - "It has not been proved that ..."

Karl Popper made the point that you cannot prove anything, only disprove it. Therefore the phrase "It has not been proved that ..." is unscientific and can mislead non-scientific people. The correct approach is to give a probability, for example, "The hypothesis that the new variant of CJD comes from Bovine Spongiform Encephalopy, BSE, has a 99% probability of being correct". If one should be opposed to this statement, it is misleading to say only "The nvCJD coming from BSE hypothesis has not been proved". Rather it would be not incorrect to write "The nvCJD from BSE hypothesis has not been proved but it is the only reasonable hypothesis and there is a 99% probability that it is correct" - for scientists should give <u>all</u> the information and not partial information.

But for scientists and for non-scientists, it is better to avoid the phrase "It has not been proved that ...", unless the probability is also given.

6.2 Agreed facts

1980 - in meat processing/rendering plants, the temperature was reduced and in many the solvent strength was also reduced. This had previously been done in the States and lowers costs.

About 1980 - sheep and cattle remains(offal) were added to animal feed which was given to cows and many other animals species. This had previously been done in the States and raises profits.

April 1985 - first cases of BSE observed.

1986 - John Wilesmith made an epidemiological study and deduced that BSE was caused by eating the changed animal feed.

6.3 "Changes in Rendering allowed Scrapie to pass" hypothesis

It was hypothesised that the old temperature and solvent used destroyed scrapie but the new lower temperature and lower solvent strength, did not kill the scrapie agent which could then infect sheep and cows.

6.4 "BSE came from Sheep" Hypothesis

Since sheep have a long-established prion disease, scrapie, and parts of sheep had been fed to cows, then it was a reasonable hypothesis that BSE came from scrapie. But should one not expect a scrapie epidemic in sheep?

6.5 "British Beef is Safe" Hypothesis

Many studies have failed to show any ill effects from eating sheep with scrapie. If BSE in cows came from them eating sheep some of whom had scrapie, then it was reasoned that this derived disease, BSE, would also not cause any illness in humans, i.e. "British Beef is Safe" for humans to eat - note beef is reasonably safe - it is the food containing offals that was dangerous.

6.6 "BSE came from Cows" Hypothesis

Humans are now starting to die from a new variant of a prion disease named nvCJD, for which the only reasonable explanation is that they caught it from eating contaminated parts of cows with BSE. It is therefore time to re-consider the hypothesis put forward earlier by some, that BSE is a new prion disease originating in cows and not in sheep.

6.6.1 Test of Rendering Hypothesis

Prosper De Mulder, the leading company that processes/renders carcass meat, helped to carry out experiments to test if the hypothesis was true that lowering the temperature and/or lowering the amount of solvent employed, then allowed scrapie to be passed whereas before it was destroyed by the old processing. They found that the scrapie prion was very resistant and that the changes made no differences. Thus in 1995, this key hypothesis was tested and found to be wrong[7].

6.6.2 Limits of Epidemiology

Epidemiology can show correlations which give possible indications of an effect. But then further scientific research is needed to establish the correctness of these indications. For example, statistics over many years showed that there was a correlation between the 11-year sunspot cycle and the number of Republicans in the US Senate and also the occurrence of influenza outbreaks in the Soviet Union - but further data showed these correlations were accidental. Thus the hypothesis that changes in rendering techniques allowed scrapie in sheep to cause BSE in cattle, needed scientific experiments to justify it - epidemiology is not enough.

6.6.3 Spontaneous Outbreaks of Prion Diseases

There is basic evidence that mutations of the prion protein are frequent and there is evidence with prions and elsewhere in medicine, that mutations can cause disease. With prions, there is strong evidence that a new disease can occur spontaneously. Kuru broke out among the Fore tribe in Papua New Guinea about 1900 - 1920 and was spread orally, but other cannibal tribes did not have kuru. Some 85% of CJD cases occur spontaneously in all parts of the world and have no correlation with food or other possible causes - the most reasonable (only) assumption is that sporadic CJD occurs by a spontaneous mutation in all parts of the World.

6.6.4 Staggers

Long before 1980, many UK farmers had observed an occasional cow to have unusual symptoms which they called "staggers". When BSE occurred, the farmers said that staggers appeared identical to BSE clinically. It is reasonable to hypothesise that BSE originated from staggers. It became an epidemic only when dead cows were fed to other cows thus causing positive feedback. Before 1980 staggers occurred rarely, but there is now evidence of vertical transmission from mother cow to calf, so once the transmutation occurred in the UK, it could have been continued by vertical transmission. From 1980 onwards, transmission of BSE was horizontal by feeding infected offal from cows to other cows - thus causing the BSE epidemic.

6.6.5 Why no BSE in the States?

Until now tests by the Animal and Plant Health Inspection Service, APHIS, have not found any BSE in the States (based mainly on the examination over a number of years, of over 3000 cows which were a sample of cows which were "downers", i.e. who fell down from an unexplained cause).

If BSE occurred as a rare spontaneous mutation in the UK, it is reasonable that it had not occurred elsewhere just as kuru occurred in one tribe but not in other cannibal tribes in PNG.

(Defenders of the BSE from sheep hypothesis say that in the States there are many more cows than sheep whereas the UK is completely different with more sheep - this is correct, but is completely irrelevant. If it were considered important, then since animals are processed locally, it could be tested by taking the different regions of the States and different regions of the UK and studying the cow/sheep ratios in each region and then comparing with the incidence of BSE in each region. It may be noted that American scrapie probably came from Britain via Canada).

There are about 100.000 "downers" per year in the USA which is only about 0.1% of the total population of 100 million cows.

6.6.6 Scrapie into Cows

The second part of the sheep causing BSE hypothesis, was that scrapie caused BSE. This was finally tested[8] experimentally in the United States when (American) scrapie from sheep's brains was injected into cows' brains. A disease was observed which was different from BSE but which resembled scrapie. The cows became sick after 14 to 18 months and died about 5 months later. (Defenders of the BSE from sheep hypothesis argue that this is one strain of scrapie and maybe the scrapie strain that caused BSE is different. Surprisingly there does not appear to have been published any experiment in the UK to test this where all known strains of scrapie were injected into cows).

6.6.7 Strain Type

The Edinburgh technique of injecting different mice with a prion disease and then studying the amount of lesions in different parts of the brain, gives a characteristic signature or "lesion profile". With sheep, some 20 different profiles have been distinguished indicating some 20 different strains of scrapie. However BSE in cows gives a lesion profile which is quite different from any of the scrapie signatures. Further when other animals which had eaten contaminated animal feed, such as cats, and various animals in zoos, were studied[9], their lesion profiles were identical to that of cows with BSE. Even stronger evidence is that when a sheep, a goat and a pig were injected with the BSE agent, they all gave signatures which strongly resembled that of BSE from cows but were quite different from scrapie signatures.

6.6.8 Conclusion

It is concluded that BSE does not come from scrapie in sheep but originated in a rare spontaneous mutation in a British cow well before 1980.

7. REPLICATION MECHANISM - WHY IS THE INCUBATION PERIOD SO LONG?

The abnormally long incubation period of prion diseases must be a major indication of the process. "Why does it take 15 years to travel from the stomach to the brain?", is a question that needs answering.

A simple mechanism in which the PrP-sen is steadily converted to PrP-res according to the amount of PrP-res available, will not reproduce the observed behaviour of prion diseases.

The most probable mechanism involves nucleation and the approach of Jarrett and Lansbury[10] will be used.

The conversion of one molecule of PrP-sen into one molecule of PrP-res is taken to be thermodynamically unfavourable so that it takes place rarely and if does take place, there is a high probability that the reverse process of changing PrP-res back into PrP-sen is favoured.

Prion proteins tend to aggregate. When several have aggregated, a seed is formed of a critical size. The process of conversion of PrP-sen into PrP-res in the presence of this seed, is now taken to be thermodynamically favourable. Then the rate of conversion increases as the seed(nucleus) grows. Thus there is a lag time until the aggregation reaches the critical seed size, and then there is a period of rapid growth followed by a period of equilibrium controlled by the rate of formation of fresh PrP-sen. Note - these thermodynamical considerations do not seem to have experimental support but are part of the hypothesis. However Jarrett and Lansbury also suggest that multiple attachment points are needed and this defines the critical size of the seed.

Thus the rate of conversion during the lag time is very sensitive to the density of PrP-res. As an example, Jarrett and Lansbury[10] quote the case where the seed is an octamer, that is, has eight proteins in it. Then for a two-fold saturation the lag time will be 45 years; for a 10-fold saturation it will be one hour: while for 100-fold saturation the lag time is only 36 microseconds. Clearly for prion proteins, the number of proteins must be less than eight but more than one. Now at this meeting Detlev Riesner reported finding insoluble spheres of 10 nm diameter with a molecular weight of 117 kDa so that they could be 4 to 6 molecules of PrP27-30 - and this might be about right for a seed. Clearly more work and confirmation should be encouraged.

Another way of considering the difficulty of forming the critical-sized seed, is to note that in some human diseases, both the paternal and the maternal alleles play a role as shown by the importance at codon 129 of whether there are Methione or Valine amino acids. Hence the seed should have a minimum of three proteins, namely the invasive PrP-res, and one prion from the maternal and one from the paternal side.

A qualitative description of this is given in the paper by Reed Wickner when discussing yeast.

It is reported that one Infectious Unit, IU, contains about 10^5 or 10^6 (or 10^4) PrP-res - the aggregation effect is well-established even though the number may vary perhaps because of the variable concentrations of the three prions. It is possible that this aggregation represents the local equilibrium state and is the basis of amyloid deposits.

When the residual prion protein, PrP-res, is attacked by proteinase K, it is cleaved and reduced to a smaller unit, e.g. PrP27-30. It is these smaller units which add to the aggregate. The cleavage would expose new surfaces to aggregation. As the PrP-res can give many forms depending on the mutations involved, the aggregate will correspondingly have many forms which will give different clinical and pathological effects accounting for different names for the prion disease.

Thus it is suggested that a threshold exists. This is good news as it would suggest that there will be fewer cases than otherwise of nvCJD in Europe where many have consumed small quantities of cow offal which has been contaminated by small amounts of BSE-infected material

An interesting and useful mathematical description of the various processes has been made by Manfred Eigen[11] based on discussions with Charles Weissmann(see ref. 4).

8. MECHANISM OF SPORADIC PRION DISEASE - WHY IS THE AVERAGE AGE SO HIGH?

The age distribution[12] of sporadic CJD, given in fig. 2, shows that the disease hardly occurs in children but increases steeply with age (the maximum near 72 years is generally considered to be an artefact due to under-reporting in the elderly as there is a tendency to ascribe the clinical symptoms to the much more frequent Alzheimer's Disease, AD).

Slowly progressing cancers have a similar age distribution different from diseases arising from a single infection such as influenza, where the age distribution is flat, that is, constant with age. For the former, the explanation is that for a cancer to develop, about four or five different things have to happen

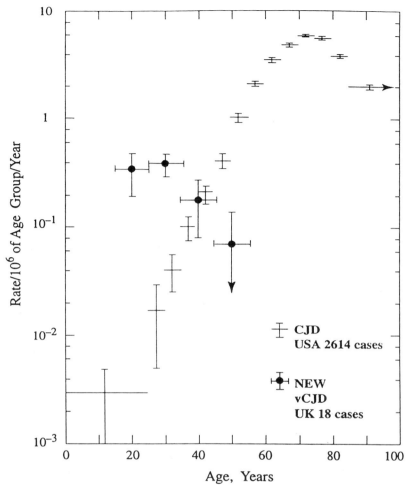

Fig. 2. The age distributions at death, rate per million of age group, are given for sporadic CJD in the USA[12] and for 18 cases as of 31 May 1997, for the new variant of CJD in the UK[14].

- since the body has several components to its defence mechanisms(immune system), these have to be knocked out before the cancer can spread. For influenza etc., only one single happening is sufficient which is why the age distribution is flat. It is concluded that for sporadic CJD to occur, several events have to happen first.

As the rate of sporadic CJD is about the same everywhere and no correlation has been found with diet, climate etc., it may be concluded that it arises from a series of spontaneous transformations. Then we make a major assumption - that these transformations are accumulative. Then the question is how many transformations are required to reach the critical level and give the "seed"?

Since these transformations are random, their number will be statistically distributed. Thus it will be very rare for sufficient transformations to occur in children for a seed to be formed. With increasing age, the probability of accumulating enough transformations to form the critical seed, increases. This mechanism will then give the observed shape of the age distribution of sporadic CJD.

An exception has been found to the rule that sporadic CJD is uniform everywhere - Libyan Jews have a higher rate than the normal one per million per year. Since this group has an unusual mutation on codon 200, it is not surprising that some special mutation favours sporadic CJD. Further this group of people have historically been isolated which would preserve the mutation among them.

The Conventional Wisdom of the Prion Only hypothesis, is that the presence of an abnormal prion, PrP^{sc}, is necessary to declench the conversion of normal prions to the abnormal form. But it has been noted that when there is a largely excessive concentration of normal prion proteins and no abnormal ones, then a prion disease does result. This is very hard to understand on the usual ideas of prions. But if one adopts the nucleation hypothesis, it becomes possible. This is because the greater concentration of PrP-sen can increase the probability of a critically-sized seed being formed by the rare abnormal PrP-res which are formed sporadically. That is, we are assuming that PrP-res are being formed by transformations, sporadically all the time, but usually they are too few in number to form a nucleus of critical size and hence no disease results. But if there is a large excess concentration of PrP-sen, then the infrequent PrP-res have a greater chance of finding and reacting with the PrP-sen and hence a greater chance of forming a seed of critical size.

The life cycle of the prion protein may be recalled - it is created in the endoplasmic reticulum inside the cell, passes through the membrane and attaches itself to the outer wall of the membrane. It then re-enters the cell where it is dissolved by detergents such as proteinase K - unless it meets a seed where it may be preserved as a PrP-res. Thus the probability of a PrP-res encountering a PrP-sen is normally low.

9. AGE DISTRIBUTION OF NEW VARIANT OF CJD CASES

It has been noted by Will et al.[13], that the age of people with the new version of Creutzfeldt-Jakob Disease, nvCJD, is much less than that of those with normal CJD. In Fig. 2 the two age distributions are compared quantitatively.

Fig. 2 shows the age distribution at death for 2614 CJD cases in the USA for the period 1979 to 1990, taken from National Center for Health Statistics by Holman et al.[12]. It may be seen that the rate, here on a log scale, is extremely low for young people but rises to a maximum of 5.9 cases per million at age of about 72 years. The apparent decline above this is possibly due to under-reporting, there being the possibility that the disease would be assigned to Alzheimer's without study.

Also shown are the rates for 18 cases[14] of the new variant in the UK as of 31st May 1997 - there is a peak near 30 years and then a sharp decline.
The two distributions are significantly different with a probability considerably greater than one in a million (calculations beyond this have little meaning).

It is tempting to interpret the nvCJD data. The minimum incubation period for human-to-human transmission in kuru is about four and a half to 5 years. For cow-to-human transmission, the "species barrier" will be expected to length this minimum incubation period. Now significant amounts of BSE-infected material from cows entered the human food chain in the period 1984-1985 according to Anderson et al.[15] and the first cases of nvCJD appeared in 1995-96. This indicates a minimum incubation period of about 10 years which is twice as long as that for kuru. Part of this longer period is due to the species barrier but part should be caused by the much bigger doses with kuru.

Apparently 49% of the clients at fast-food restaurants are in the 16 to 24 year age group. Adding the above minimum incubation period of about 10 years, the two age distributions may not be inconsistent with a broad peak near 30 years of age. It is important to note that the nvCJD cases occurring now correspond to the minimum incubation period and not to the average which must be much longer. This suggests that a serious statistical study of dietary habits as a function of age, may be justified.

10. NEW VARIANT OF CJD - POSSIBLE FUTURE

To estimate the likely number of cases of nvCJD, several pieces of information are needed, but they are missing or there are only indications.

As nvCJD following oral contamination, involves a change of species from cow-to-human, the average incubation period is expected to be longer than with human-to-human contamination but the extra period cannot be estimated reliably.

Whether there is a threshold of infectious prion material is unsure but possible. It has been shown that the incubation period is reduced when the dose of infectious material is increased but the amount is not known for humans. An example[15] was found for cows when going in steps from one gram to 300 grams, the incubation period decreased from about 52 months to about 40 months (about 23%). For the nvCJD case, the input amounts involved were probably small since normally the material from many cows were mixed together diluting the concentration of infective agent - except in some gourmet foods.

10.1 Average Incubation Period of Human-to -Human Prion Diseases

Carleton Gajdusek kindly sent me all his papers with annotations on kuru which might have some bearing on estimating the average incubation period of kuru. The statistics have not been collected in the most complete form, not surprisingly, in view of conditions in the Highlands of Papua New Guinea at that time.

Ritual cannibalism ended the early 1950's in the North Fore tribe and in about 1957-1959 in the South Fore tribe. The peak of the epidemic was about 1961 with some 200 deaths per year. This has fallen steadily to about 15 deaths per year in 1975. The decrease continued and in 1995 there were 3 deaths, probably about 35 to 40 years after infection. A precise statistical analysis is not justified in view of the uncertainties, in particular one does not know how many feasts were attended and when, but this would suggest an average incubation period of the order of 15 to 20 years for this human-to-human disease with high doses.

The incubation period has been found to vary with the mode of entry, being very short for entry into the brain or via the eye or ear, longer according to the place of injection and longest for oral entry. Thus iatrogenic cases following brain surgery cannot usefully be included for estimating times following oral infection.

Initially human growth hormone, HGH, injections were made into young people far from the brain, often in the leg. After a case of CJD was found in 1985, abnormally in a young person, the human growth hormone was no longer taken from the pituitary gland in the brain of cadavers, instead artificial hormones are now being used. The latest statistics[16] for average incubation periods are;

France (about 50 cases) 8 years
UK (24 cases)   19 years

USA (26 cases)   19 years.

These average incubation periods have tended to increase with time as more cases of CJD arise.

It is known that the dose was greater in France because of the method of selection of pituitaries. This is evidence that the incubation period varies greatly with the dose and/or there is a threshold. So for the majority of people who have only eaten small doses of BSE-infected food, there may be no illness (threshold effect) or the incubation period may be very long.

Combining these data, an incubation period for low doses of oral infection, would be about 15 to 20 years for human-to-human transmission.

10.2 Incubation Period of nvCJD

Since transmission between two different species has always been found to increase the average incubation period, then for nvCJD with cow-to-human transmission, the period might be expected to be more than 20 years, but more precise estimates cannot be given with confidence now. However the evidence presented above that the minimum incubation period is about 5 years for high dose human-to-human transmission (kuru) and about 10 years for low dose cow-to-human transmission, nvCJD, does suggest that the "species barrier'" in this case, does increase the incubation period substantially. Hence the average length of the incubation period of nvCJD could be more than 20 years, perhaps even 25 or 30 years.

From Anderson et al.[15] the consumption of BSE-infected food began slowly about 1984 and had its peak about 1990/92. So the peak of the human nvCJD illness will be approximately about 2010 - 2020 but maybe earlier in countries with gourmet eating customs where larger concentrations of infected food are eaten.

10.3 Threshold Effect and Dose Rate Effect

There are diverse indications that small doses of infectious agent produce no disease - threshold effect - but this is not well quantised and cannot easily be applied to nvCJD. The data for human growth hormone suggest there could be a threshold - some 10^4 children received frequent injections often for several years[17] of hormones extracted from a mixture of thousands of pituitaries and possibly fewer than 1% will develop CJD.

Another indicative experiment has been carried out by Clouscard et al.[18] with sheep. Normally about 1% of sheep get scrapie. The mode of transmission in sheep appears not to have been systematically studied but is partly horizontally and partly vertically (involving the mother, but not necessarily genetically). It might be expected that when taken orally, the infected material would pass through the gastrointestinal tract and exit. However at this meeting Heino Diringer showed that infection entered the spinal chord at the same place as nerves enter the gastrointestinal tract. Clouscard et al. fed sheep with parasites which attack the walls of the tract. Some 80% of the sheep got scrapie relatively quickly. When the walls of the gastrointestinal tract were examined, numerous lesions were observed. Again this can be considered weak evidence in favour of there being a threshold since the rate went from 1% to 80%, and also some evidence for a concentration dependent rate.

Other arguments in favour of there being an effective threshold(not an absolute one), are given above when discussing the reaction mechanism.

10.4 Transmission from Mother to Child

Since nvCJD affects primarily young people, there is a worry that a mother could pass the disease on to her children. In PNG, there have been many cases of mothers with kuru giving birth but in no case has the child been infected. Carleton is quite emphatic; there is no transmission from mother to child.

With animals the situation is different. There are indications that sheep can pass scrapie to their young and an experiment[19, 20] in the UK with cows, showed that vertical transmission definitely does occur. It is suspected that this may be from contamination of walls or fences, or the eating of placenta, but is probably not maternal (that is, not genetic).

11. IS nvCJD AN APPROPRIATE NAME FOR THE NEW HUMAN BOVINE DISEASE?

There are a number of differences between the now well-known sporadic Creutzfeldt-Jakob Disease, CJD, and the new human illness which the evidence suggests is caused from eating parts of cows infected by Bovine Spongiform Encephalopathy, BSE.

Clinically, there are a number of differences from CJD but these are not considered to be decisive. The ECG tests tend to be normal whereas with CJD generally characteristic waves are observed. However the most distinctive differences are seen in neuropathological examination where characteristic floral patterns are seen which are usually absent with CJD though not always[21].

Since the cautious and euphemistical name, "new variant of CJD" and written vCJD or nvCJD, is not only cumbersome but also inappropriate, a shorter and more indicative name is preferable. Perhaps the first people to identify this important new disease should follow custom, and chose a more suitable name for it.

12. DRUGS AND THE NEED FOR SIMPLE EARLY DETECTION

The nature of prion diseases is such that a vaccine is unlikely.

However as more is learnt, it is possible that a drug can be found which would attenuate or delay the disease, possibly for a considerable time. This might be possible, for example, by reducing the rate of producing prion proteins or by altering the anchor binding the prion protein to the outer membrane of the cell. First results have been presented at this meeting which show the incubation period can be lengthened. The best results are when the drug is given at the same time as the infectious prion enters the animal, however this is not practicable for humans who have already been infected by BSE.

What is needed is for infected persons to be able to take a drug when the disease can be detected but before onset of symptoms. This requires a simple reliable detection system for a small concentration of infective agent. Further one would wish a detection system which is sensitive to small amounts of infectious prions so that the disease can be detected early. No such system exists. The lesion profile system in mice takes more than a year. The Western Blot method is a major improvement but does not at present, match the requirements, taking several days and requiring significant amounts of infectious agent. However this is a new technique so one expects that it will be developed further, for example, it would be good if quantitative measurements could made giving the precise numbers of kDaltons of the various residues and the intensity of the blot by measuring the image; plus errors on these quantities found by double blind controls etc.. The Western Blot technique looks like the future.

A concerted effort is needed to develop new detection systems and to test new drugs.

It may be important to note that some breeds of sheep are immune to scrapie. It might be useful to study their prion proteins, e.g. their amino acid sequences, to investigate the explanation of their immunity.

13. PRION PROTEINS PURPOSE? THE BRAIN

The Conventional Wisdom is that there are 10^{11} neurons and 10^4 synaptic functions and these 10^{15} elements are enough for all brain and memory operations. It seems a large number but our brain plus memory is so fantastic that I have always doubted it. Hence welcomed the paper by Bray[22] that many proteins appear to be involved in brain operations as this would greatly increase the number of elements employed. Bray says that such proteins have as their primary function, the transfer and processing of information and they are capable of amplification, integration and storing information. He does not specifically mention prion proteins but it is difficult not to note that transgenic mice with no prion proteins, have some problems, for example, with circadian rhythm, that may suggest their function. But for me, the great fundamental mystery is how can you store memory using organic elements which have a lifetime much shorter than memory?

14. SOME HIGHLIGHTS

There were many highlights; these are just three which were the most striking to one physicist.

14.1 Prion Structure

Rudi Glockshuber presented wonderful slides showing for the first time the three-dimensional structure of the most important part of a prion - codons 121 to 231. The three helixes were clearly separated and the β-flat was seen to be two rows of amino acids joined in an anti-parallel fashion, giving an almost flat surface. The remarkable stability of the prion could be understood as the weakest parts were well-protected. Although it appeared tightly folded, there were still some central voids - for any purpose?

The addition of electrons gave further thought-inspiring images. Some surface regions were positive and some negative. Also the surface was extremely indented with hollows and promontories - very suggestive of a surface which lent itself to multiple regions of attachment. In particular, it provided the possibility of requiring several precise criteria before another prion could be attached. And this is what would be required for very selective mutations - the fit has to be just right before aggregates can be formed - this would be consistent with Jarrett and Lansbury's suggestion[10] that the requirements of multiple attachments might explain the occurrence of a seed of minimum size. These striking images are published in ref. 23.

Recently the Zurich group's study has been extended[24, 25] to the full-length mouse prion protein from codon 23 to 231. The additional N-terminal segment from 23 to 120, looks like a flexible 'random coil-like' polypeptide chain. The tightly bound segment, 121 to 231, would appear to contain most of the critical sites. In a further paper[26] by Billeter et al., several surface recognition sites are identified and species barriers for prion diseases are discussed. For the bovine protein, a unique isolated surface with a negative charge has been identified.

The complex structure with prominences and hollows and areas of positive and negative charges, supports the concept that aggregation can proceed by multiple attachments and the hypothesis that there is a seed of critical size.

14.2 Yeast Better than Mammalian Prions for Studies?

Reed Wickner gave an impressive talk where he showed firstly, that two proteins in yeast behave like prions and secondly, since the yeast system is much simpler and cleaner, it is possible to get information and do experiments better than in the case of prions in mammals. Thus the evidence for the Protein Only hypothesis is strong since there are probably no undiscovered viruses in yeast.

The concept of reverse curability was strongly presented - when the material is cured to remove all the abnormal prions, they can be formed again, though with low frequency. This supports the model of the replication mechanism suggested above, as does the evidence that over-production of the normal form increases the rate of production of PrP-res. It would be good to have quantitative measurements of the concentrations of pure PrP-sen which can give a prion disease.

14.3 Use of Transgenes

A feature of the meeting was the number of important experiments which could only be made by using transgenic mice introduced by Charles Weissmann's group. Laboratory experiments, in vitro, are important and give new indications, but there are so many factors involved with real live subjects that only in vivo experiments can deal with the complications of real living mammals.

15. CLAIMED LINK OF FARMERS WITH CJD

In 1995 it was claimed[27] that four cattle farmers had CJD and this was "more than happenstance" as the odds were 10,000 to one. However it was pointed out[28] that these high odds were obtained by taking a very restricted sample of farmers and that if a more general sample were taken, then the odds became not unusual. Now in 1997, it has again been claimed[29] that farmers in the UK have too high a rate of CJD on the bases of six cases. However there are problems of under-reporting. The most reliable statistics may be the American ones given by Holman et al.[12] who find a CJD rate of 0.9 per million per year - reliable because the rate is essentially constant over the period 1979 to 1990. On the other hand, the UK rate increases by a factor of four over the period 1970-1996 (indicating some under-reporting) and also the rates are lower than the USA rate.

A further indication that these farmers do not have nvCJD, has been given recently by Bruce et al.[30] who have taken lesion profiles of two of the farmers who died of CJD and who had cows with BSE in their herds. They found that these two farmers had normal sporadic CJD.

16. FINAL DISCUSSION AND CONCLUSIONS

16.1 The "Protein Only" hypothesis has now reached a stage where the extent and quality of the evidence is considered sufficient so that it is best to assume it is correct and to work on this basis. If claims are made to have found a virus or virino which can cause a TSE, then this evidence needs to be very strong and needs confirmation.

16.2 The creation of the infectious agent, PrP-res, from the normal prion protein, PrP-sen, is not really understood. However some things are strongly indicated;

16.2.1 Aggregates are necessary

16.2.2 There is a threshold

16.2.3 Mutations and polymorphisms are very important and are correlated with the characteristics of the TSE

16.2.4 A new TSE disease can arise spontaneously (kuru, staggers), probably as a series of mutations which slowly accumulate to give a seed.

16.2.5 Sporadic diseases, e.g. sporadic CJD, could occur as a result of a series of transformations which accumulate to give a seed

16.2.6 The values of the concentration of PrP-res and PrP-sen are important (especially since high concentrations of PrP-sen with an absence of PrP-res, can alone give a TSE).

16.2 7 The nucleation theory is preferred - a seed of a critical size is needed.

16.2.8 The long incubation period is explainable in terms of the lag time to create the critical seed and of the low concentrations of the molecules that form this seed.

16.3 Probably all mammals have prion proteins and hence can have TSE diseases.

16.4 The fact that yeast has two proteins which appear to replicate as prions, suggests that a new method of replication, independent of nucleic acids (DNA, RNA) exists and further examples of it may be expected.

16.5 It is still not clear what is the role or use of the prion protein.

16.6 It is highly probable that a spontaneous mutation produced staggers as a fairly rare disease which was not formally recognised. When cows were fed the offal from other cows, then a positive feedback mechanism was produced giving the BSE epidemic. BSE did not come from scrapie in sheep.

(Note added. In Nature of 2nd October 1997, two papers confirm the above conclusion. Moira Bruce et al.[30] show that the lesion profile of nvCJD is the same as that of BSE observed in several species. John Collinge et al.[31] used a biochemical analysis, Western Blot, to show that nvCJD is similar to BSE but different from other human prion diseases.)

16.7 The minimum incubation period of nvCJD appears to be about 10 years, i.e. twice as long as the 5 years for kuru, showing the difference, due to the 'species barrier', between human-to-human and cow-to-human transmission plus some dose effect.

16.8 The average incubation period of nvCJD is not known but presumably because of the 'species barrier', is longer than for kuru and for human growth hormone infection. This might indicate a value of about 20 to 30 years.

16.9 The extent of the coming human nvCJD epidemic is not known. A first calculation[32] has courageously been published giving a range from 151 to 80,000 cases. The higher numbers were obtained with longer incubation periods. However these calculations did not impose the condition that the minimum incubation period should be about 10 years which might change the calculation and explain some of the difficulty they had in fitting data. Fortunately there appears to be a threshold.

16.10 David Westaway gave an extensive comparison of Alzheimer's Disease, AD, and prion diseases. While there are some clinical similarities between them, there is a fundamental difference in that AD is not transmissible unlike prion disease. Also prion diseases are being interpreted in terms of mutations and polymorphisms while the causes of AD are being found in other molecules. Thus while the causes of the two diseases appear fundamentally different, it may be that the progress of the clinical symptoms have some similarities - GSS being closest to Alzheimer's.

17. OVERALL CONCLUSION

The overall conclusion is that it was a very successful meeting with many new and interesting papers and results.

Another very important feature was that it was possible for most of the world's experts on prions to meet in a secluded spot and discuss freely (once they had escaped the TV cameras).

For me personally, the greatest pleasure was to meet a new community of scientists and discover a very fine group who although disagreeing at times, kept their friendly relations, and who shared a common enthusiasm for Science and for Humanity.

References

1. Alper T., et al., Nature 214(1967)764-766.
2. Giffith, J.S., Nature 215(1967)1043-1045.
3. Prusiner, S.B., Science 216(1982)16-144.
4. Weissmann, C., et al., Cold Spring Harbour Symposia on Quantitative Biology, Vol. LXI, 1996.
5. Bueler, H., et al. Nature 356(1992)577 and Cell73(1993)1339.
6. Collinge, J. et al., Nature 383(1996)685-690 .
7. Nature, 382(1996)106-107.
8. The New Scientist, 13 April 1996, p4 and J. Infectious Diseases, April 1996.
9. Bruce, ME, Br. Med. Bull. 49(1993)822.
10. Jarrett, J.T., and Lansbury, P.T., Cell 73(1993)1055-1058.
11. Eigen, M. "Prionics or the Kinetic Basis of Prion Diseases". Biophys. Chem. 69(1996)A1-A18.
12. Holman, R.C. et al., Neuroepidemiology 14(1995) 174-181.
13. Will, R.G, et al., Lancet, 347(1996)921.
14. Data from the UK National CJD Surveillance Unit.
15. Anderson, R.M., et al. Nature 382(1996)779-788.
16. Brown, P., at Erice meeting on Planetary Emergencies, August 1997.
17. Coste, J., et al., BMJ 315(1997)708-713.
18. Clouscard, C. et al., J. Gen. Vir. 76(1995)2097-2101.
19. Ridley, R.M. and Baker, HF, Nature 384(1996) 17.
20. Hoinville LJ, Wilesmith JW, and Richards, MS., Vet. Rec., 136(1995)312-318.
21 Takashima, S. et al., Lancet, 350(1997)865-866 and Kopp, N. et al., Lancet 348(1996)1239-40.
22. Bray, D., Nature 376(1995) 307-312.
23. Riek, R. et al., Nature, 382(1996)180-182.
24. Hornemann, S. et al, FEBS Lett. 413(1997)271-281.
25. Riek, R. et al., FEBS Lett. 413(1997)282-288.
26. Billeter, M. et al., Proc. Natl. Aca. Sci. 94(1997)7281-7285.
27. Gore, S.M., Brit. Med. J. 311(1995) 1416-18.
28. Morrison, D.R.O., Brit. Med. J. 313(1996)560.
29. Cousens, S.N., et al., Brit. Med. J., 315(1997)389-397.
30. Bruce, M.E. et al. Nature 389(1997)498.
31. Collinge, J. et al., Nature 389(1997)448.
32. Cousens, S.N., et al., Nature 385(1997)197-198.

LIST OF SPEAKERS

Chairman; Douglas Morrison
- it is the Erice tradition to have one chairman

Session 1
Introduction - John Collinge and Antonio Zichichi
Moira Bruce, Edinburgh
Paul Brown, Bethseda, Maryland
Andrea LeBlanc, McGill
Elio Lugaresi, Bologna
Katie Sidle, London
Pierluigi Gambetti, Cleveland

Session II
John Collinge , London
Bruce Chesebro, Hamilton, Montana
Stanley Prusiner, San Francisco
Charles Weissmann, Zurich
David Westaway, Toronto

Session III
Pavel Liberski. Lodz
David Harris, St. Louis
Lajos Laszlo, Budapest
Sergei Inge-Vechtomov, St. Petersburg
Reed Wickner, Bethseda, Maryland
Rudi Glockshuber, Zurich

Session IV - Special Joint Session with Workshop on Planetary Emergencies
Stanley Prusiner, San Francisco
Paul Brown, Bethseda, Maryland
John Wilesmith, Addlestone, Surrey
Robert Will, Edinburgh
John Collinge, London
Luc Montagnier, Paris

Session V
Luc Montagnier, Paris
Randell Nixon, San Francisco
David Westaway, Toronto
Victor Zouev, Moscow
Dominique Dormont, Fontenay-aux-Roses
Adriano Aguzzi, Zurich

Session VI
Anthony Clarke, Bristol
Graham Jackson, London
Byron Caughey, Hamilton, Montana
Boris F. Semenov, Moscow
Detlev H. Riesner, Dusseldorf
Maurizio Pocchiari, Rome
Eva Mitrova, Bratislava/Budapest
Session VII
Hans Kretzschmar, Gottingen
Fabrizio Tagliavini, Milan
Heino Diringer, Berlin
Bernadino Ghetti and Pedro Piccardo, Indianapolis
Ruth Gabizon, Jerusalem
Albert Taraboulos, Jerusalem
Douglas Morrison, Geneva - Chairman; John Collinge

NAMES AND ADDRESSES OF PARTICIPANTS

Prof. Adriano Aguzzi
Dept. of Pathologie der Universitat   Tel. 41 1 255 28 69
Universitatsspital   21 07
Schelzbergstrasse 12
CH-8091 Zurich   Fax 41 1 255 44 02
Switzerland   Email; adriano@pathol.unizh.ch

Prof. Paul Brown
Lab. of Central Nervous System Studies   Tel. 1 301 496 52 92
NINDS/NIH   1 301 496 32 81(Sec)
9000 Rockville Pike
Building 36 Room 5B21   Fax 1 301 846 15 69
Bethseda
Maryland 20892-0036   Email; PWB@codon.nih.gov
USA

Prof. Moira E. Bruce
Inst. for Animal Health   Tel. 44 131 667 52 04
BBSRC/MRC Neuropathogenisis Unit,
Ogston Building   Fax 44 131 668 38 72
West Mains Road
Edinburgh EH9 3JF   Email   Moira.Bruce@bbsrc.ac.uk
UK

Prof. Harald Bruessow   Tel. 41 21 785 86 76(dir)
Nestec Ltd. Research Centre   81 11 (gen)
P.O. Box 44   Fax 41 21 785 89 25
1000 Lausanne 26
Switzerland

Prof. Byron Caughey
Lab. of Persistent Viral Diseases   Tel. 1 406 363 92 64 (dir)
NIH/NIAID   32 11(s/b)
Rocky Mountain Labs   92 91(lab)

Hamilton
Montana 59840
USA

FTS (700)   322 84 00
Fax 1 406 363 93 71
Email; Byron.Caughey@nih.gov
BCaughey@atlas.niaid.nih.gov

Prof. Bruce Chesebro
Lab. of Persistent Viral Diseases
NIH/NIAID
Rocky Mountain Labs
Hamilton
Montana 59840-2999
USA

Tel. 1 406 363 92 91(lab)
32 11(s/b)
94 00 (let)

Fax 1 406 363 92 04
Email; B.Chesebro@atlas.niaid.nih.gov

Prof. Anthony Clarke
Molecular Recognition Centre
Dept. of Biochemistry
School of Medical Sciences
University of Bristol
University Walk
Bristol BS8 1TD
UK

Tel. 44 117 928 8665

Fax 44 117 928 8274

Prof. Harriet Coles
Nature
Macmillan Magazines
4 Little Essex St.
London WC2R 3LF
UK

Tel. 44 171 836 6633

Fax 44 171 843 45 96

Prof. John Collinge
Neurogenetics Unit
Imperial College School of Medicine at St. Mary's
Norfolk Place
London W2 1PG
UK

Tel. 44 171 594 37 69 (dir)
723 12 52, ext 5469
594 37 60
Fax 44 171 706 70 94
F.Oliver@ic.ac.uk
J. Collinge@ic.ac.uk

Prof. Heino Diringer
Robert Koch-Institut
Bundesinstitut fur Infektionskrankheiten
und nicht Ubertragbare Krankheiter
Nordufer 20
Berlin D-13353
Germany

Tel. 49 30 45 47 22 30
or 2359 or 2526

Fax 49 30 45 47 26 09

Prof. Dominique Dormont
Direction des Sciences du Vivant
Dept. de Recherche Medicale
Service de Neurologie
Centres d'Etudes Nucleaires, CEA
60-68 Ave. General Leclerc
BP 6
92265 Fontenay-aux-Rose Cedex
France

Tel. 33 1 46 54 81 22

Fax 33 1 46 54 77 26

Prof. Ruth Gabizon
Dept. of Neurology
Hadassah University Hospital
Ein Karem
91120 Jerusalem
Israel

Tel. 972 2 77 78 58

Fax 972 2 43 77 82

Prof. Pierluigi Gambetti
Institute of Pathology
University School of Medecine
Case Western Reserve University
2085 Adelbert Rd.
Cleveland
Ohio 44106
USA

Tel. 1 216 368 05 86(office)
05 87(main)

Fax 1 216 844 18 10
36 27(dir)

Prof. Bernardino Ghetti
Div. of Neuropathology
Medical Science Building A142
Indiana University
635 Barnhill Drive MS A142
Indianapolis
IN 46202-5120
USA

Tel. 1 317 274 78 18

Fax 1 317 274 48 82

Prof. Rudi Glockshuber
Institut fur Molekularbiologie und Biophysik
ETH - Hoennggerberg
CH-8093 Zurich
Switzerland

Tel. 41 1 633 68 19

Fax 41 1 633 10 36

Email rudi@mol.biol.ethz.ch

Prof. David Harris
Dept of Cell Biology and Physiology
Washington University Medical Centre
Campus Box 8228
660 S. Euclid Ave.
St. Louis
Mo 63110-1093
USA

Tel. 1 314 362 46 90
69 45 (gen)

Fax 1 314 362 74 63

Prof. Sergei G. Inge-Vechtomov
Dept. of Genetics and Breeding
St. Petersburg State University
Universitetskaya Nab. 7/9
St. Petersburg 199034
Russia

Tel. 7 812 218 15 90
7 812 428 71 41(home)
Fax 7 812 218 96 34

Email inge@genet.bio.lgu.spb.su
Secr. (Mila Mirova) Email miron@btc.bio,pu.ru

Prof. Graham Jackson
Prion Disease Group
Dept. of Biochemistry
St. Mary's Hospital Medical School
Norfolk Place
London W2 1PG
UK

Tel. 44 171 594 37 69 (dir)
723 12 52, ext 5469

Fax 44 171 706 70 94

f.oliver@ic.ac.uk

Prof. Hans A. Kretzschmar
Dept of Neuropathology
Georg-August University of Gottingen
Robert Koch Str. 40
37075 Gottingen
Germany

Tel. 49 551 39 27 00
84 72 (dir?)

Fax 49 551 39 84 72

Prof. Lajos Laszlo
Dept. of General Zoology
Eotvos Lorand University
Puskin u. 3
P.O. Box 330
Budapest
H-1445 Hungary

Tel. 36 1 266 98 33 (ext 3002)
Ans. mach, 36 1 160 64 91

Fax 36 1 266 78 84

Email Lazlo@ludens.elte.hu

Prof. Pawel Liberski
Electron Microscopy Lab
Dept of Oncology
Medical Academy of Lodz

Tel. 48 42 81 88 40 Ext 460
Tel (Sec.) 81 11 17

Paderewskiego 4
Lodz
Poland

Fax 48 42 32 23 47

Prof. Andrea C. LeBlanc
Dept. of Neurology and Neurosurgery
McGill University
Montreal
Canada H3T 1E2
or
Lady Davis Inst.
Jewish General Hospital
3755 Ch. cote Ste Catherine
Montreal, H3T IE2
Canada

Tel. 1 514 340 82 22
49 76 (off)
4837 (lab)
Fax 1 514 340 82 95
Email mdal@musica.mcgill.ca

Prof. Elio Lugaresi
Istituto di Clinica Neurologica
Via Ugo Foscolo, 7 (porto Saragozza)
Universita di Bologna
Bologna 40123
 Italy

Tel. 39 51 58 50 51/53/58

Fax 39 51 64 42 165

Prof. Paul M. Mathews
Lab for Molecular Neuroscience
Mailman Research Centre
Harvard Medical School
McLean Hospital
115 Mill St.
Belmont
MA 02178-9106
USA

Tel. 1 617 855 35 65

Fax 1 617 855 31 98

Email   mathews@helix.mgh.harvard.edu

Prof. Eva Mitrova
Ambassador Extraordinary and Plenipotentiary of the Slovak Republic
Stefania ut 22-24
H-1143 Budapest
Hungary
Also
Research Institute of Preventive Medecine
Limbova 14
83301 Bratislava- Kramare
Slovakia

Tel. 36 1 251 25 68
18 60
Fax 36 1 251 14 60

Tel. 42 7 37 35 60

Fax 42 7 37 39 06

Prof. Luc Montagnier
Fondation Mondiale Recherche et Prevention de Sida   Tel. 33 1 45 68 45 20
1 rue Miollis   38 41
F-75015Paris   Fax 33 1 42 73 37 45
France
Institut Pasteur   Tel. 33 1 45 68 80 00
28 Rue du Dr. Roux
F75724 Paris Cedex 15   Fax 33 1 43 06 98 35
France

Prof. Douglas R.O. Morrison   Tel. 41 22 767 35 32 (dir)

CH-1296 Coppet
Switzerland   Fax 41 22 767 90 75
                                     E mail; douglas.morrison@cern.ch

Prof. Randal Nixon
Dept. of Pathology   Tel. 1 415 476 52 36(dir)
HSW 430   17 01(s/b)
University of California, San Francisco   Fax 1 415 476 79 63
San Francisco
CA 94143- 0506   Email; randal_nixon.pathmail@quickmail.ucsf.edu
USA   (Sec. Marlyna Stewart)

Prof. John Pattison
Dean of the Medical School Administration   Tel. 44 171 209 63 03
UCL Medical School
Gower Street
London WC1E 6BT   Fax 44 171 383 24 62
UK   Email; n.dyke@ucl.ac.uk

Prof. Pedro Piccardo
Div. of Neuropathology MS A142   Tel. 1 317 274 01 07
Medical Science Building A142
Indiana University
635 Barnhill Drive
Indianapolis   Fax 1 317 278 20 18
IN 46202-5120   Email; ppiccard@indyvax.iupui.edu
USA

Prof. Maurizio Pocchiari
Lab. of Virology   Tel. 39 6 49 90 32 03
Inst. Superio de Sanita   44 62 548
Viale Regina Elena 299   Fax 39 6 49 38 71 99
I-00161 Rome   pocchiar@viral.iss.infn.it
Italy   pocchiar@viral1.iss.infn.it

Prof. Stanley B. Prusiner
Dept. of Neurobiology   Tel. 1 415 476 44 82
University of California San Fransisco
San Francisco   Fax 1 415 476 83 86
CA 94143-0518
USA

Prof. Detlev H. Riesner
Institut fur Physikalsche Biologie   Tel. 49 211 811 48 40
Heinrich Heine University
Universitatsstr 1   Fax 49 211 811 51 67
D-40225 Dusseldorf
Germany   Email; riesner@biophys.uni-duesseldorf,de

Prof. Mark Rogers
Biotechnology Centre   Tel. 353 1 706 28 06
University College Dublin
Belfield   Fax 353 1 269 20 16
Dublin 4   Email; mrogers@acadamh.ucd.ie
Ireland   Mark.Rogers@ucd.ie

Prof. Acad. Boris F. Semenov
Mechnikov Research Institute on Vacines and Sera   Tel. 7 095 917 49 00
Russian Academy of Medical Sciences
Malyl Kazennyl Pereulok. 5a
Moscow 103064   (10.00 to 16.00 hours)   Fax 7 095 917 54 60
Russian Federation

Prof. Katie Sidle   Tel. 44 171 594 37 69 (dir)
Prion Disease Group   723 12 52, ext 5469
Dept. of Biochemistry
St. Mary's Hospital Medical School   Fax 44 171 706 70 94
Norfolk Place
London W2 1PG   f.oliver@ic.ac.uk
UK

Prof. Fabrizio Tagliavini
National Institute of Neurology   Tel. 39 2 23 94 260
Carlo Besta
Via Celoria 11
20133 Milano   Fax 39 2 70 63 82 17
Italy

Prof. Albert Taraboulos
Dept. of Neurology   Tel. 972 2 75 70 86

Hadassah University Hospital
Ein Karem
91120 Jerusalem   Fax 972 2 78 40 10
Israel

Prof. Charles Weissmann
Neurologische Klinik   Tel. 41 1 633 11 11(s/b)
Institut fur Molekularbiologie I   24 90/1(dir)
Universitat Zurich
CH 8093 Zurich   Fax 41 1 371 72 05
Switzerland   Email; Weissma@molbio1.unizh.ch

Prof. David Westaway
University of Toronto   Tel. 1 416 978 15 56
Centre for Research in Neurodegenerative Diseases
Tanz Neuroscience Building   Fax 1 416 978 18 78
6 Queens Park Cresent West
Toronto   Email; david.westaway%utoronto.ca@bureau-de-poste.utcc.utoronto.ca
Ontario M5S 1A8   Email; david.westaway@utoronto.ca
Canada

Prof. Reed Wickner
Section on Genetics of Simple Eukaryotes   Tel. 1 301 496 34 52
National Institute of Diabetes and Digestive and Kidney Diseases
Bldg. 8, Room 225
National Institutes of Health
Center Dr. MSC 0830   Fax 1 301 402 02 40
Bethseda
MD 20892-0830   Email wickner@helix.nih.gov
USA

Prof. John Wilesmith
Epidemiology Dept   Tel. 44 1932 34 11 11
Central Veterinary Lab   35 76 18(dir)
New Haw   Fax 44 1932 34 99 83
Addlestone
Surrey KT15 3NB   Email; j.wilesmith@cvl.maff.gov.uk
UK

Prof. Robert G. Will
National CJD Surveillance Unit   Tel. 44 131 332 25 25 (?)
Dept. of Clinical Neuroscience   21 17(dir)

Western General Hospital
Edinburgh EH4 2XU
UK

537 19 80(lab)
Fax 44 131 343 14 04(dir)
537 10 13(lab)
Email; RGW@srv0.med.ed.ac.uk

Prof. Viktor A. Zuyev
The Gamaleya Inst. for Epidemology and Microbiology
Russian Acad. of Medical Sciences
Gamaleya St. 18
Moscow 123098
Home
12 Osennig Bul'var
Building 3, Appt 355
Moscow 121360
Russia

Tel. 7 095 190 43 73
76 11
Fax 7 095 95 47 305
Email   postmaster@niiem.msk.ru

Tel. 7 095 413 22 55

INDEX